CONTEMPORARY MATHEMATICS

393

Recent Advances in Operator-Related Function Theory

Conference on Recent Advances
in Operator-Related Function Theory
Trinity College
Dublin, Ireland
August 4–6, 2004

Alec L. Matheson
Michael I. Stessin
Richard M. Timoney
Editors

American Mathematical Society
Providence, Rhode Island

2000 *Mathematics Subject Classification.* Primary 46E15, 46E20, 47B33, 47B38, 47B32, 30E20, 30E25.

1004740285

Library of Congress Cataloging-in-Publication Data
Conference on Recent Advances in Operator-Related Function Theory (2004 : Dublin, Ireland)
    Recent advances in operator-related function theory : Conference on Recent Advances in
Operator-Related Function Theory, Trinity College, Dublin, Ireland, August 4–6, 2004 /
Alec L. Matheson, Michael I. Stessin, Richard M. Timoney, editors.
        p. cm. — (Contemporary mathematics (American Mathematical Society) ; v. 393)
    Includes bibliographical references.
    ISBN 0-8218-3925-X (alk. paper)
    1. Functional analysis—Congresses. 2. Linear operators—Congresses. 3. Functions of complex
variables—Congresses. I. Matheson, Alec L., 1946– II. Stessin, Michael I., 1953– III. Timoney,
Richard Martin. IV. Title. V. Series.

QA319.C66   2004
515'.7—dc22
                                                                            2005053622

This volume is graciously dedicated to Prof. Joseph A. Cima on the occasion of his seventieth birthday. Over the past 45 years Prof. Cima has dedicated himself to mathematics as a researcher, with over 100 published papers; a devoted teacher; and a sage colleague. From his very humble beginnings in a hardscrabble Pennsylvania mining town, Joseph Cima learned the value of hard work and kept a Franklinesque ethic of "early to bed, early to rise" throughout his entire career. This is evident in his research program, which is just as active now as it was when he started—with no signs of slowing down.

After earning his degree from Penn State under Josephine Mitchell, Professor Cima started working in functional analysis and then moved to modern complex analysis and further to several complex variables. He keeps his hand and active mind in several other fields of analysis as well. Though he has held visiting positions and guest lectureships at various universities throughout the world, his first love is the University of North Carolina at Chapel Hill, where he has been a professor and a thesis advisor for over 45 years. He is a true believer in the egalitarian nature of public education.

As a person, Professor Cima is the epitome of professionalism and integrity. His standards are high, and he demands just as much from himself as he does from the rest of us. He is always generous with his time and is known in the mathematics community as willing to seek out and develop young talent. His unending energy is an inspiration to all those who get the chance to meet him.

We hope this humble offering, a collection of papers from some of Professor Cima's favorite fields of analysis, begins to show how much we appreciate all that he has done for us individually, for analysis, and for the profession in general.

Contents

Preface

This volume comprises the conference proceedings of a conference entitled "Recent advances in operator-related function theory", held at Trinity College, Dublin, August 4–6, 2004. The main theme of the conference was the role of Aleksandrov-Clark measures in modern operator theory and its attendant function theory. An introduction to these measures and their context for operator theory and function theory can be found in the paper of Poltoratski and Sarason. Further applications are found in the paper of Matheson and Stessin.

Garcia's paper provides a survey of recent results related to conjugation operators, Clark's operators and related topics. Ross's paper provides an introduction to the classical Dirichlet space, with an exposition of known results and a list of open problems.

The remaining papers deal with recent results in specific directions related to the theme of the conference. We believe they serve to illustrate the richness of the subject and the wealth of recent progress. We hope also that this collection will help inspire further work in the near future.

List of Participants

Alexandru Aleman
University of Lund

John T. Anderson
College of the Holy Cross

Catherine Beneteau
Seton Hall University

Christopher Boyd
University College Dublin

Patrick A. Brown
SUNY Albany

Stephen Buckley
National University of Ireland
Maynooth

Tom Carroll
University College Cork

Joseph Cima
University of North Carolina

Olivia Constantin
University of Lund

Seán Dineen
University College Dublin

Ron Douglas
Texas A&M University

Konstantin Dyakonov
Universitat de Barcelona

Eva A. Gallardo-Gutierrez
Universidad de Zaragoza

Stephan R. Garcia
University of California, Santa Barbara

Daniel Girela
Universidad de Málaga

Mark Halinan
Trinity College Dublin

Mary Hanley
University College Dublin

Remo Hügli
Trinity College Dublin

James Jamison
University of Memphis

Hans Jarchow
Universität Zürich

Siarhei Katsialouski
Belarus

Angeliki Kazas
SUNY Oneonta

Dmitry Khavinson
University of Arkansas

Greg Knese
Washington University, St. Louis

Urs Kollbrunner
Universität Zürich

Jussi Laitila
University of Helsinki

Wayne Lawton
National University of Singapore

David Malone
Communications Network Research
Institute

María José Martín Gómez
Universidad Autónoma de Madrid

Alec Matheson
Lamar University

John McCarthy
Washington University, St. Louis

Alfonso Montes-Rodríguez
Universidad de Sevilla

Jennifer Moorhouse
Colgate University

Raymond Mortini
Université de Metz

Don Moynihan
Trinity College Dublin

Artur Nicolau
Universitat Autònoma de Barcelona

Pekka Nieminen
University of Helsinki

Maria Nowak
Lublin, Poland

Donal O'Donovan
Trinity College Dublin

Jordi Pau
Universitat Autònoma de Barcelona

Serpil Pehlivan
Suleyman Demirel University, Turkey

Alexei Poltoratski
Texas A&M University

Manuel Ponce Escudero
Universidad de Sevilla

Arthur Prendergast
Trinity College Dublin

Stefan Richter
University of Tennessee

Alejandro Rodriguez Martinez
Universidad de Sevilla

William Ross
University of Richmond, Virginia

Ray Ryan
National University of Ireland, Galway

Eero Saksman
University of Jyväskylä

Rebecca Schmitz
University of Virginia

António Serra
Instituto Superior Técnico, Lisbon

Aristomenis G. Siskakis
Aristotle University of Thessaloniki

Wayne Smith
University of Hawaii

Michael Stessin
SUNY Albany

Karel Stroethoff
University of Montana

Carl Sundberg
University of Tennessee

Richard M. Timoney
Trinity College Dublin

Hans-Olav Tylli
University of Helsinki

Mitsuru Uchiyama
Fukuoka University of Education

Prasada R. Vegulla
Washington University, St. Louis

Dragan Vukotić
Universidad Autónoma de Madrid

David Walsh
National University of Ireland,
Maynooth

Warren R. Wogen
University of North Carolina

Lawrence Zalcman
Bar-Ilan University

Contemporary Mathematics
Volume **393**, 2006

Aleksandrov–Clark Measures

Alexei Poltoratski and Donald Sarason

To Joe Cima, on the occasion of his 70th birthday

ABSTRACT. This is a survey of the many roles Aleksandrov–Clark measures play in questions of function theory and operator theory.

1. Introduction

Associated with a holomorphic self-map of the unit disk $\mathbb{D}$ is a family of Borel measures on the unit circle $\mathbb{T}$, indexed by the points of $\mathbb{T}$. The measures have been referred to of late both as Clark measures and (by some of the same people) as Aleksandrov measures. Both names are appropriate: D. N. Clark's seminal paper [**C**] first called attention to them, while many of their deepest properties were discovered by A. B. Aleksandrov [**A1**]–[**A5**]. We shall call them AC measures.

This article reviews the substantial theory that has developed around AC measures in the past 20 years or so. In Section 2 we introduce a family of unitary transformations discovered by Clark. We shall present some of the results spawned by this basic discovery, including a surprising refinement of P. Fatou's classical theorem on existence of nontangential boundary values. In Section 3 we discuss further applications of AC measures in the study of boundary behavior of Cauchy-Stieltjes integrals. Section 4 describes the role of AC measures in perturbation theory, the topic of Clark's original paper. Section 5 deals with a different topic in operator theory, holomorphic composition operators, where AC measures also arise. Section 6 discusses a circle of ideas centered on the notion of a rigid function in the Hardy space H^1. That story is continued in Section 7, where it will be seen how AC measures arose, pre-Clark, in the Nehari interpolation problem.

In the remainder of this introduction we shall explain what AC measures are and derive a few of their properties. For a holomorphic self-map φ of $\mathbb{D}$ and a point α on $\mathbb{T}$, the function $\frac{\alpha+\varphi}{\alpha-\varphi}$ is holomorphic in $\mathbb{D}$ and has a positive real part. Its real part, the function $\frac{1-|\varphi|^2}{|\alpha-\varphi|^2}$, is thus the Poisson integral of a positive measure μ_α on

---

2000 *Mathematics Subject Classification.* 30D50, 30D55, 30E05, 30E20, 47A45, 47A55, 47B15, 47B33.

The first author is supported by N.S.F. Grant No. 0200699.

$\mathbb{T}$:

(1.1) $$\frac{1 - |\varphi(z)|^2}{|\alpha - \varphi(z)|^2} = \int \frac{1 - |z|^2}{|\xi - z|^2} \, d\mu_\alpha(\xi) \qquad (z \in \mathbb{D}).$$

We let $M_\varphi = \{\mu_\alpha : \alpha \in \mathbb{T}\}$, the family of AC measures associated with φ.

Setting $z = 0$ in (1.1), one finds that $\|\mu_\alpha\| = (1 - |\varphi(0)|^2)/|\alpha - \varphi(0)|^2$. Thus

$$\frac{1 - |\varphi(0)|}{1 + |\varphi(0)|} \leq \|\mu_\alpha\| \leq \frac{1 + |\varphi(0)|}{1 - |\varphi(0)|}.$$

If $\varphi(0) = 0$, then every μ_α is a probability measure.

We let m denote normalized Lebesgue measure on $\mathbb{T}$. In a sense to be explained, one has the equality

(1.2) $$\int \mu_\alpha \, dm(\alpha) = m.$$

Namely, if f is a function in $C(\mathbb{T})$ then the function $\alpha \mapsto \int f \, d\mu_\alpha$ is also in $C(\mathbb{T})$. Indeed, this is clear by (1.1) if f is the Poisson kernel of a point z in $\mathbb{D}$, hence if f is a finite linear combination of Poisson kernels. Such linear combinations are norm dense in $C(\mathbb{T})$, which, together with the boundedness of the measures μ_α, enables one to pass to the general f in $C(\mathbb{T})$ by a limit argument. One can thus define a (clearly positive) linear functional Λ on $C(\mathbb{T})$ by

$$\Lambda(f) = \int \left(\int f \, d\mu_\alpha \right) dm(\alpha).$$

It is asserted that Λ is the functional induced by m: $\Lambda(f) = \int f \, dm$. In fact, if f is a Poisson kernel this follows easily from (1.1), from which the assertion in general follows by the density mentioned earlier. Thus (1.2) holds in the sense that

(1.3) $$\int f \, dm = \int \left(\int f \, d\mu_\alpha \right) dm(\alpha)$$

for all f in $C(\mathbb{T})$.

As Aleksandrov pointed out in [**A1**], a limit argument enables one to strengthen the preceding conclusion: If f is a Borel function in $L^1(m)$, then f is in $L^1(\mu_\alpha)$ for almost every α, and (1.3) holds.

Atoms of the measures μ_α are related to angular derivatives of the function φ. One says φ has an angular derivative in the sense of Carathéodory (an ADC) at the point ζ of $\mathbb{T}$ if φ has a nontangential limit $\varphi(\zeta)$ of modulus 1 at ζ, and the difference quotient $\frac{\varphi(z) - \varphi(\zeta)}{z - \zeta}$ has a limit, denoted $\varphi'(\zeta)$, as $z \to \zeta$ nontangentially. A discussion of this notion and references can be found, for example, in [**S4**], where in particular one finds the following result, the original version of which is due to R. Nevanlinna [**Ne**]: The function φ has an ADC at the point ζ of $\mathbb{T}$ if and only if there is an α in $\mathbb{T}$ such that μ_α has an atom at ζ. In that case $\varphi(\zeta) = \alpha$, and $\mu_\alpha(\{\zeta\}) = 1/|\varphi'(\zeta)|$.

A related result (see [**S4**]) goes back to M. Riesz [**Ri**]: μ_α has a point mass at ζ if and only if φ has the nontangential limit α at ζ and for all (or one) β in $\mathbb{T}$ different from α,

(1.4) $$\int \frac{d\mu_\beta(\xi)}{|\xi - \zeta|^2} < \infty.$$

In their papers [**C**] and [**A1**], Clark and Aleksandrov were concerned only with the case where φ is an inner function, i.e. $|\varphi(\zeta)| = 1$ for almost all ζ in $\mathbb{T}$. If φ is inner then, for α in $\mathbb{T}$, the function $\frac{1-|\varphi|^2}{|\alpha-\varphi|^2}$ has a vanishing nontangential limit almost everywhere on $\mathbb{T}$, implying that the measure μ_α is singular (with respect to m). Moreover, because the Poisson integral of a positive singular measure has the nontangential limit ∞ almost everywhere with respect to the measure, one sees that μ_α must be carried by the subset $\{\varphi = \alpha\}$ of $\mathbb{T}$. When φ is inner, therefore, the measures in M_φ are mutually singular; they are carried by mutually disjoint sets. The family M_φ is in this case a so-called canonical system of measures (see [**P**], [**R**]).

The most elementary example of an AC family for an inner φ is given by $\varphi(z) = z^n$. In this case every μ_α in M_φ consists of n point masses, each of size $1/n$, placed at the nth roots of α.

The inner function $\varphi(z) = \exp\left(\frac{z+1}{z-1}\right)$ (the atomic inner function) affords another simple example. It continues analytically across $\mathbb{T}\backslash\{1\}$, at each point of which it thus has not only an angular derivative but an ordinary derivative. For α in $\mathbb{T}$ the set $\{\varphi = \alpha\}$ is countable and clusters only at the point 1. The measure μ_α is thus discrete. Its mass at a point ζ where $\varphi(\zeta) = \alpha$ equals $|\zeta - 1|^2/2$.

To conclude this introduction we note that Clark measures can also be associated, much as above, with holomorphic maps of the upper half-plane $\mathbb{C}_+$ into $\mathbb{D}$. The measures live on the extended boundary $\widehat{\mathbb{R}} = \mathbb{R} \cup \{\infty\}$ of $\mathbb{C}_+$ and are indexed by the points of $\widehat{\mathbb{R}}$ (see [**P**]).

2. Clark's construction and boundary behavior of pseudocontinuable functions

As usual, we denote by S the shift operator, the operator of multiplication by z on the Hardy space H^2 in the unit disk. A famous theorem of A. Beurling, published in 1949, says that the general nontrivial invariant subspace of S has the form IH^2 where I is an inner function. It follows that for the adjoint operator, the backward shift S^*, given by

$$S^* f(z) = \frac{f(z) - f(0)}{z},$$

the set of all proper invariant subspaces consists of the spaces K_I defined as orthogonal differences $K_I = H^2 \ominus IH^2$. A well-known result of R. G. Douglas, H. S. Shapiro and A. L. Shields [**DSS**] says that the spaces K_I are populated by so-called pseudocontinuable functions. These spaces have many important operator and function theoretic properties, see for instance [**Ni**].

The model operator S_I on K_I is defined by $S_I = P_I S$, where P_I is the orthogonal projection from H^2 onto K_I. A basic fact of operator model theory is that all contractive operators on Hilbert spaces, satisfying some natural conditions, can be realized as (are unitarily equivalent to) model operators on properly chosen spaces K_I or their vector-valued analogues.

AC measures were first introduced in [**C**] as part of the following construction. To simplify the formulas we will assume that the inner function I satisfies $I(0) = 0$. Consider the family of operators U_α on K_I defined, for α in $\mathbb{T}$, by

$$U_\alpha f = S_I f + \alpha \langle f, S^* I \rangle.$$

The deficiency indices of the model contraction S_I are $(1,1)$: the kernel of the operator is generated by the function S^*I and the orthogonal complement of its range is one-dimensional, generated by the constant function 1. From this it is not difficult to see that the operators U_α are unitary. Each of them differs from S_I by a one-dimensional operator. Moreover, $\{U_\alpha\}_{\alpha \in \mathbb{T}}$ is the family of all possible unitary rank-one perturbations of S_I.

The family $\{U_\alpha\}_{\alpha \in \mathbb{T}}$ contains full information about S_I and, at the same time, is much easier to study. All its operators are unitary and cyclic, and therefore, by the Spectral Theorem, each of them can be realized as multiplication by z in L^2 with respect to its (scalar) spectral measure. The main point of Clark's construction is that these spectral measures can be easily found from I. Namely, the spectral measure of U_α is μ_α defined by (1.1) with $\varphi = I$. The family $\{\mu_\alpha\}_{\alpha \in \mathbb{T}}$ of these spectral measures is the family of AC measures M_I.

In Section 4 we discuss the importance of Clark's construction in perturbation theory of unitary and self-adjoint operators. Until then we will be more concerned with relevant questions from function theory.

In most applications of the Spectral Theorem it is useful to understand the mechanism of the action of the unitary mapping that sends the original Hilbert space into L^2 over the spectral measure. Such a mapping is often called the generalized Fourier transform. This term is justified because the classical Fourier transform can be viewed as the particular example of such a mapping for the case of the operator of differentiation.

In Clark's situation the generalized Fourier transform is the unitary operator Φ_α from K_I to $L^2(\mu_\alpha)$ that "sends" U_α into the operator of multiplication by z on $L^2(\mu_\alpha)$. The natural question about the mechanism of the action of Φ_α was raised by Sarason in the early 1990's. It was answered in [**P1**], where it was shown that for α in $\mathbb{T}$, any f in K_I has a nontangential boundary value $f^*(\xi)$ for μ_α-almost every point ξ in $\mathbb{T}$, and $\Phi_\alpha f = f^*$. Note that in particular the $L^2(\mu_\alpha)$-norm of the boundary values of f coincides with the H^2-norm of f. This beautiful fact can be viewed as an analog of the Plancherel equality.

Further progress in this area is due to Aleksandrov [**A3**]. Using a theorem by Nikishin, he obtained an extension of the convergence result from [**P1**]. It was shown that, for any measure μ for which every function f in K_I can be defined μ-almost everywhere on $\mathbb{T}$ in a certain (weak) natural sense, every function f in K_I has finite nontangential boundary values μ-almost everywhere on $\mathbb{T}$. A similar result is true for the L^p-analogue of K_I $(p > 0)$.

In regard to the unitarity of the Clark operator, Aleksandrov proved [**A4**] that a positive μ such that $\|f\|_{H^2} = \|f^*\|_{L^2(\mu)}$ for all functions f in K_I that are continuous up to the boundary must be an AC measure corresponding to I or a "larger" function: μ must belong to $M_{I\varphi}$ for some φ from the unit ball of H^∞. Here one should mention that a similar question of the description of those μ such that the mapping $f \mapsto f^*$ is bounded as an operator from K_I to $L^2(\mu)$ is equivalent to the famous two-weight Hilbert transform problem; see [**NV**].

An analogue of the maximal principle for the Clark construction was proved by Aleksandrov in [**A5**]. The result says that if f is in K_I and f is in both $L^p(\mu_\alpha)$ and $L^p(\mu_\beta)$, where $2 \le p \le \infty$ and $\alpha \ne \beta$, then f is in H^p.

It follows from the discussion above that for every α in $\mathbb{T}$ the subset $\{I = \alpha\}$ of $\mathbb{T}$ is a uniqueness set for K_I in the following sense: if for the function f in K_I the nontangential boundary values $f^*(\xi)$ are equal to 0 for all ξ in $\{I = \alpha\}$ for which $f^*(\xi)$ exists, then $f \equiv 0$. Actually, even if $f^*(\xi) = 0$ only at μ_α-almost every point ξ, we can still conclude that $f \equiv 0$. This is no longer true in the L^p-analogues of K_I for $p < 2$; see [**A2**]. Further discussion of uniqueness sets on $\mathbb{T}$ and separating measures for K_I^p, including some open problems, can be found in [**P2**] .

3. Estimates of Cauchy Transforms

The inverse to the Clark operator Φ_α is the unitary operator $\Phi_\alpha^* : L^2(\mu_\alpha) \to K_I$ that recovers a function f in K_I from its boundary values f^* in $L^2(\mu_\alpha)$. The operator Φ_α^* can be defined in a very natural way. If μ is a finite complex measure on $\mathbb{T}$ we will denote by $K\mu$ the Cauchy integral in $\mathbb{D}$ of the measure μ:

$$K\mu(z) = \int_{\mathbb{T}} \frac{d\mu(\xi)}{1 - \bar{\xi}z}.$$

If h is in $L^1(|\mu|)$ we will write $K(h\mu)(z)$ for the Cauchy integral

$$\int_{\mathbb{T}} \frac{h(\xi)d\mu(\xi)}{1 - \bar{\xi}z}.$$

We will use the notation $\mathcal{C}_\mu h$ for the normalized Cauchy transform of h:

$$\mathcal{C}_\mu h = \frac{K(h\mu)}{K\mu}.$$

In general $\mathcal{C}_\mu h$ is a meromorphic function in $\mathbb{D}$, but in the case of positive μ it is analytic for any h. It is not difficult to show that

(3.1) $$\Phi_\alpha^* = \mathcal{C}_{\mu_\alpha}.$$

To prove this formula one need only verify it on the set of reproducing kernels for K_I and use the density of their linear combinations in $L^2(\mu_\alpha)$.

The normalized Cauchy transform $\mathcal{C}_\mu$ has many intriguing properties. It can be viewed as a generalization of the usual weighted Cauchy and Hilbert transforms to the case of singular weights; see [**P7**]. The usual Cauchy transform is the particular case of $\mathcal{C}_\mu$ with $\mu = m$.

The relation between $\mathcal{C}_\mu$ and the Clark model parallels the relation between the canonical Cauchy and Fourier transforms. The study of $\mathcal{C}_\mu$ gives analogues of classical theorems of P. Fatou, M. Riesz, R. A. Hunt–B. Muckenhoupt–R. L. Wheeden, etc., for the singular situation. For all the known results in this direction, some of which are discussed below, the main tools were provided via the connection with the Clark model given by (3.1).

Let μ be an arbitrary probability measure on $\mathbb{T}$. Then μ can be realized as $\mu = \mu_1$ in M_φ for a properly chosen function φ. To find such a φ one simply needs to solve the equation (1.1). If one adds the restriction $\varphi(0) = 0$, then such a φ can be found uniquely. As was mentioned in the introduction, singular μ correspond to inner φ. In view of this, the existence for any function f in K_I of nontangential boundary values μ_1–a.e. translates into the statement that the normalized Cauchy transform $\mathcal{C}_\mu h(z)$ converges to $h(\xi)$ nontangentially at μ-almost every point ξ, for any h in $L^2(\mu)$. The result follows immediately from (3.1) for singular measures μ and the general statement can be obtained with a little additional effort. It is also easy to extend the result to h in $L^1(\mu)$; see [**P1**]. This can be considered an extension of the classical Fatou Theorem to the case of weights with nontrivial singular parts. At the moment, all the known proofs of this general fact rely on the complex structure of AC families. In addition to [**P1**], another version of the proof was recently published in [**JL**]. It utilizes the same key ideas based on AC measures, which appear there in the form of spectral measures of rank-one perturbations of self-adjoint operators.

The convergence result allows one to view $\mathcal{C}_\mu$ as a linear transform on $L^p(\mu)$ in the usual sense: the image of a function h in $L^p(\mu)$, $p \geq 1$, under $\mathcal{C}_\mu$ is the boundary function of $\mathcal{C}_\mu h$. The first natural question is whether $\mathcal{C}_\mu$ is bounded in $L^p(\mu)$. The same convergence result states that the boundary values of $\mathcal{C}_\mu h$ are equal to h almost everywhere with respect to the singular component of μ. Thus, for purely singular μ, $\mathcal{C}_\mu$ is not only bounded in all $L^p(\mu)$, $p \geq 1$, it is actually the identity map. This reflects the fact that for singular μ the measure $h\mu$ is never "antianalytic." The general situation is not as nice. For an arbitrary positive μ, the operator $\mathcal{C}_\mu$ is still bounded in $L^p(\mu)$ for $1 < p \leq 2$. But $\mathcal{C}_\mu$ is unbounded in general for $p > 2$ [**P7**]. Unlike the case of singular μ, the operator is not, generally, bounded for $p = 1$ either: the simplest counterexample is $\mu = m$.

Another natural question is whether $\mathcal{C}_\mu$ is bounded as an operator from $L^p(\mu)$ to H^p. As follows from (3.1) and the discussion above, when μ is singular, $\mathcal{C}_\mu$ not only maps $L^2(\mu)$ into H^2, it actually preserves norms. For an arbitrary positive μ and other p, the operator $\mathcal{C}_\mu$ is bounded as an operator from $L^p(\mu)$ to H^p for $1 < p \leq 2$ and unbounded, in general, for $p > 2$. In particular, if μ is singular and $\mathcal{C}_\mu : L^p(\mu) \to H^p$ is bounded for some $p > 2$ then μ is discrete; see [**A2**].

Any convergence result in analytic function theory is closely associated with a result on the corresponding nontangential maximal function. For any function g on $\mathbb{D}$ we define its nontangential maximal function Mg on $\mathbb{T}$ by

$$(3.2) \qquad Mg(\xi) = \sup_{z \in \Gamma_\xi} |g(z)|$$

where Γ_ξ is the sector $\left\{ |z - \xi| \leq \sqrt{2}\Re(1 - \bar{\xi}z), |z| > \frac{1}{\sqrt{2}} \right\}$. The study of the maximal operator $M\mathcal{C}_\mu : f \mapsto M\mathcal{C}_\mu f$ in $L^p(\mu)$ includes a number of interesting questions. Many of them are still open.

As usual, we say that an operator is of weak type (p, p) if it acts from $L^p(\mu)$ into "weak $L^p(\mu)$", defined by

$$L^{p,\infty}(\mu) = \left\{ f : \mu(\{|f| > t\}) < \frac{C}{t^p}, \text{ for some } C < \infty \right\}$$

(with the infimum of such C raised to the power $1/p$ viewed as a norm).

Results of [**P1**] show that for any positive μ the operator $M\mathcal{C}_\mu$ is of weak type $(2,2)$ and is bounded in $L^p(\mu)$ for $1 < p < 2$. It is, generally, unbounded for $p > 2$. There is an example of a singular μ and an f in $L^\infty(\mu)$ such that $M\mathcal{C}_\mu f$ is not in $L^p(\mu)$ for any $p > 2$ [**P7**]. The operator is obviously unbounded for $p = 1$: just put $\mu = m$. Whether weak type $(2,2)$ can be improved to strong type remains unknown. It is also unknown if the operator is of weak type $(1,1)$. Once again, all the known proofs in this area are based on the structure of AC families.

One of the intriguing features of the normalized Cauchy transform is the contrast with the classical situation. The Cauchy transform in L^p spaces with absolutely continuous weights and the corresponding maximal operator are bounded under the same conditions on the weight. For arbitrary weights and the normalized Cauchy transform the situation is different. For instance, as we saw above, if we restrict our attention to the case of singular measures μ, then $\mathcal{C}_\mu$ is bounded for *all*

$p \geq 1$. The maximal operator however will still be, generally, unbounded for $p > 2$, as follows from the example mentioned above.

Another important area in the study of the boundary behavior of Cauchy transforms concerns distributions of their boundary values. Stemming from the remarkable work of G. Boole published in 1857, this area includes the classical weak-type estimates by A. Kolmogorov and later results by B. Davis, M. Goluzina, O. Tsereteli, S. Vinogradov, S. Hruschev and others. Using the tools provided by AC measures one can obtain shorter proofs for many of these results and develop a new understanding of their nature; see [**P6**].

4. Perturbations of self-adjoint and unitary operators

In this section we discuss perturbations of spectra of unitary and self-adjoint operators. Perturbation theory, even in its finite-dimensional form, is a vast area of spectral theory, consisting of many different subareas. Here we only present a small collection of results most closely related to the Clark construction.

If U_1 is an arbitrary unitary operator with simple spectrum and v, $\|v\| = 1$, is one of its cyclic vectors, then one can consider the standard family of rank-one unitary perturbations of U_1 generated by v:

$$(4.1) \qquad U_\alpha = U_1 + (1 - \alpha)\langle \cdot, U_1^{-1}v\rangle v, \qquad \alpha \in \mathbb{T}.$$

Let the inner function I be the characteristic function of the contraction $U_1 - \langle \cdot, U_1^{-1}v\rangle v$ in the sense of operator model theory. Then the spectral measures of U_α are immediately produced by (1.1) with $\varphi = I$. If we replace the unitary U_1 with a self-adjoint A_0 and consider the family of self-adjoint rank-one perturbations

$$A_\lambda = A_0 + \lambda\langle \cdot, v\rangle v, \qquad \lambda \in \mathbb{R},$$

which is a more traditional setting for perturbation theory, then the same scheme will give the spectral measures of A_λ: one will just have to write down the upper half-plane analogue of (1.1); see [**P4**]. Since such a correspondence between cyclic unitary (self-adjoint) operators and inner functions is bijective, problems in rank-one perturbation theory become, via the Clark construction, problems on the boundary behavior of inner functions. From the rank-one situation one can, with some additional effort, pass to the case of finite rank; see [**P5**]. The case of infinite-dimensional perturbations in the Clark model is still under development; see [**KP**].

Many of the standard notions of perturbation theory translate via the Clark construction into objects more familiar to function theorists. For instance, the Kreĭn-Lifshits shift function becomes $\arg(1 + I)/(1 - I)$, which immediately implies a number of its classical properties; see [**P4**].

A well-known paper by W. F. Donoghue [**D**] presented examples of families of rank-one perturbations $\{A_\lambda\}_{\lambda \in \mathbb{R}}$ with unstable spectral type. Even though Donoghue's paper did not mention inner functions, his approach was very close to Clark's. One of his examples is of a family in which A_0 has a pure point spectrum on $[0, 1]$ but all other A_λ have singular continuous spectrum on the same interval. This amounts to constructing a family of AC measures M_I on $\mathbb{T}$ such that one of the measures is discrete but the rest of them are singular continuous. To do this one can start with any dense countable subset of $\mathbb{T}$ and place point masses at the points of the set so that the resulting discrete measure μ_1 satisfies

$$\int \frac{d\mu_1(\xi)}{|w - \xi|^2} = \infty$$

for all w in $\mathbb{T}$. In view of the discussion of point masses of μ_α and formula (1.4) in the introduction, all other measures in the family M_I that includes μ_1 will automatically be singular continuous. Several more examples of this sort are contained in [**D**]. Before [**D**] it was unclear if the spectral type of an operator can change under a rank-one perturbation. Several recent papers contain similar examples displaying the lack of stability of various properties of the spectra under finite rank perturbations. Some of them use Clark's construction as their main tool; see for instance [**P8**] for further references.

The integral characterization of point masses of AC measures (1.4) together with Aleksandrov's formula (1.2) immediately gives the following property. The measures $\mu_\alpha \in M_I$ are discrete for almost every α in $\mathbb{T}$ if and only if for one (or all) of these measures the integral

$$\int \frac{d\mu_\alpha(\xi)}{|w - \xi|^2}$$

is finite for almost every w in $\mathbb{T}$. Independently of Clark's construction and subsequent studies by function theorists, this result was found by B. Simon and T. Wolff in their important paper [**SW**] in the setting of rank-one perturbations of self-adjoint operators. Many mathematical physicists now call this property the Simon-Wolff criterion. The criterion was used in [**SW**] to prove the so-called Anderson localization for the discrete Schroedinger operator in one dimension.

After the lack of stability of the spectral type in rank-one problems was established in [**D**], the next natural step was to check the stability of other spectral properties. Some of the questions in this direction were answered by a result from [**P9**], which says that any two cyclic unitary operators U and V whose spectral measures are singular with respect to Lebesgue measure and with respect to each other, and having spectrum equal to $\mathbb{T}$, are equivalent modulo a rank-one perturbation, i.e., there exists a unitary operator W such that $WUW^* - V$ is a rank-one operator. In terms of AC families this means that for any two singular and mutually singular positive measures μ and ν on $\mathbb{T}$ satisfying supp μ = supp ν = $\mathbb{T}$, there exist positive functions f in $L^1(\mu)$ and g in $L^1(\nu)$ such that the measures $f\mu$ and $g\nu$ belong to the same family M_I for some inner I.

In the opposite direction the most significant result was proved by Aleksandrov (unpublished) and independently by R. Del Rio, N. Makarov and Simon [**DMS**], and by A. Ya. Gordon [**Go**]. In these papers it was formulated in terms of spectral perturbations of self-adjoint operators. The AC-version of the result can be formulated as follows: If for some inner I one of the measures μ_α is densely supported on $\mathbb{T}$ then for a dense G_δ-set $B \subset \mathbb{T}$ all the measures μ_β for β in B are singular continuous. (An equivalent way of saying that one of the μ_α's is densely supported is to say that all μ_α's are densely supported or to say that the spectrum of I contains $\mathbb{T}$.)

Many questions in this area stem from mathematical physics. They are related to mathematical models of solid state physics and quantum dynamics, in particular to the problem of Anderson localization, which remains largely open in dimensions higher than 1. The translation of such questions into the language of AC families provides a variety of open problems for specialists in analytic function theory. For some such questions and discussions of further directions in this area see, for instance, [**P4, P5, P8, Si**].

5. Holomorphic Composition Operators

A holomorphic self-map φ of $\mathbb{D}$ induces a composition operator C_φ, which transforms a function f on $\mathbb{D}$ according to $C_\varphi f = f \circ \varphi$. The operator C_φ is known to act boundedly on many spaces of functions on $\mathbb{D}$, in particular, on the Hardy spaces H^p. It can also be defined, via composition with Poisson integrals, on the L^p spaces of $\mathbb{T}$, as well as on the space $M(\mathbb{T})$ of finite Borel measures on $\mathbb{T}$. This lets one, in particular, interpret the AC measure μ_α associated with φ as $C_\varphi \delta_\alpha$, where δ_α is the Dirac measure at α.

Here we shall limit our attention to C_φ as an operator on H^2. In a paper that stimulated much of the subsequent interest in composition operators, J. H. Shapiro [**Sh**] characterized the compactness of C_φ in terms of the Nevanlinna counting function of φ. The counting function is defined by

$$ N_\varphi(w) = \sum_{z \in \varphi^{-1}(w)} \log \frac{1}{|z|} \qquad (w \in \mathbb{D}), $$

where the sum on the right is understood to take account of multiplicities and to have the value 0 if $\varphi^{-1}(w)$ is empty. Its finiteness (for $w \neq \varphi(0)$) follows from the Blaschke condition.

Shapiro's result is that C_φ is compact if and only if

$$ \text{(5.1)} \qquad \lim_{|w| \to 1} \frac{N_\varphi(w)}{\log \frac{1}{|w|}} = 0; $$

the condition says, roughly speaking, that φ does not assume values near $\mathbb{T}$ too often. Shapiro in fact proved that the essential norm $\|C_\varphi\|_e$, the distance of C_φ from the compact operators, is given by

$$ \text{(5.2)} \qquad \|C_\varphi\|_e^2 = \limsup_{|w| \to 1} \frac{N_\varphi(w)}{\log \frac{1}{|w|}}. $$

A different necessary and sufficient condition for C_φ to be compact came from work of Sarason [**S2**] and Shapiro and C. Sundberg [**SS**]. It is simply the condition that all the AC measures μ_α in M_φ be absolutely continuous. This condition is therefore equivalent to (5.1); there is a close connection between AC measures and the Nevanlinna counting function, a connection that is not at all apparent on the surface. The question was raised of understanding the connection from a function-theoretic viewpoint.

Progress was made by J. A. Cima and A. L. Matheson [**CM**]. They proved in particular that $\|C_\varphi\|_e^2$ equals the supremum of the norms of the singular parts of the measures μ_α, thus, in view of (5.2), enhancing the connection between M_φ and N_φ.

The definitive answer is due to P. Nieminen and E. Saksman [**NS1**]. They convert the counting function N_φ into a measure-valued function $\mathcal{M}^\varphi$, whose value at a point w of $\mathbb{D}$ is given by

$$\mathcal{M}^\varphi(w) = \frac{1}{\log \frac{1}{|w|}} \sum_{z \in \varphi^{-1}(w)} \left(\log \frac{1}{|z|}\right) \delta_z.$$

As in the definition of N_φ, the sum on the right is understood to take account of multiplicities, and it is interpreted to be 0 if $\varphi^{-1}(w)$ is empty. The norm of $\mathcal{M}^\varphi(w)$ equals $N_\varphi(w)/\log \frac{1}{|w|}$. The beautiful theorem of Nieminen and Saksman states that, for α in $\mathbb{T}$, the measures $\mathcal{M}^\varphi(w)$ converge in the weak-star topology of $M(\overline{\mathbb{D}})$ (as the dual of $C(\overline{\mathbb{D}})$) to the singular part of μ_α as w converges nontangentially to α off a small exceptional set (technically, a set of finite Green capacity). Nieminen and Saksman show also that this result is optimal in a certain natural sense.

AC measures arise also in the problem of determining when the difference of two composition operators is compact. The question has been studied by J. E. Shapiro [**Sha**] and by Nieminen and Saksman [**NS2**].

6. Rigid Functions

A nonzero function in the Hardy space H^1 is said to be rigid if it is determined by its argument on $\mathbb{T}$, in the sense that the only other functions in H^1 with the same argument as it (almost everywhere) on $\mathbb{T}$ are the positive scalar multiples of itself. It is easy to see that rigid functions are outer functions. First of all, if u is a nonconstant inner function then it is not rigid, for it has the same argument on $\mathbb{T}$ as $(1+u)^2$: $\bar{u}(1+u)^2 = |1+u|^2$. Thus, if the nonzero function g in H^1 is divisible by the nonconstant inner function u, then g is not rigid, for its argument on $\mathbb{T}$ agrees with that of $(1+u)^2 g/u$. By the same token, if the nonconstant function g in H^1 is divisible in H^1 by $(1+u)^2$ for some nonconstant inner function u, then g is not rigid, for its argument on $\mathbb{T}$ agrees with that of $ug/(1+u)^2$.

If u is a nonconstant inner function then $(1+u)^2$ is an outer function. Thus, in view of the preceding remarks, not all outer functions are rigid. AC measures arise when one pursues the question: Which outer functions are rigid? There is as yet no satisfying answer to this question; no one has found a structural characterization of rigidity.

There are two simple and easily proved characterizations of rigid functions, which, however, do not reveal much about their internal structure. One is in terms of Toeplitz operators on H^2: The nonzero function g in H^1 is rigid if and only if the Toeplitz operator $T_{\bar{g}/|g|}$ has a trivial kernel (see for example [**S4**]). The other is in terms of the geometry of H^1: The nonzero function g in H^1 is rigid if and only if $g/\|g\|_1$ is an exposed point of the unit ball of H^1 (see for example [**BJH**], [**S4**]).

Rigid functions appear in a characterization of discrete completely nondeterministic stationary Gaussian processes (those for which the past and the future have a trivial intersection): A stationary Gaussian process is completely nondeterministic if and only if its spectral measure has the form $|g|\, m$ with g a rigid function in H^1 (see for example [**BJH**]). They play a key role in E. Hayashi's characterization of the kernels of Toeplitz operators [**H**], [**S5**], and they arise in the linear-fractional parameterizations of the solutions of Nehari interpolation problems, to be discussed in Section 7.

To bring AC measures into the picture we consider a nonconstant outer function g satisfying the harmless normalizations $\|g\|_1 = 1$ and $g(0) > 0$. Let the holomorphic self-map b of $\mathbb{D}$ be defined by

$$\frac{1 + b(z)}{1 - b(z)} = \int_{\mathbb{T}} \frac{\xi + z}{\xi - z} |g(\xi)| \, dm(\xi).$$

Thus, the AC measure μ_1 in M_b equals $|g| \, m$. (We are using "b" here instead of "φ" for the sake of harmony with other literature on rigid functions.)

The function g can be written as $g = f^2$ where f is an outer function satisfying $\|f\|_2 = 1$ and $f(0) > 0$. Define the function a by $a = (1 - b)f$. We note the following properties for reference in Section 7 (see [S1]):

(6.1) $\qquad\qquad$ a and b are in the unit ball of H^∞;

(6.2) $\qquad\qquad$ a is an outer function, and $a(0) > 0$;

(6.3) $\qquad\qquad$ $b(0) = 0$;

(6.4) $\qquad\qquad$ $|a|^2 + |b|^2 = 1 \qquad$ almost everywhere on $\mathbb{T}$.

In [S1] properties of the de Branges–Rovnyak space $\mathcal{H}(b)$ were exploited to show that if $g \; (= a^2/(1 - b)^2)$ is a rigid function then $a^2/(1 - \bar{\alpha}b)^2$ is a rigid function of unit H^1-norm for every α in $\mathbb{T}$. A consequence is that the rigidity of g implies that every measure in M_b is absolutely continuous.

In [S1] the conjecture was made (in a slightly different form) that the following conditions are equivalent:

$\quad$ (i) g is rigid.
$\quad$ (ii) There is no nonconstant inner function u such that $g/(1 + u)^2$ is in H^1.
$\quad$ (iii) Every measure in M_b is absolutely continuous.

The implication (i) $\Rightarrow$ (ii) was explained at the beginning of the section and the implication (i) $\Rightarrow$ (iii) just now. The implication (ii) $\Rightarrow$ (iii) is proved in [S3]. Although a bit of evidence had accumulated in favor of the conjecture, it was not long before J. Inoue [I] refuted it by constructing a nonrigid function g that satisfies (ii).

Further light on the subject was shed in [P3] where each nonrigid outer function was assigned an order, a natural number or ∞. If the implication (ii) $\Rightarrow$ (i) were true it would say that every nonrigid outer function has order 1. Inoue's counterexample has order ∞. One of the results of [P3] is that there exist nonrigid outer functions of all possible orders. The same paper contains a counterexample to the implication (iii) $\Rightarrow$ (ii). Thus, each of the implications (i) $\Rightarrow$ (ii) and (ii) $\Rightarrow$ (iii) is strict.

There is much here that remains to be understood.

7. Nehari Interpolation Problem

In the Nehari problem one is given a sequence $(c_n)_1^\infty$ of complex numbers. The question is whether there is a function f in the unit ball of L^∞ of $\mathbb{T}$ having the numbers in the sequence as its negatively indexed Fourier coefficients: $\hat{f}(-n) = c_n$ ($n = 1, 2, \ldots$). Z. Nehari [N] proved that such an f exists if and only if the Hankel matrix formed from the sequence, the matrix whose $(j, k)^{\text{th}}$ entry is c_{j+k-1} ($j, k = 1, 2, \ldots$), acts contractively on the space ℓ^2.

V. M. Adamyan, D. Z. Arov and M. G. Kreĭn (hereafter referred to as AAK) obtained a description of the solution set of a Nehari problem in the case a solution exists and is not unique [AAK1]. They proved that in this (the indeterminate) case there is a pair of functions a, b satisfying conditions (6.1)–(6.4) (and uniquely determined by the data of the problem) such that the solution set is given by

$$\left\{ \frac{a}{\bar{a}} \left(\frac{\omega - \bar{b}}{1 - b\omega} \right) : \ \omega \in H^\infty, \|\omega\|_\infty \leq 1 \right\}.$$

The problem is then suggested of determining which pairs a, b satisfying (6.1)–(6.4) arise this way from an indeterminate Nehari problem. We shall call such pairs Nehari pairs.

AAK noted that a Nehari pair a, b has a property beyond (6.1)–(6.4), namely, all the measures in M_b must be absolutely continuous. They asked whether (6.1)–(6.4) together with this property are adequate to characterize Nehari pairs. In [**HHN**] they formulated a conjecture, in a more general context, which would if correct provide an affirmative answer to their question.

J. B. Garnett [**G**] found an alternative approach to the AAK linear-fractional parametrization. From his analysis one can deduce that a pair a, b satisfying (6.1)–(6.4) is a Nehari pair if and only if the function $a^2/(1-b)^2$ has unit norm in H^1 and is rigid (details are in [**S1**]). Thus, AAK's question is equivalent to the question of whether condition (iii) from the preceding section implies condition (i), and so in fact it has a negative answer, thanks to the examples from [**I**] and [**P3**].

A. Ya. Kheifets [**K**] found an alternative and very interesting way of answering AAK's question. Using an abstract interpolation problem he had developed in collaboration with V. E. Katsnelson and P. M. Yuditskii, he proved that every indeterminate Hamburger moment problem produces a counterexample to the implication (iii) $\Rightarrow$ (i).

References

[AAK1] V. M. Adamyan, D. Z. Arov and M. G. Kreĭn, Infinite Hankel matrices and generalized Carathéodory–Fejér and I. Schur problems, *Funkcional. Anal. i Prĭlozhem.* **2** (1968), no. 4, 1–17 (MR 58#30446).

[A1]   A. B. Aleksandrov, Multiplicity of boundary values of inner functions, *Izv. Akad. Nauk Armyan. SSR Ser. Math.* **22** (1987), no. 5, 490–503, 515 (MR 89e:30058).

[A2]   A. B. Aleksandrov, Inner functions and related spaces of pseudocontinuable functions, *Zap. Nauchn. Sem. Leningrad. Otdel. Mat. Inst. Steklov. (LOMI)* **170** (1989), *Issled. Linein Oper. Teorii Funktsii.* 17, 7–33, 321; translations in *J. Soviet Math.* **63** (1993), no. 2, 115–159 (MR 910:30063).

[A3]   A. B. Aleksandrov, On the existence of angular boundary values for pseudocontinuable functions, *Zap. Nauchn. Sem. S.-Peterburg. Otdel. Mat. Inst. Steklov. (POMI)* **222** (1995), *Issled. po Linein Oper. i Teor. Funktsii.* 23, 5–17, 307; translation in *J. Math. Soc. New York* **87** (1997), no. 5, 3781–3787 (MR 97a:30046).

[A4]   A. B. Aleksandrov, Isometric embeddings of coinvariant subspaces of the shift operator, *Zap. Nauchn. Sem. S.-Peterburg. Otdel. Mat. Inst. Steklov. (POMI)* **232** (1996), *Issled. po Linein Oper. i Teor. Funktsii.* 24, 5–15, 213; translation in *J. Math. Sci. (New York)* **92** (1998), no. 1, 3543–3549 (MR 98k:30050).

[A5]   A. B. Aleksandrov, *On the maximum principle for pseudocontinuable functions.* (Russian) Zap. Nauchn. Sem. S.-Peterburg. Otdel. Mat. Inst. Steklov. (POMI) 217 (1994), Issled. po Linein. Oper. i Teor. Funktsii. 22, 16–25, 218; translation in J. Math. Sci. (New York) 85 (1997), no. 2, 1767–1772 (MR 96c:30033).

[BJH] P. Bloomfield, N. P. Jewell and E. Hayashi, Characterizations of completely nondeterministic stochastic processes, *Pacific J. Math.* **107** (1983), no. 2, 307–317 (MR 85g:60043).

[CM]   J. A. Cima and A. L. Matheson, Essential norms of composition operators and Aleksandrov measures, *Pacific J. Math.* **179** (1997), no. 1, 59–64 (MR 98e:47047).

[C]   D. N. Clark, One dimensional perturbations of restricted shifts, *J. Analyse Math.* **25** (1972), 169–191 (MR 46#692).

[DMS]   R. Del Rio, N. Makarov, B. Simon, *Operators with singular continuous spectrum. II. Rank one operators.* Comm. Math. Phys. 165 (1994), no. 1, 59–67 (MR 97a:47002).

[D]   W. F. Donoghue, *On the perturbation of spectra*, Comm. Pure Appl. Math., 18, 1965, 559–576 (MR 32#8171).

[DSS]   R. G. Douglas, H. S. Shapiro and A. L. Shields, *Cyclic vectors and invariant subspaces for the backward shift operator,* Ann. Inst. Fourier (Grenoble), 20 (1970), 37–76 (MR 42#5088).

[G]   J. B. Garnett, *Bounded Analytic Functions*, Pure and Applied Mathematics, 96, Academic Press, Inc., New York-London, 1981 (MR 83g:30037).

[Go]   A. Ya. Gordon, *Instability of dense point spectrum under finite rank perturbations.* Comm. Math. Phys. 187 (1997), no. 3, 583–595 (MR 98k:47026).

[HHN]   V. P. Havin, S. V. Hruščëv and N. K. Nikolskii (editors), *Linear and Complex Analysis Problem Book*, 199, Research Problems, Lecture Notes in Mathematics, 1043, Springer-Verlag, Berlin, 1984, pp. 160–163 (MR 85k:46001).

[H]   E. Hayashi, Classification of nearly invariant subspaces of the backward shift, *Proc. Amer. Math. Soc.* **110** (1990), no. 2, 441–448 (MR 90m:47013).

[I]   J. Inoue, *An example of a non-exposed extreme function in the unit ball of H^1.* Proc. Edinburgh Math Soc., 37 (1993), 47–51 (MR 95d:30d55).

[JL]   V. Jakšić and Y. Last, *A new proof of Poltoratskii's theorem.* J. Funct. Anal. 215 (2004), no. 1, 103–110.

[KP]   V. V. Kapustin and A. G. Poltoratski, *Boundary convergence of vector-valued pseudo-continuable functions*, Preprint.

[K]   A. Ya. Kheifets, Hamburger moment problem: Parseval equality and A-singularity, *J. Funct. Anal.* **141** (1996), no. 2, 374–420 (MR 97k:47014).

[NV]   F. Nazarov and A. Volberg, Bellman function, two-weight Hilbert transform, and embeddings of the model space K_θ, J. d'Analyse Math. **87** (2002), 385–414 (MR 2003j:30081).

[N]   Z. Nehari, On bounded bilinear forms, *Ann. of Math.* **(2)05** (1977), 153–162 (MR 18,602g).

[Ne]   R. Nevanlinna, Remarques sur le lemme de Schwarz, *Comptes Rendus Acad. Sci. Paris* **188** (1929), 1027–1029.

[NS1]   P. J. Nieminen and E. Saksman, Boundary correspondence of Nevanlinna counting functions for self-maps of the unit disc, *Trans. Amer. Math. Soc.* **356** (2004), no. 8, 3167–3187.

[NS2]   P. J. Nieminen and E. Saksman, On compactness of the difference of composition operators, *J. Math. Anal. Appl.* **298** (2004), no. 2, 501–522.

[Ni]   N. K. Nikolski, *Treatise on the shift operator*, Springer-Verlag, New York, 1986 (MR 87i:47042).

[P]   A. G. Poltoratski, Canonical systems and finite rank perturbations of spectra, MSRI preprint, 1995 (available at http://www.msri.org)

[P1]   A. G. Poltoratski, Boundary behavior of pseudocontinuable functions, *Algebra i Analiz* **5** (1993), no. 2, 189–210; translation in *St. Petersburg Math. J.* **5** (1994), no. 4, 389–408 (MR 94k:30090).

[P2]   A. G. Poltoratski, Integral representations and uniqueness sets for star-invariant subspaces, Systems, approximations, singular integral operators, and related topics (Bordeaux 2000), 425–443, *Oper. Theory Adv. Appl.* 129, Birkhäuser, Basel, 2001 (MR 2003F:30043).

[P3]   A. G. Poltoratski, Properties of exposed points in the unit ball of H^1, *Indiana Univ. Math. J.* **50** (2001), no. 4, 1789–1806 (MR 2003a:30039).

[P4]   A. G. Poltoratski, Kreĭn's spectral shift and perturbations of spectra of rank one, *Algebra i Analiz* **10** (1998), no. 5, 143–183; translation in *St. Petersburg Math. J.* **10** (1999), no. 5, 833–859 (MR 2000d:47028).

[P5]   A. G. Poltoratski, Finite rank perturbations of singular spectra, *Internat. Math. Res. Notices* (1997), no. 9, 421–436 (MR 98d:47035).

[P6]   A. G. Poltoratski, On the distribution of boundary values of Cauchy integrals, *Proc. Amer. Math. Soc.* **124** (1996), no. 8, 2455–2463 (MR 96j:30057).

[P7]   A. G. Poltoratski, *Maximal properties of the normalized Cauchy transform*, J. Amer. Math. Soc. 16 (2003), no. 1, 1–17.

[P8]   A. G. Poltoratski, *Survival probability in rank-one perturbation problems.* Comm. Math. Phys. 223 (2001), no. 1, 205–222.

[P9]   A. G. Poltoratski, *Equivalence up to a rank one perturbation.* Pacific J. Math. 194 (2000), no. 1, 175–188 (MR 2001j:47013).

[Ri]   M. Riesz, Sur certaines inégalités dans la théorie des fonctions, *Fysiogr. Sällsk. Lund Förh.* **1** (1931), Nr. 4, 18–38.

[R]   V. A. Rokhlin, On the fundamental ideas of measure theory, *Mat. Sbornik N.S.* **25(67)** (1949), 107–150 (MR 11,40).

[S1]   D. Sarason, Exposed points in H^1, I, *The Gohberg Anniversary Collection, Vol. II (Calgary, AB, 1986)*, 485–496, Oper. Theory Adv. Appl., 41, Birkhäuser, Basel, 1989 (MR 91h:46043).

[S2]   D. Sarason, Composition operators as integral operators, Analysis and Partial Differential Equations, 545–565, Lecture Notes in Pure and Appl. Math., 122, Decker, New York, 1990 (MR 92a:47040).

[S3]   D. Sarason, Exposed points in H^1, II, *Topics in Operator Theory: Ernst D. Hellinger Memorial Volume*, 333–347, Oper. Theory Adv. Appl. 48, Birkhäuser, Basel, 1990 (MR 94a:46031).

[S4]   D. Sarason, *Sub-Hardy Hilbert Spaces in the Unit Disk*, University of Arkansas Lecture Notes in the Mathematical Sciences, 10, A Wiley-Interscience Publication, John Wiley and Sons, Inc., New York, 1994 (MR 96k:46039).

[S5]   D. Sarason, Kernels of Toeplitz operators, *Toeplitz Operators and Related Topics (Santa Cruz, CA, 1992)*, 153–164, Oper. Theory Adv. Appl., 71, Birkhäuser, Basel, 1994 (MR 95k:47039).

[Sha]   J. E. Shapiro, Aleksandrov measures used in essential norm inequalities for composition operators, *J. Operator Theory* **40** (1998), no. 1, 133–146 (MR 99i:47062).

[Sh]   J. H. Shapiro, The essential norm of a composition operator, *Ann. of Math.* **(2)125** (1987), no. 2, 375–404 (MR 88c:47052).

[SS]   J. H. Shapiro and C. Sundberg, Compact composition operators on L^p, *Proc. Amer. Math. Soc.* **108** (1990), no. 2, 443–449 (MR 90d:47035).

[Si]   B. Simon, *Spectral analysis of rank one perturbations and applications.* Mathematical quantum theory. II. Schroedinger operators (Vancouver, BC, 1993), 109–149, CRM Proc. Lecture Notes, 8, Amer. Math. Soc., Providence, RI, 1995 (MR 97c:47008).

[SW]   B. Simon and T. Wolff, *Singular continuous spectrum under rank one perturbations and localization for random Hamiltonians.* Comm. Pure Appl. Math. 39 (1986), no. 1, 75–90 (MR 87k:470932).

TEXAS A&M UNIVERSITY, DEPARTMENT OF MATHEMATICS, COLLEGE STATION, TX 77843, USA

E-mail address: alexeip@math.tamu.edu

UNIVERSITY OF CALIFORNIA, DEPARTMENT OF MATHEMATICS, BERKELEY, CA 94720-3840, USA

E-mail address: sarason@math.berkeley.edu

Contemporary Mathematics
Volume **393**, 2006

Applications of spectral measures

Alec Matheson and Michael Stessin

Dedicated to Professor Joseph A. Cima on the occasion of his seventieth birthday.

ABSTRACT. The spectral measures arising from a holomorphic function ϕ mapping the unit disk into itself have found application in a number of areas of analysis. We survey applications to the theory of composition operators, conditional expectations, generalized factorizations associated with Wold decompositions, and similarity problems for operators on Hilbert space.

1. Introduction

A holomorphic function ϕ mapping the unit disk $\mathbb{D}$ into itself determines a family of measures $\{\,\tau_\alpha \colon \alpha \in \mathbb{T}\,\}$ as follows. For each α on the unit circle $\mathbb{T}$, the function $\frac{\alpha+\phi}{\alpha-\phi}$ has positive real part, and so, by Herglotz's theorem there is a positive measure τ_α on $\mathbb{T}$ such that

$$\Re\frac{\alpha + \phi}{\alpha - \phi} = \Re \int_{\mathbb{T}} \frac{\zeta + z}{\zeta - z}\, d\tau_\alpha(\zeta).$$

The family of all such measures is the family of spectral measures associated with the self-map ϕ of the unit disk. In particular each inner function θ generates such a family of measures and in this case each of the measures τ_α is singular with respect to Lebesgue measure. On the other hand, any positive measure μ on the unit circle is τ_1 for some ϕ, and it is not difficult to see that ϕ is inner if and only if μ is singular. In the case of inner functions τ_α is carried on the set on $\mathbb{T}$ where ϕ takes the value α.

Although the first use of spectral measures was made by Nevanlinna [29] in his proof of the Julia-Carathéodory theorem, the first systematic investigation was made by D. N. Clark [11] in his study of perturbations of restricted shifts on star-invariant subspaces. It was Aleksandrov who discovered that these spectral measures can be considered in the context of classical harmonic analysis. In his 1987 paper [2] he found a natural approach to the study of inner functions using spectral measures. In particular he explained the phenomenon of uncountable boundary multiplicity for inner functions. The existence of such functions was discovered in

---

2000 *Mathematics Subject Classification.* Primary 47B38; Secondary 30E20, 30E25.

1962 by MacLane and Ryan [**23**] in a completely nonconstructive way. Aleksandrov showed that every continuous singular measure is associated with an inner function of uncountable boundary multiplicity, and described an explicit method for constructing such functions.

Aleksandrov's work initiated extensive studies on boundary behavior of Cauchy transforms of measures supported on the unit circle [**3, 35, 34**], pseudocontinuations [**4**], and relationships with composition operators [**36, 37, 9**].

Another extremely important breakthrough was made by Sarason [**36, 37, 38**] who showed explicitly how the spectral measures associated with ϕ interact with the operator C_ϕ of composition with ϕ. For example, he showed that composition acts compactly on the space M of measures on $\mathbb{T}$ precisely when each spectral measure τ_α is absolutely continuous with respect to Lebesgue measure. Subsequently Shapiro and Sundberg [**41**] showed that this condition is equivalent to Shapiro's compactness condition on H^2 [**40**], Shapiro also calculates the essential norm of C_ϕ acting on H^2, and subsequently Cima and Matheson [**10**] showed how to compute this explicitly in terms of the spectral measures of ϕ.

In 1949, Beurling showed that a subspace of the Hardy space H^2 is invariant for the shift operator S if and only if it has the form θH^2, where θ is some inner function. The invariant subspaces for the backward shift B are thus the orthogonal complements of these subspaces, denoted by K_θ^2. Douglas, Shapiro, and Shields [**13**] showed that a function f belongs to K_θ^2 if and only if f/θ has a pseudocontinuation, that is, a function g in $H^2(\mathbb{D}_e)$ such that $g(\infty) = 0$ and f/θ and g have nontangential limits on the unit circle which coincide almost everywhere. There is a similar definition of K_θ^p for $0 < p < \infty$. A number of deep and interesting problems are related to the spaces K_θ^p.

2. Composition operators

If ϕ is a holomorphic map of the unit disk $\mathbb{D}$ into itself, it is a consequence of Littlewood's subordination principle [**22**] that composition with ϕ induces a bounded operator C_ϕ on each Hardy space H^p. A recurring theme in the study of composition operators has been the search for function theoretic conditions on ϕ which guarantee the compactness of the operator C_ϕ.

When $p = 2$, J. H. Shapiro [**40**] gave a definitive answer to this problem in terms of the Nevanlinna counting function, given by

$$(1) \qquad N_\phi(z) = \sum_{\phi(\zeta)=z} \log \frac{1}{|\zeta|}.$$

Shapiro showed that C_ϕ is compact on H^2, (and hence on H^p, for any p with $0 < p < \infty$) if and only if $\limsup_{|z|\to 1} \frac{N_\phi(z)}{\log \frac{1}{|z|}} = 0$. In the course of his proof, Shapiro established the inequality

$$(2) \qquad \|C_\phi\|_e^2 \geq \limsup_{|a|\to 1} \|C_\phi f_a\|^2 \geq \limsup_{|a|\to 1} \frac{N_\phi(a)}{\log \frac{1}{|a|}}.$$

where $f_a(z) = \frac{\sqrt{1-|a|^2}}{1-\bar{a}z}$, is the normalized kernel function for $a \in \mathbb{D}$, and where $\|C_\phi\|_e$ is the essential norm of C_ϕ acting on H^2, that is, $\|C_\pi\|_e$ is the distance in the operator norm from C_ϕ to the compact operators. We recall that the essential norm $\|T\|_e$ of the operator T is the distance in operator norm from T to the compact operators. In fact Shapiro actually proved that the three quantities in (2) are equal, thus obtaining the following theorem.

THEOREM 1 (Shapiro). *If ϕ is a holomorphic mapping of the unit disk into itself, then the composition operator C_ϕ acting on the Hardy space H^2 has essential norm $\|C_\phi\|_e$ given by*

$$\|C_\phi\|_e^2 = \limsup_{|a| \to 1} \frac{N_\phi(a)}{\log \frac{1}{|a|}}.$$

Remark. Recently Shapiro's theorem was extended to composition operators acting from the Hardy space in the unit disk into the Hardy space in either a polydisk or a ball in $\mathbb{C}^n$ which are induced by holomorphic mappings of these domains into the unit disk (see [47] for details).

For C_ϕ acting on Bergman spaces, again boundedness follows from Littlewood's principle. Shapiro showed in this case that C_ϕ is compact if and only if ϕ has no angular derivative at any point of the unit circle $\mathbb{T}$. The angular derivative $\phi'(\zeta)$ for $\zeta \in T$ with $|\phi(\zeta)| = 1$ is the limit $\lim_{z \to \zeta} \frac{\phi(z) - \phi(\zeta)}{z - \zeta}$, provided the limit exists nontangentially. Let $\Sigma_\alpha = \{ \zeta \in T \mid \phi(\zeta) = \alpha \}$. The Julia-Carathéodory theorem asserts that for each $\zeta \in \Sigma_\alpha$, this limit exists or is infinite. Nevanlinna's proof [29] shows in fact that $\tau_\alpha(\{ \zeta \}) = \frac{1}{|\phi'(\zeta)|}$ for each $\zeta \in \Sigma_\alpha$.

If μ is a positive measure on $\mathbb{T}$, its Poisson integral U_μ is a positive harmonic function on $\mathbb{D}$, and, conversely, every positive harmonic function arises in this way. In particular, the positive harmonic function $U_\mu \circ \phi$ is the Poisson integral of some positive measure ν, and, following Sarason [36], we define $C_\phi \mu = \nu$. Since $\|\mu\| = U_\mu$, Harnack's inequality shows that

$$\|\nu\| \leq \frac{1 + |\phi(0)|}{1 - |\phi(0)|} \|\mu\|.$$

Applying C_ϕ to each component of the Jordan decomposition of the complex measure μ extends the composition operator C_ϕ to a bounded operator on the space M of complex Borel measures on $\mathbb{T}$. With an application of the Schur test, Sarason shows directly that C_ϕ is a bounded operator acting on each L^p space ($1 \leq p < \infty$), and that this operator has norm 1 in case $\phi(0) = 0$, thus obtaining Littlewood's subordination principle as a corollary. Turning to questions of compactness, Sarason obtains the following theorem.

THEOREM 2 (Sarason). *The following are equivalent.*

(i) *C_ϕ is compact on L^1.*
(ii) *C_ϕ is compact on M.*
(iii) *C_ϕ maps M into L^1.*

Since it is easy to see that the spectral measures τ_α for ϕ are just the measures $C_\phi \delta_\alpha$, the third condition implies that all the measures τ_α are absolutely continuous with respect to Lebesgue measure. We shall call this the *absolute continuity* condition. It is not hard to see that this condition is equivalent to the conditions in Sarason's theorem, and hence is equivalent to compactness of C_ϕ on L^1 or on M.

Since H^1 is a subspace of L^1 and C_ϕ restricted to H^1 is the usual composition operator, it follows immediately that the absolute continuity condition implies the compactness of C_ϕ on H^1, and, hence, by a theorem of Shapiro and Taylor [39], on H^p for $0 < p < \infty$. In particular, the absolute continuity condition implies

Shapiro's compactness condition. Sarason asked whether the converse was true. This question was answered affirmatively by Shapiro and Sundberg [**41**]. Since both the absolute continuity condition and Shapiro's condition are function-theoretic in nature, Sarason [**38**] asked for a direct function-theoretic proof of this equivalence. This was provided in the proof of the following theorem of Cima and Matheson [**10**]

THEOREM 3 (Cima and Matheson). *If* σ_α *denotes the singular part of the Aleksandrov measure* τ_α *for* ϕ, *then the essential norm of* C_ϕ *acting on* H^2 *is given by*

$$(3) \qquad \|C_\phi\|_e^2 = \sup_{\alpha \in \mathbb{T}} \|\sigma_\alpha\|.$$

In another direction spectral measures have been used to investigate compactness of differences of composition operators. The primary motive for these studies has come from the desire to understand the topological structure of the set of composition operators acting on H^2. Perhaps the most significant results to date are the following theorems of Nieminen and Saksman [**30**]. In what follows, $T = C_\phi - C_\psi$, where ϕ and ψ are holomorphic functions mapping the unit disk into itself.

THEOREM 4 (Nieminen and Saksman). *The following are equivalent:*

   (i) *T is compact on* H^p *for all* $1 \le p < \infty$,
   (ii) *T is compact on* H^p *for some* $1 \le p < \infty$,
   (iii) *T is weakly compact on* H^1.

This theorem parallels the known results for a single composition operator. In the next theorem, M denotes the space of Borel measures on $\mathbb{T}$, $\{\mu_\alpha\}$ and $\{\nu_\alpha\}$ are the families of spectral measures for ϕ and ψ, respectively, and the measures admit Lebesgue decompositions $\mu_\alpha = \mu_\alpha^a + \mu_\alpha^s$ and $\nu_\alpha = \nu_\alpha^a + \nu_\alpha^s$, respectively.

THEOREM 5 (Nieminen and Saksman). *The following are equivalent:*

   (i) *T is weakly compact on* M,
   (ii) *T is compact on* M,
   (iii) *T is weakly compact on* L^1,
   (iv) *T is compact on* L^1,
   (v) $\mu_\alpha^s = \nu_\alpha^s$ *for each* $\alpha \in \mathbb{T}$ *and the set* $\{\mu_\alpha^a - \nu_\alpha^a : \alpha \in \mathbb{T}\}$ *is uniformly integrable.*

In the case of composition operators, compactness on H^1 is equivalent to compactness on L^1. This is no longer the case for differences of composition operators. In the same work, Nieminen and Saksman construct two analytic functions $\phi, \psi \colon \mathbb{D} \to \mathbb{D}$ such that the difference operator $C_\phi - C_\psi$ is compact on H^1 but not compact on L^1.

These results extend a program begun by J. E. Shapiro [**42**]. Related work can be found in [**27, 28, 19**].

3. Conditional expectation and the decomposition of measures

In 1987 Aleksandrov [**2**] established a theorem on the disintegration of Lebesgue measure with respect to the spectral measures τ_α determined by a holomorphic function ϕ mapping the unit disk into itself. As we shall see, this decomposition underlies most of the deeper applications of spectral measures to function theory and to related parts of operator theory.

THEOREM 6 (Aleksandrov). *Let ϕ be a holomorphic function mapping the unit disk into itself, and let $\{\tau_\alpha\}$ be the associated family of spectral measures. Then we have the weak-$*$ decomposition of Lebesgue measure*

$$m = \int_{\mathbb{T}} d\tau_\alpha,$$

which means that

(4) $$\int_{\mathbb{T}} f(\zeta)\, dm(\zeta) = \int_{\mathbb{T}} \int_{\mathbb{T}} f(\zeta)\, d\tau_\alpha\, dm(\alpha),$$

for every $f \in C$. Moreover, the formula (4) holds for every $f \in L^1$, where the integrals with respect to the measures τ_α exist for almost every $\alpha \in \mathbb{T}$.

Furthermore, if Σ is the subset of the unit circle where ϕ has nontangential limits of modulus one, and if σ_α denotes the singular part of τ_α, then the formula

(5) $$\int_{\Sigma} f(\zeta)\, dm(\zeta) = \int_{\mathbb{T}} \int_{\mathbb{T}} f(\zeta)\, d\sigma_\alpha(\zeta)\, dm(\alpha),$$

holds for all $f \in L^1$ as in Aleksandrov's theorem.

When ϕ is an inner function, the spectral measures τ_α are all singular with respect to Lebesgue measure, and, moreover, are pairwise singular. In fact, if Σ_α denotes the subset of $\mathbb{T}$ where ϕ has nontangential limit equal to α for each $\alpha \in T$, then the measure τ_α is carried by Σ_α in the sense that $\tau_\alpha(E) = \tau_\alpha(E \cap \Sigma_\alpha)$ for each Borel measurable set E. Since the sets E_α are mutually disjoint, the pairwise singularity of the measures τ_α follows.

When ϕ is an inner function with $\phi(0) = 0$, Aleksandrov introduces the operator

$$Tf(\alpha) = \int_{\mathbb{T}} f(\zeta)\, d\tau_\alpha(\zeta),$$

and notes that it is an absolute contraction, i.e., it is a contraction on each L^p, $1 \le p \le \infty$. This operator (for general ϕ) was studied by Matheson in [**26**]. As we shall see, this operator is related to the conditional expectation operator that arises by considering ϕ as a measurable map of the unit circle.

First we recall some properties of conditional expectation. Let $(X, \mathcal{F}, \mu)$ be a measure space with μ a positive measure. and let $\mathcal{G}$ be a sub-σ-algebra of $\mathcal{F}$. The conditional expectation of a function $f \in L^1(X, \mathcal{F}, \mu)$ is a $\mathcal{G}$-measurable function $\mathbb{E}(f|\mathcal{G})$ which satisfies

$$\int_{E} \mathbb{E}(f|\mathcal{G})\, d\mu = \int_{E} f\, d\mu \quad \text{for all } E \in \mathcal{G}.$$

The existence of $\mathbb{E}(f|\mathcal{G})$ follows by applying the Radon-Nikodým theorem to the measure $\nu(E) = \int_E f\, d\mu$ restricted to the σ-algebra $\mathcal{G}$, or, alternatively, by considering the orthogonal projection of $L^2(X, \mathcal{F}, \mu)$ onto $L^2(X, \mathcal{G}, \mu)$ Clearly, $\mathbb{E}(\cdot|\mathcal{G})$ is a positive linear operator of norm 1 on $L^1(X, \mathcal{F}, \mu)$ which has the following properties.

    (i) $\mathbb{E}(\mathbb{E}(f|\mathcal{G})|\mathcal{G}) = \mathbb{E}(f|\mathcal{G})$.
    (ii) $\mathbb{E}(fg|\mathcal{G}) = f\mathbb{E}(g|\mathcal{G})$ if f is $\mathcal{G}$-measurable.
    (iii) $\mathbb{E}(\mathbb{E}(f|\mathcal{G})|\mathcal{H}) = \mathbb{E}(f|\mathcal{H})$ if $\mathcal{H}$ is a sub-σ-algebra of $\mathcal{G}$.
    (iv) $\|\mathbb{E}(f|\mathcal{G})\|_p \le \|f\|_p$ for all $f \in L^p(X, \mathcal{F}, \mu)$, $1 \le p \le \infty$.

The last property asserts that conditional expectation is an absolute contraction (see [**12**, p. 184]).

When ϕ is an inner function with $\phi(0) = 0$, let $\mathcal{F}_\phi$ be the σ-algebra generated by ϕ, i.e., $\mathcal{F}_\phi$ is the smallest complete sub-σ-algebra of the σ-algebra of Lebesgue measurable sets with respect to which ϕ is measurable. Then, as Aleksandrov notes, the operator $T(f) \circ \phi$ coincides with the conditional expectation operator $\mathbb{E}(f|\mathcal{F}_\phi)$. It follows from (4), that this conditional expectation preserves analyticity in the sense that $\mathbb{E}(f|\mathcal{F}_\phi) \in H^1$ whenever $f \in H^1$. Indeed, since the conditional expectation is bounded on L^1, if p_n is a sequence of polynomials converging to $f \in H^1$, it follows that $\mathbb{E}(p_n|\mathcal{F}_\phi) \to \mathbb{E}(f|\mathcal{F}_\phi)$ in L^1. But

$$\mathbb{E}(p_n|\mathcal{F}_\phi)(\zeta) = \int_{\mathbb{T}} p_n(z)\, d\sigma_\zeta(z),$$

and it follows from (4) that the last integral is an analytic polynomial in ζ of the same degree as p_n. This condition is equivalent to the assertion that the conditional expectation commutes with the Riesz projection. Of course the preservation of analyticity follows easily from the fact that conditional expectation is the orthogonal projection of L^2 onto the linear span of the powers of ϕ (both positive and negative), but we emphasize the use of spectral measures as this point of view leads to interesting consequences (see Section 5). In 1986 Aleksandrov [1] had proved the remarkable result that all such conditional expectations arise in this way.

THEOREM 7 (Aleksandrov). *If $\mathcal{F}$ is a complete sub-σ-algebra of the σ-algebra of Lebesgue measurable sets on $\mathbb{T}$, then $\mathcal{F} = \mathcal{F}_\phi$ for some inner function ϕ if and only if the corresponding conditional expectation operator $\mathbb{E}(\cdot|\mathcal{F})$ commutes with the Riesz projection.*

We note the following related observation, whose proof is an easy observation based on the above theorem. In the statement of the theorem, the third condition means that if $f(\alpha) \in L^1$ is the boundary function of an analytic function then so is the function $Tf(\alpha) = \int f(\zeta)\, d\tau_\alpha(\zeta)$.

THEOREM 8. *Let $\{\tau_\alpha\}_{\alpha \in \mathbb{T}}$ be a family of Borel probability measures on $\mathbb{T}$ satisfying the following conditions.*

    (i) *the measures τ_α have pairwise disjoint carriers,*
    (ii) *Aleksandrov's decomposition formula (4) holds,*
    (iii) *the family $\{\tau_\alpha\}_{\alpha \in \mathbb{T}}$ preserves analyticity.*

Then all the measures τ_α are singular, and there is an inner function ϕ vanishing at the origin such that $\{\tau_\alpha\}_{\alpha \in \mathbb{T}}$ is the family of spectral measures for ϕ.

PROOF. Let $\{E_\alpha\}_{\alpha \in T}$ be a pairwise disjoint set of carriers for the measures τ_α, and let $\mathcal{F}$ be the smallest σ-algebra with respect to which all the E_α's are measurable. $\square$

PROBLEM. *Can the condition that the measures have pairwise disjoint carriers be eliminated in the above theorem?*

4. Generalized factorization

Spectral measures present an effective tool for describing a certain type of factorization of Hardy functions associated with a given inner function. In this section we present the problems which lead to this factorization and state appropriate results.

Let φ be a bounded analytic function. We want to describe the lattice of closed subspaces of H^p invariant under the multiplication by φ. When $\varphi(z) = z$ and $p = 2$, this lattice was described in a famous 1949 work of Beurling [6]. This was extended to the case where φ is a finite Blaschke product by Lax [21] in 1957. In 1965 Srinivasan and Wang extended Beurling's result to the H^p spaces with $p \neq 2$. In 1961 Halmos [15] proved a general theorem about isometries in a Hilbert space, which leads to an answer to the multiplication problem in the case when $p = 2$ and φ is an arbitrary inner function.

Halmos's theorem (called the Wold decomposition theorem) states the following. Let X be a Hilbert space and $T : X \to X$ be an isometry. The subspace $X_1 = X \ominus TX$ is called the wandering subspace, and the space $X_0 = \bigcap_{n=1}^{\infty} T^n X$ is called the stable subspace. Then

$$X = X_0 \oplus \bigoplus_{n=0}^{\infty} T^n X_1.$$

In the special case when $X = H^2$ and T is multiplication by an inner function φ, the stable subspace is obviously trivial, and Halmos's theorem implies that every closed φ-invariant subspace M of H^2 is generated by $M \ominus TM$, and each generator t satisfies the condition

$$(6) \qquad \int_{\mathbb{T}} |t(z)|^2 \varphi(z)^k \, dm(z) = 0, \quad k = 1, 2, ...,$$

where $dm(z)$ is normalized Lebesgue measure on the unit circle $\mathbb{T}$. Functions of norm one in H^2 which satisfy (6) are called φ-2-inner. It also follows that if φ is a finite Blaschke product, then the number of generators does not exceed the order of φ (this is Lax's theorem).

For $p \neq 2$ the Hardy space H^p does not have a Hilbert space structure, and, therefore, Halmos's theorem is not applicable. Still it is natural to conjecture that some analogs of φ-2-inner functions generate every closed φ-invariant subspace of H^p. For the case of finite Blaschke products this conjecture was proved by Lance and Stessin [20]. The result uses the following factorization which generalizes the classical Riesz inner-outer factorization of Hardy functions.

We call a function h of H^p-norm one φ-p inner if

$$\int_{\mathbb{T}} |h(z)|^p \varphi(z)^k \, dm(z) = 0, \quad k = 1, 2, ...$$

It was proved in [20] that every H^p-function f is uniquely (up to a unimodular constant factor) represented in the form

$$f = h_p \, F \circ \varphi,$$

where h_p is φ-p inner and F a classical outer function in H^p.

The following result holds [20].

THEOREM 9. *If φ is a finite Blaschke product of order n and $p \geq 1$, then any φ-invariant subspace M has a basis over $H^p[\varphi]$ consisting of at most n φ-p-inner functions. That is there are $k \leq n$ φ-p inner functions $h_1, ..., h_k$ in M such that every $f \in M$ can be uniquely expressed as*

$$f(z) = \sum_{l=1}^{k} h_l(z) f_l \circ \varphi(z),$$

where $f_1, ..., f_k \in H^p$.

We note in passing that no similar result is known for infinite Blaschke products, or, more generally, for arbitrary inner functions. The general problem of finding the right form of the characterization of φ-invariant subspaces of H^p remains open.

We also remark that generalized factorization naturally appears in the problem of characterizing the commutant of an analytic Toeplitz operator (see [**45**]).

Both the φ-p inner and outer parts of a function can be characterized in terms of spectral measures associated with φ.

THEOREM 10 (Stessin, Zhu [**45**]). *Let* $0 < p < \infty$. *Then* $h \in H^p$ *is* φ-p *inner if and only if*

$$\int_{\mathbb{T}} |h(z)|^p \, d\sigma_w(z) = 1$$

for almost all $w \in \mathbb{T}$.

PROOF. By Aleksandrov's theorem

$$\int_{\mathbb{T}} |h(z)|^p \varphi(z)^k \, dm(z) = \int_{\mathbb{T}} w^k \left(\int_{\mathbb{T}} |h(z)|^p \, d\sigma_w(z) \right) dm(w).$$

Now the result follows from the uniqueness theorem for Fourier coefficients. □

THEOREM 11 (Stessin, Zhu [**45**]). *Let* $0 < p < \infty$, $f \in H^p$, *and* $f = h\, F \circ \varphi$ *be the* φ-p *factorization of* f. *Then for all* $z \in \mathbb{T}$

$$(7) \qquad F(z) = \exp \left\{ \frac{1}{p} \int_{\mathbb{T}} \frac{\zeta + z}{\zeta - z} \log \left(\int_{\mathbb{T}} |f(\tau)|^p \, d\sigma_\zeta(\tau) \right) dm(\zeta) \right\}.$$

We remark that when $\varphi(z) = z$, σ_w is a point mass at w of mass 1 and the expression in the last theorem is just the classical Riesz formula.

Further investigation of generalized factorization was made in [**8**]. It follows from (11) that for every $f \in H^p$ ($1 \le p \le \infty$) and for every $r \le p$ the φ-r outer part of f is in H^p. It turned out that the φ-p inner part of the factorization does not enjoy this regularity. Indeed, the following result holds.

THEOREM 12 (Cima, Kazas, Stessin [**8**]). *Let* φ *be an inner function such that for every* $w \in \mathbb{T}$ *the spectral measure* σ_w *is continuous (that is the measure of each single point is zero). Then there exists an* H^∞ *function* f *such that for every* $0 < p \le \infty$ *the* φ-p *inner part of* f *is not in* L^q *for any* $q > p$.

5. Operators similar to a contraction

Let H be a Hilbert space. An operator $T \colon H \to H$ is a *contraction* if $\|T\| \le 1$, and it is *similar to a contraction* if there is a linear isomorphism $C \colon H \to H$ such that $\|C^{-1}TC\| \le 1$. The celebrated inequality of von Neumann [**48**] says that if T is a contraction, then $\|p(T)\| \le \|p\|_\infty$ for every polynomial p, where $\|p\|_\infty = \sup\{ |p(z)| \colon z \in \mathbb{D} \}$. This shows that $p(T)$ is a contraction for every p such that $\|p\|_\infty \le 1$.

An operator $T \colon H \to H$ is *polynomially bounded* if there is a constant C such that $\|p(T)\| \le C\|p\|_\infty$. Halmos raised the question as to whether every polynomially bounded operator is similar to a contraction. An operator T is *completely polynomially bounded* if there is a constant C such that for every $n \times n$ matrix $P(z)$ with polynomial entries, we have $\|P(T)\| \le C \sup_{z \in \mathbb{D}} \|P(z)\|$, where the norm on the left is the norm of $P(T)$ as an operator on H^n, the direct sum of n copies of H, and the norm on the right is the norm of the matrix $P(z)$ acting on $\mathbb{C}^n$. Arveson [5] generalized von Neumann's inequality by showing that that if T is a contraction, then $\|P(T)\| \leq \sup_{z \in \mathbb{D}} \|P(z)\|$. In 1984, Paulsen [32] showed that every completely polynomially bounded operator is similar to a contraction. In 1996 Pisier [33] produced a polynomially bounded operator which is not similar to a contraction, thus answering Halmos's question in the negative. A good source for all of this is the recent book of Vern Paulsen [31].

Halmos [14] and others showed that T is similar to a contraction whenever T^n is similar to a contraction for some positive integer n. If ϕ is an inner function which is analytic on an open set containing the spectrum of T, then the functional calculus produces the operator $\phi(T)$. R. G. Douglas asked the question as to whether T is similar to a contraction if $\phi(T)$ is similar to a contraction for some inner function. Mascioni [24] proved that the result holds if ϕ is a finite Blaschke product. An alternate proof of Mascioni's result was given by Lance and Stessin [20] who used estimates related to the Wold decomposition of H^2 for the multiplication operator M_ϕ. Neither of these proofs works for an infinite Blaschke product. Douglas had suggested the use of the Wold decomposition for this problem and had conjectured that the result should hold for a wide class of infinite Blaschke products. Stessin [44] established the conjecture for a large class of infinite Blaschke products with the following theorem.

THEOREM 13 (Stessin). *Let $\phi(z)$ be a Blaschke product whose zeros (a_n) satisfy*

$$(8) \qquad \sum_{n=1}^{\infty} (1 - |a_n|^2)^{1/2} < \infty.$$

Then T is similar to a contraction whenever $\phi(T)$ is similar to a contraction.

The use of spectral measures is essential for the proof of this theorem. Let $H^2[\phi]$ denote the closure of the algebra of polynomials in ϕ, where ϕ is a Blaschke product with zeros (a_n). The Wold decomposition of H^2 with respect to the isometry of multiplication by ϕ shows that any $f \in H^2$ can be written in the form

$$(9) \qquad f = \sum_{n=1}^{\infty} s_n f_n \circ \phi,$$

where $f_n \circ \phi \in H^2[\phi]$, and the functions s_n, given by

$$s_n(z) = \frac{\overline{a_n}}{|a_n|} \cdot \frac{\sqrt{1 - |a_n|^2}}{1 - \overline{a_n}z} \prod_{k=1}^{n-1} \frac{\overline{a_k}}{|a_k|} \frac{a_k - z}{1 - \overline{a_k}z}$$

form a special orthonormal basis for the wandering subspace $H^2 \ominus \phi H^2$. This decomposition gives rise to the component operators $Q_n \colon H^2 \to H^2$, where $f_n = Q_n(f)$ in the decomposition.

In order to show that T is similar to a contraction, it is enough, by Paulsen's theorem, to show that T is completely polynomially bounded. To this end, the above decomposition can be applied to the matrix function $P(z)$ to obtain

$$P(T) = \sum_{n=1}^{\infty} s_n(T) P_k(\phi(T)),$$

where now the P_k's are matrix functions. This immediately yields the estimate

$$\|P(T)\| = \sum_{n=1}^{\infty} \|s_n(T)\| \, \|P_k(\phi(T))\|.$$

Since the spectrum of T is at a positive distance from the poles of ϕ, it follows from the special form of the s_n's that $s_n(T)$ satisfies an estimate of the form

$$(10) \qquad\qquad \|s_n(T)\| \leq C\sqrt{1 - |a_n|^2}$$

On the other hand, since $\phi(T)$ is similar to a contraction, Arveson's theorem shows that

$$\|P_k(\phi(T))\| \leq c\sup_{z \in \mathbb{D}} \|P_k\|,$$

so it is important to estimate the last quantity. As it is easy to reduce this to the one-dimensional case, it will be enough to estimate the norms of the component operators Q_n in the supremum norm. Stessin provides the following estimates.

THEOREM 14 (Stessin). *Let* $2 \leq p \leq \infty$. *Then*

$$\|Q_n\|_{p \to p} \leq \left(\frac{1 + |\phi(0)|}{1 - |\phi(0)|}\right)^{5/2}$$

for each positive integer n.

We outline the proof of this theorem since it depends fundamentally on Aleksandrov's decomposition theorem. Let $\mathbb{E}(\cdot|\phi)$ denote the conditional expectation with respect to the σ-algebra generated by ϕ. If $\{\sigma_\alpha\}$ is the family of spectral measures for ϕ, it follows from Aleksandrov's theorem that

$$(11) \qquad\qquad \frac{1 - |\phi(0)|^2}{|\alpha - \phi(0)|^2} f_0(\alpha) = \int_{\mathbb{T}} f(\zeta)\, d\sigma_\alpha(\zeta)$$

for $f \in L^2$, where f_0 is given by $f_0 \circ \phi = \mathbb{E}(f|\phi)$. For each $a \in \mathbb{D}$, define the operator $R_a \colon H^1 \to H^1$ by

$$R_a f(z) = (1 - |a|^2)\frac{\dfrac{f(z)}{1 - \overline{a}z} - \dfrac{f(a)}{1 - |a|^2}}{z - a}.$$

It is straightforward to estimate

$$\|R_a f\|_p \leq \frac{(1 + |a|)^{2 - 1/p}}{(1 - |a|)^{1 + 1/p}} \|f\|_p$$

for $f \in H^p$ with $f(0) = 0$. Also let t_n be given by

$$t_n(z) = \frac{\overline{a_n}}{|a_n|}\frac{z\sqrt{1 - |a_n|^2}}{1 - \overline{a_n}z}\prod_{k=n+1}^{\infty}\frac{\overline{a_k}}{|a_k|}\frac{a_k - z}{1 - \overline{a_k}z}$$

On $\mathbb{T}$ the function t_n coincides with $-\phi\overline{s_n}$. If $f \in H^2$ and $f = \sum_{n=1}^{\infty} s_n f_n \circ \phi$, then $f_n = -R_{\phi(0)}\hat{f}_n$, where $\hat{f}_n \circ \phi = \mathbb{E}(f_n|\phi)$. This formula is established through an application of Aleksandrov's theorem. Now an application of (11) and Hölder's inequality shows that

$$|\hat{f}_n(\alpha)|^p \leq \left(\frac{1 + |\phi(0)|}{1 - |\phi(0)|}\right)^{1 + p/2}\int_{\mathbb{T}} |f(\zeta)|^p\, d\sigma_\alpha(\zeta).$$

Another application of Aleksandrov's theorem completes the proof.

Recently, Kazas and Kelley [**18**] have found a variation on Theorem 13. They showed that the conclusion holds for a Blaschke product ϕ whose zeros (a_n) lie in a Stolz angle and satisfy a growth estimate of the form

$$1 - |a_n| \sim \frac{1}{n^\beta},$$

where $1 < \beta < 2$. They also showed that the conclusion holds for a singular inner function ϕ given by a measure which is a finite combination of point masses.

References

1. A. B. Aleksandrov, *Measurable partitions of the circle induced by inner functions*, Zap. Nauchn. Sem. Leningrad. Otdel. Mat. Inst. Steklov. (LOMI) **149** (1986), no. Issled. Linein. Teor. Funktsii. XV, 103–106, 188. MR **0849298 (87i:**30065)

2. _____, *Multiplicity of boundary values of inner functions*, Izv. Akad. Nauk Armyan. SSR Ser. Mat. **22** (1987), no. 5, 490–503, 515. MR **0931885 (89e:**30058)

3. _____, *Inner functions and related spaces of pseudocontinuable functions*, Zap. Nauchn. Sem. Leningrad. Otdel. Mat. Inst. Steklov. (LOMI) **170** (1989), no. Issled. Linein. Oper. Teorii Funktsii. 17, 7–33, 321. MR **1039571 (91c:**30063)

4. _____, *On the existence of angular boundary values of pseudocontinuable functions*, Zap. Nauchn. Sem. S.-Peterburg. Otdel. Mat. Inst. Steklov. (POMI) **222** (1995), no. Issled. po Linein. Oper. i Teor. Funktsii. 23, 5–17, 307. MR **1359992 (97a:**30046)

5. William B. Arveson, *Subalgebras of C^*-algebras*, Acta Math. **123** (1969), 141–224. MR 0253059 (40 #6274)

6. Arne Beurling, *On two problems concerning linear transformations in Hilbert space*, Acta Math. **81** (1948), 17. MR 0027954 (10,381e)

7. Robert George Blumenthal, *Holomorphically closed algebras of analytic functions*, Math. Scand. **34** (1974), 84–90. MR 0380423 (52 #1323)

8. Joseph A. Cima, Angeliki Kazas, and Michael I. Stessin, *Growth estimates for generalized factors of H^p spaces*, Studia Math. **158** (2003), no. 1, 19–38. MR 2014549

9. Joseph A. Cima and Alec Matheson, *Cauchy transforms and composition operators*, Illinois J. Math. **42** (1998), no. 1, 58–69. MR **1492039 (98k:**42028)

10. Joseph A. Cima and Alec L. Matheson, *Essential norms of composition operators and Aleksandrov measures*, Pacific J. Math. **179** (1997), no. 1, 59–64. MR **1452525 (98e:**47047)

11. Douglas N. Clark, *One dimensional perturbations of restricted shifts*, J. Analyse Math. **25** (1972), 169–191. MR 0301534 (46 #692)

12. J. L. Doob, *Measure theory*, Graduate Texts in Mathematics, vol. 143, Springer-Verlag, New York, 1994. MR **1253752 (95c:**28001)

13. R. G. Douglas, H. S. Shapiro, and A. L. Shields, *Cyclic vectors and invariant subspaces for the backward shift operator.*, Ann. Inst. Fourier (Grenoble) **20** (1970), no. fasc. 1, 37–76. MR 0270196 (42 #5088)

14. P. R. Halmos, *Ten problems in Hilbert space*, Bull. Amer. Math. Soc. **76** (1970), 887–933. MR 0270173 (42 #5066)

15. Paul R. Halmos, *Shifts on Hilbert spaces*, J. Reine Angew. Math. **208** (1961), 102–112. MR 0152896 (27 #2868)

16. Marc J. Jaffrey, Timothy L. Lance, and Michael I. Stessin, *Submodules of the Hardy space over polynomial algebras*, Pacific J. Math. **194** (2000), no. 2, 373–392. MR **1760788 (2001e:**46097)

17. J. Jones, *Generators of the disk algebra*, Ph.D. thesis, Brown University, 1977.

18. Angeliki Kazas and Amy Kelley, *Clark measures and operators similar to a contraction*, preprint.

19. Thomas Kriete and Jennifer Moorhouse, *Linear relations in the calkin algebra for composition operators*, preprint.

20. T. L. Lance and M. I. Stessin, *Multiplication invariant subspaces of Hardy spaces*, Canad. J. Math. **49** (1997), no. 1, 100–118. MR **1437202 (97m:**30048)

21. Peter D. Lax, *Translation invariant spaces*, Acta Math. **101** (1959), 163–178. MR 0105620 (21 #4359)

22. J. E. Littlewood, *On inequalities in the theory of functions*, Proc. London Math. Soc. **23** (1925), 481–519.

23. G. R. MacLane and F. B. Ryan, *On the radial limits of Blaschke products*, Pacific J. Math. **12** (1962), 993–998. MR 0145080 (26 #2615)

24. Vania Mascioni, *Ideals of the disc algebra, operators related to Hilbert space contractions and complete boundedness*, Houston J. Math. **20** (1994), no. 2, 299–311. MR **1283278** **(95e:47023)**

25. Alec Matheson and Michael I. Stessin, *Cauchy transforms of characteristic functions and algebras generated by inner functions*, Proc. Amer. Math. Soc. **133** (2005), no. 11, 3361–3370. MR 2161161 to appear.

26. Alec L. Matheson, *Aleksandrov operators as smoothing operators*, Illinois J. Math. **45** (2001), no. 3, 981–998. MR **1879248** **(2002j:47059)**

27. Jennifer Moorhouse, *Compact differences of composition operators*, J. Funct. Anal. **219** (2005), no. 1, 70–92. MR 2108359

28. Jennifer Moorhouse and Carl Toews, *Differences of composition operators*, Trends in Banach spaces and operator theory (Memphis, TN, 2001), Contemp. Math., vol. 321, Amer. Math. Soc., Providence, RI, 2003, pp. 207–213. MR **1978818** **(2004b:47046)**

29. Rolf Nevanlinna, *Remarques sur le lemme de Schwarz*, C. R. Acad. Sci. Paris **188** (1929), 1027–1029.

30. Pekka J. Nieminen and Eero Saksman, *On compactness of the difference of composition operators*, J. Math. Anal. Appl. **298** (2004), no. 2, 501–522. MR **2086972** **(2005e:30065)**

31. Vern Paulsen, *Completely bounded maps and operator algebras*, Cambridge Studies in Advanced Mathematics, vol. 78, Cambridge University Press, Cambridge, 2002. MR **1976867** **(2004c:46118)**

32. Vern I. Paulsen, *Every completely polynomially bounded operator is similar to a contraction*, J. Funct. Anal. **55** (1984), no. 1, 1–17. MR **0733029** **(86c:47021)**

33. Gilles Pisier, *A polynomially bounded operator on Hilbert space which is not similar to a contraction*, J. Amer. Math. Soc. **10** (1997), no. 2, 351–369. MR **1415321** **(97f:47002)**

34. Alexei G. Poltoratski, *On the distributions of boundary values of Cauchy integrals*, Proc. Amer. Math. Soc. **124** (1996), no. 8, 2455–2463. MR **1327037** **(96j:30057)**

35. A. G. Poltoratskiĭ, *Boundary behavior of pseudocontinuable functions*, Algebra i Analiz **5** (1993), no. 2, 189–210. MR **1223178** **(94k:30090)**

36. Donald Sarason, *Composition operators as integral operators*, Analysis and partial differential equations, Lecture Notes in Pure and Appl. Math., vol. 122, Dekker, New York, 1990, pp. 545–565. MR **1044808** **(92a:47040)**

37. _____, *Weak compactness of holomorphic composition operators on H^1*, Functional analysis and operator theory (New Delhi, 1990), Lecture Notes in Math., vol. 1511, Springer, Berlin, 1992, pp. 75–79. MR **1180749** **(93h:47041)**

38. _____, *Sub-Hardy Hilbert spaces in the unit disk*, University of Arkansas Lecture Notes in the Mathematical Sciences, 10, John Wiley & Sons Inc., New York, 1994. MR **1289670** **(96k:46039)**

39. J. H. Shapiro and P. D. Taylor, *Compact, nuclear, and Hilbert-Schmidt composition operators on H^2*, Indiana Univ. Math. J. **23** (1973/74), 471–496. MR 0326472 (48 #4816)

40. Joel H. Shapiro, *The essential norm of a composition operator*, Ann. of Math. (2) **125** (1987), no. 2, 375–404. MR **0881273** **(88c:47058)**

41. Joel H. Shapiro and Carl Sundberg, *Compact composition operators on L^1*, Proc. Amer. Math. Soc. **108** (1990), no. 2, 443–449. MR **994787** **(90d:47035)**

42. Jonathan E. Shapiro, *Aleksandrov measures used in essential norm inequalities for composition operators*, J. Operator Theory **40** (1998), no. 1, 133–146. MR **1642538** **(99i:47062)**

43. N. Sibony and J. Wermer, *Generators for $A(\Omega)$*, Trans. Amer. Math. Soc. **194** (1974), 103–114. MR 0419838 (54 #7856)

44. M. I. Stessin, *Wold decomposition of the Hardy space and Blaschke products similar to a contraction*, Colloq. Math. **81** (1999), no. 2, 271–284. MR **1715351** **(2000g:47036)**

45. Michael Stessin and Kehe Zhu, *Generalized factorization in Hardy spaces and the commutant of Toeplitz operators*, Canad. J. Math. **55** (2003), no. 2, 379–400. MR **1969797** **(2004a:30033)**

46. Michael I. Stessin and Pascal J. Thomas, *Algebras generated by two bounded holomorphic functions*, J. Anal. Math. **90** (2003), 89–114. MR **2001066** **(2004g:46070)**

47. Michael I. Stessin and Kehe Zhu, *Composition operators induced by symbols defined on a polydisk*, (2005), preprint.
48. Johann von Neumann, *Eine Spektraltheorie für allgemeine Operatoren eines unitären Raumes*, Math. Nachr. **4** (1951), 258–281. MR 0043386 (13,254a)
49. J. Wermer, *Subalgebras of the disk algebra*, Colloque d'Analyse Harmonique et Complexe, Univ. Aix-Marseille I, Marseille, 1977, p. 7 pp. (not consecutively paged). MR **0565008 (81e:**46035)
50. John Wermer, *Rings of analytic functions*, Ann. of Math. (2) **67** (1958), 497–516. MR 0096817 (20 #3299)

DEPARTMENT OF MATHEMATICS, LAMAR UNIVERSITY, BEAUMONT, TX 77710
E-mail address: matheson@math.lamar.edu

DEPARTMENT OF MATHEMATICS AND STATISTICS, UNIVERSITY AT ALBANY, ALBANY, NY 12222
E-mail address: stessin@math.albany.edu

Contemporary Mathematics
Volume **393**, 2006

Parametrizing Distinguished Varieties

Jim Agler and John E. McCarthy

This paper is dedicated to Joseph Cima on the occasion of his 70th birthday.

ABSTRACT. A distinguished variety is a variety that exits the bidisk through the distinguished boundary. We look at the moduli space for distinguished varieties of rank (2,2).

0. Introduction

In this paper, we shall be looking at a special class of bordered algebraic varieties that are contained in the bidisk $\mathbb{D}^2$ in $\mathbb{C}^2$.

DEFINITION 0.1. *A non-empty set V in $\mathbb{C}^2$ is a* distinguished variety *if there is a polynomial p in $\mathbb{C}[z, w]$ such that*

$$V = \{(z, w) \in \mathbb{D}^2 : p(z, w) = 0\}$$

and such that

$$(0.2) \qquad \overline{V} \cap \partial(\mathbb{D}^2) = \overline{V} \cap (\partial\mathbb{D})^2.$$

Condition (0.2) means that the variety exits the bidisk through the distinguished boundary of the bidisk, the torus. We shall use ∂V to denote the set given by (0.2): topologically, it is the boundary of V within the zero set of p, rather than in all of $\mathbb{C}^2$.

In [**1**], the authors studied distinguished varieties, which we considered interesting because of the following two theorems:

THEOREM 0.3. *Let T_1 and T_2 be commuting contractive matrices, neither of which has eigenvalues of modulus 1. Then there is a distinguished variety V such that, for any polynomial p in two variables, the inequality*

$$\|p(T_1, T_2)\| \leq \|p\|_V$$

holds.

2000 *Mathematics Subject Classification.* Primary 14H15.

The first author was partially supported by National Science Foundation Grant DMS 0400826.

The second author was partially supported by National Science Foundation Grant DMS 0501079.

THEOREM 0.4. *The uniqueness variety for a minimal extremal Pick problem on the bidisk contains a distinguished variety V that contains each of the nodes.*

It is the goal of this paper to examine the geometry of distinguished varieties more closely, and in particular to parametrize the space of all distinguished varieties of rank $(2,2)$ (see Definition 0.5 below).

Notice that if V is a distinguished variety, for each z in the unit disk $\mathbb{D}$, the number of points w satisfying $(z,w) \in V$ is constant (except perhaps at a finite number of multiple points, where the w's must be counted with multiplicity). So the following definition makes sense:

DEFINITION 0.5. *A distinguished variety is of rank (m,n) if there are generically m sheets above every first coordinate and n above every second coordinate.*

The principal result of this paper, Theorem 2.1, is a parametrization of distinguished varieties of rank $(2,2)$.

1. Structure theory

For positive integers m and n, let

$$(1.1) \qquad U = \begin{pmatrix} A & B \\ C & D \end{pmatrix} \; : \; \mathbb{C}^m \oplus \mathbb{C}^n \; \to \; \mathbb{C}^m \oplus \mathbb{C}^n$$

be an $(m+n)$-by-$(m+n)$ unitary matrix. Let

$$(1.2) \qquad \Psi(z) = A + zB(I - zD)^{-1}C$$

be the m-by-m matrix valued function defined on the unit disk $\mathbb{D}$ by the entries of U. This is called the *transfer function* of U. Let

$$U' = \begin{pmatrix} D^* & B^* \\ C^* & A^* \end{pmatrix} \; : \; \mathbb{C}^n \oplus \mathbb{C}^m \; \to \; \mathbb{C}^n \oplus \mathbb{C}^m,$$

and let

$$\Psi'(w) = D^* + wB^*(I - wA^*)^{-1}C^*.$$

Because $U^*U = I$, a calculation yields

$$(1.3) \qquad I - \Psi(z)^*\Psi(z) = (1 - |z|^2)\, C^*(I - \bar{z}D^*)^{-1}(I - zD)^{-1}C,$$

so $\Psi(z)$ is a rational matrix-valued function that is unitary on the unit circle. Such functions are called rational matrix inner functions, and it is well-known that all rational matrix inner functions have the form (1.2) for some unitary matrix decomposed as in (1.1) — see *e.g.* [**2**] for a proof. The set

$$(1.4) \qquad V = \{(z,w) \in \mathbb{D}^2 \; : \; \det(\Psi(z) - wI) = 0\}$$

$$(1.5) \qquad = \{(z,w) \in \mathbb{D}^2 \; : \; \det(\Psi'(w) - zI) = 0\}$$

$$(1.6) \qquad = \{(z,w) \in \mathbb{D}^2 \; : \; \det\begin{pmatrix} A - wI & zB \\ C & zD - I \end{pmatrix} = 0\}$$

is a distinguished variety, because when $|z| = 1$, the eigenvalues of $\Psi(z)$ are unimodular (and a similar statement holds for Ψ'). The converse was proved in [**1**]: all distinguished varieties of rank (m,n) can be represented in this way. So the moduli space for distinguished varieties of rank (m,n) is a quotient of the space of $(m+n)$-by-$(m+n)$ unitaries. Let us write $\mathcal{U}_n^m$ to denote the set of $(m+n)$-by-$(m+n)$ unitaries decomposed as in (1.1). The following result is well-known.

PROPOSITION 1.7. *Let U and U_1 be in $\mathcal{U}_n^m$, with respective decompositions*

$$U = \begin{pmatrix} A & B \\ C & D \end{pmatrix} \qquad U_1 = \begin{pmatrix} A_1 & B_1 \\ C_1 & D_1 \end{pmatrix}.$$

Then they give rise to the same transfer function if and only if there is an n-by-n unitary W such that

(1.8) $$\begin{pmatrix} A_1 & B_1 \\ C_1 & D_1 \end{pmatrix} = \begin{pmatrix} I & 0 \\ 0 & W^* \end{pmatrix} \begin{pmatrix} A & B \\ C & D \end{pmatrix} \begin{pmatrix} I & 0 \\ 0 & W \end{pmatrix}.$$

PROOF: By looking at the coefficients of powers of z in the transfer function, we see that U and U_1 have the same transfer function if and only if

(1.9) $$\begin{aligned} A &= A_1 \\ BD^nC &= B_1 D_1^n C_1 \quad \forall\, n \in \mathbb{N}. \end{aligned}$$

Equation (1.8) is equivalent to

$$\begin{aligned} A_1 &= A \\ B_1 &= BW \\ C_1 &= W^*C \\ D_1 &= W^*DW. \end{aligned}$$

Clearly these equations imply (1.9).

To see the converse, note that the fact that U and U_1 are unitaries and $A = A_1$ means $BB^* = B_1 B_1^*$. If B is invertible, define W to be $B^{-1}B_1$. This is unitary since B and B_1 have the same absolute values, and then the equations $BC = B_1 C_1$ and $BDC = B_1 D_1 C_1$ yield (1.8).

If B is not invertible, then A has norm one. Decompose

$$\begin{pmatrix} A & B \\ C & D \end{pmatrix} = \begin{pmatrix} \begin{pmatrix} A' & 0 \\ 0 & A'' \end{pmatrix} & \begin{pmatrix} 0 \\ B'' \end{pmatrix} \\ \begin{pmatrix} 0 & C'' \end{pmatrix} & D \end{pmatrix},$$

and apply the same argument to B'' and C''. $\qquad\square$

Remark 1.10 Note that W is unique unless $\|A\| = 1$.

2. Parametrizing distinguished varieties of rank $(2,2)$

In this section, we address the question of when two different unitaries in U_2^2 give rise to the same distinguished variety. From the previous section we see that this is equivalent to asking when two rational matrix inner functions are isospectral.

THEOREM 2.1. *Let U, Ψ and V be as in (1.1), (1.2) and (1.4), with U in U_2^2. Let*

$$U_0 = \begin{pmatrix} A_0 & B_0 \\ C_0 & D_0 \end{pmatrix}$$

be another unitary in U_2^2. Then U and U_0 give rise to the same distinguished variety iff

(i) A and A_0 have the same eigenvalues.
(ii) D and D_0 have the same eigenvalues.
(iii) BC and B_0C_0 have the same trace.

PROOF: For simplicity in the proof we will assume that $\det(A) \neq 0$ and that A and D both have two eigenvalues. (We can attain the remaining cases as a limit of these.)

Let

$$
\begin{aligned}
(2.2) \qquad Q(z,w) &= \det \begin{pmatrix} A - wI & zB \\ C & zD - I \end{pmatrix} \\
(2.3) \qquad &= \frac{\det D}{\det A^*} \det \begin{pmatrix} D^* - z & wB^* \\ C^* & wA^* - I \end{pmatrix} \\
(2.4) \qquad &= p_2(z)w^2 + p_1(z)w + p_0(z) \\
(2.5) \qquad &= q_2(w)z^2 + q_1(w)z + q_0(w),
\end{aligned}
$$

where p_i and q_j are polynomials of degree at most 2. As V is the zero set of Q, it is sufficient to prove that conditions (i) — (iii) completely determine Q. Let μ_1 and μ_2 be the eigenvalues of D and λ_1 and λ_2 be the eigenvalues of A.

We have

$$
\begin{aligned}
p_2(z) &= \det(zD - I) \\
&= (z\mu_1 - 1)(z\mu_2 - 1),
\end{aligned}
$$

so is determined by (ii), the eigenvalues of D. Similarly $q_2(w)$ is determined by (i), the eigenvalues of A.

From (2.3) we see that the coefficient of z^2 in Q is $(\det D / \det A^*)$. Dividing (2.4) by p_2, we get

$$
(2.6) \qquad \det(\Psi(z) - wI) = w^2 + \frac{p_1(z)}{p_2(z)}w + \frac{p_0(z)}{p_2(z)}.
$$

As Ψ is a matrix inner function, we must have that the last term in (2.6), which is the product of the eigenvalues of Ψ, is inner. Therefore

$$
p_0(z) = e^{i\theta} (z - \overline{\mu_1})(z - \overline{\mu_2}),
$$

where

$$
e^{i\theta} = (\det D / \det A^*).
$$

It remains to determine p_1.

LEMMA 2.7. *With notation as above, let*

$$
(2.8) \qquad \det(\Psi(z) - wI) = w^2 - a_1(z)w + a_0(z).
$$

Then

$$
(2.9) \qquad a_1(z) = a_0(z)\overline{a_1(\tfrac{1}{z})}.
$$

PROOF: For any fixed z in $\mathbb{D}$, there are two w's with (z,w) in V. The function $a_0(z)$ is the product of these w's, and $a_1(z)$ is their sum. Labelling them (locally) as $w_1(z)$ and $w_2(z)$, the right-hand side of (2.9) is

$$
(w_1(z)w_2(z)) \, (\overline{w_1(\tfrac{1}{z})} + \overline{w_2(\tfrac{1}{z})}).
$$

When the modulus of z is 1, because the variety is distinguished, the right-hand side of (2.9) equals the left-hand side. By analytic continuation, they must be equal everywhere.   □

Applying the lemma to $-p_1/p_2$ and p_0/p_2, we get

(2.10)
$$p_1(z) = e^{i\theta} z^2 \overline{p_1(\frac{1}{\bar{z}})}.$$

Writing

$$p_1(z) = b_2 z^2 + b_1 z + b_0,$$

(2.10) gives the two equations

$$e^{i\theta} \overline{b_2} = b_0$$
$$e^{i\theta} \overline{b_1} = b_1.$$

Comparing (2.4) and (2.5), the coefficient of $z^2 w$ gives us b_2 (since we know q_2), and hence we also know b_0.

Finally, if we know the coefficient of zw in the power series expansion of (2.6), we will know

$$b_1 p_2(0) - b_0 p_2'(0),$$

and be done. But

$$\Psi(z) - wI = A - wI + zBC + O(z^2),$$

so the coefficient of zw is $-\mathrm{tr}(BC)$, which is given by (iii).   □

3. Open problems

Two distinguished varieties are *geometrically equivalent* if there is a biholomorphic bijection between them.

Question 3.1 When do two unitaries give rise to geometrically equivalent distinguished varieties?

Notice that all distinguished varieties of rank $(1, n)$ or $(m, 1)$ are geometrically equivalent, since they are all biholomorphic to the unit disk.

Question 3.2 When are two distinguished varieties of rank $(2, 2)$ geometrically equivalent?

Question 3.3 What is the generalization of Theorem 2.1 to distinguished varieties of rank $(2, 3)$ or $(3, 3)$?

W. Rudin showed that smoothly bounded planar domains are geometrically equivalent to distinguished varieties iff their connectivity is 0 or 1 [3].

Question 3.4 Which distinguished varieties are geometrically equivalent to planar domains?

Question 3.5 How can one read the topology of a distinguished variety from a unitary that determines it as in Section 1?

References

[1] J. Agler and J.E. MᶜCarthy. Distinguished varieties. *Acta Math.* To appear.

[2] J. Agler and J.E. MᶜCarthy. *Pick Interpolation and Hilbert Function Spaces.* American Mathematical Society, Providence, 2002.

[3] W. Rudin. Pairs of inner functions on finite Riemann surfaces. *Trans. Amer. Math. Soc.*, 140:423–434, 1969.

(J. AGLER) UNIVERSITY OF CALIFORNIA, SAN DIEGO, LA JOLLA, CALIFORNIA 92093

(J.E. MᶜCARTHY) WASHINGTON UNIVERSITY, ST. LOUIS, MISSOURI 63130

Contemporary Mathematics
Volume **393**, 2006

Approximation by CR Functions on the Unit Sphere in $\mathbb{C}^2$

John T. Anderson and John Wermer

Dedicated to Joe Cima on the occasion of his seventieth birthday

ABSTRACT. For a smoothly bounded relatively open subset Ω of the unit sphere in $\mathbb{C}^2$ we derive, using a kernel $H(\zeta, z)$ introduced by G. Henkin, an analogue of the Cauchy-Green formula in the plane:

$$\phi(z) = A(z) + \frac{1}{4\pi^2} \int_\Omega \overline{\partial}\phi(\zeta) H(\zeta, z) \omega(\zeta) - \frac{1}{4\pi^2} \int_{\partial\Omega} \phi(\zeta) H(\zeta, z) \omega(\zeta), \ z \in \Omega$$

valid for $\phi \in C^1(\overline{\Omega})$, where A is a CR function on Ω and $\omega(\zeta) = d\zeta_1 \wedge d\zeta_2$. We employ this formula to study rational approximation on compact subsets K of S, by using it to estimate the distance in $C(K)$ of ϕ to the CR functions on a neighborhood Ω of K. This requires an examination of the integral over $\partial\Omega$ appearing in the above formula, which we denote by $F_\Omega(z)$; in some circumstances we can show that F_Ω also defines a CR function on Ω, and thereby estimate the distance of ϕ to the CR functions on Ω in terms of $X(\phi)$, where X is the tangential Cauchy-Riemann operator on S. For certain K we can show that $R(K) = C(K)$ by showing that the distance of $\overline{z}_j$ to the CR functions on Ω tends to zero as Ω shrinks to K.

1. Introduction

Let D be a smoothly bounded domain in the complex plane and let f be a function in $C^1(\overline{D})$. The *Cauchy-Green formula* for D allows us to represent the value of f at a point $z \in D$ in terms of the values of f on ∂D and of the one-form $\overline{\partial} f$ on D:

$$(1.1) \qquad f(z) = \frac{1}{2\pi i} \int_{\partial D} \frac{f(\zeta)}{\zeta - z} d\zeta - \frac{1}{2\pi i} \int_D \overline{\partial} f(\zeta) \wedge \frac{d\zeta}{\zeta - z}.$$

We note the following consequences of (1.1). Let μ be a finite complex measure on $\mathbb{C}$, of compact support. The *Cauchy transform* $\hat{\mu}$ of μ is defined by

$$\hat{\mu}(\zeta) = \int \frac{d\mu(z)}{z - \zeta}, \ z \in \mathbb{C}.$$

Given a function $f \in C_0^1(\mathbb{C})$, we then have

$$(1.2) \qquad \int_{\mathbb{C}} f d\mu = \frac{1}{2\pi i} \int_{\mathbb{C}} \hat{\mu}(\zeta) \overline{\partial} f(\zeta) \wedge d\zeta.$$

2000 *Mathematics Subject Classification.* Primary 32A25, Secondary 32E30.

Equation (1.2) follows from (1.1) by letting D be a large disk, multiplying by μ and integrating over $\mathbb{C}$.

Given a compact set $K \subset \mathbb{C}$, $C(K)$ is the Banach algebra of continuous functions on K with norm $\|f\|_K = \max\{|f(z)| : z \in K\}$. For a smoothly bounded plane domain D, let $A(D)$ denote the space of functions $g \in C(\overline{D})$ with g holomorphic on D. For $f \in C^1(\partial D)$, set

$$F(z) = \frac{1}{2\pi i} \int_{\partial D} \frac{f(\zeta)}{\zeta - z} d\zeta, \quad z \in D.$$

Then F extends to $\overline{D}$ (see [**3**]) with $F \in C^1(\overline{D})$, and in particular $F \in A(D)$. The formula (1.1) gives, for $z \in D$,

$$(1.3) \quad |f(z) - F(z)| = \left| \frac{1}{2\pi i} \int_D \frac{\partial f}{\partial \overline{\zeta}}(\zeta) \frac{d\overline{\zeta} \wedge d\zeta}{\zeta - z} \right| \leq \frac{1}{\pi} \max_{\overline{D}} \left| \frac{\partial f}{\partial \overline{\zeta}} \right| \cdot \left[\int_D \frac{dm_2(\zeta)}{|\zeta - z|} \right],$$

where $m_2(\zeta)$ is two-dimensional Lebesgue measure. An inequality of Mergelyan (see [**5**]) states that

$$(1.4) \qquad\qquad \int_D \frac{dm_2(\zeta)}{|\zeta - z|} \leq 2\sqrt{\pi} \cdot \sqrt{m_2(D)}.$$

It follows by continuity of f and F on $\overline{D}$ that

$$(1.5) \quad \text{dist}(f, A(D)) \equiv \inf\{\|f - \Psi\|_{\overline{D}} : \psi \in A(D)\} \leq \frac{2}{\sqrt{\pi}} \cdot \max_{\overline{D}} \left| \frac{\partial f}{\partial \overline{\zeta}} \right| \sqrt{m_2(D)}.$$

For a compact set $K \subset \mathbb{C}$, let $R(K)$ be the closure in $C(K)$ of rational functions holomorphic in a neighborhood of K. As a corollary of (1.5) we obtain the following classical result.

THEOREM 1.1 (The Hartogs-Rosenthal Theorem). *Let K be a compact set in $\mathbb{C}$ with $m_2(K) = 0$. Then $R(K) = C(K)$.*

PROOF. Fix $f \in C^1(\mathbb{C})$. Choose a sequence of open, smoothly bounded sets $\{D_n\}$ decreasing to K. Fix $\epsilon > 0$. By (1.5), for each n there exists $F_n \in A(D_n)$ with

$$\|f - F_n\|_{\overline{D_n}} < \frac{2}{\sqrt{\pi}} \cdot \max_{\overline{D_n}} \left| \frac{\partial f}{\partial \overline{\zeta}} \right| \sqrt{m_2(D_n)} + \epsilon.$$

By Runge's Theorem, there exists $r_n \in R(K)$ such that $\|F_n - r_n\|_K < \epsilon$. It follows that

$$\|f - r_n\|_K < \frac{2}{\sqrt{\pi}} \cdot \max_{\overline{D_1}} \left| \frac{\partial f}{\partial \overline{\zeta}} \right| \sqrt{m_2(D_n)} + 2\epsilon.$$

Since $m_2(D_n) \to m_2(K) = 0$ as $n \to \infty$, we get $\|f - r_n\|_K < 3\epsilon$ for n sufficiently large. As ϵ was arbitrary, $f|_K \in R(K)$. Restrictions of functions in $C^1(\mathbb{C})$ to K are dense in $C(K)$, so $R(K) = C(K)$. $\qquad\square$

Our goal in this paper is to study the above situation when the complex plane is replaced by the unit sphere S in $\mathbb{C}^2$, and analytic functions on a domain in $\mathbb{C}$ are replaced by CR functions on a domain on S. Our work uses the kernel introduced by G. Henkin in [**6**]. Related integral formulas are given by Chen and Shaw in [**4**].

In section 2 we describe Henkin's kernel and the analogues of formulas (1.1) and (1.2) on the sphere S. In section 3 we give a Cauchy-Green formula for a smoothly bounded domain on S. The remainder of the paper is devoted to a study of the integrals appearing in this formula and applications to approximation results.

2. Henkin's kernel for S

We denote by $\mathbb{B}$ the open unit ball in $\mathbb{C}^2$, and let $S = \partial\mathbb{B}$ be the unit sphere. Points in $\mathbb{C}^2$ will normally be written as $z = (z_1, z_2)$ or $\zeta = (\zeta_1, \zeta_2)$, and the Hermitian inner product as $\langle \zeta, z \rangle \equiv \zeta_1 \bar{z}_1 + \zeta_2 \bar{z}_2$. The standard invariant three-dimensional measure on S is written σ (note: σ is not normalized). We will also frequently make use of the 2-form $\omega(\zeta) \equiv d\zeta_1 \wedge d\zeta_2$. We denote by $A(\mathbb{B})$ the *ball algebra* consisting of all functions continuous on the closure of $\mathbb{B}$ and holomorphic on $\mathbb{B}$. If Ω is an open subset of S, we say $g \in C^1(\Omega)$ is a *smooth CR function* on Ω, if $Xg = 0$ on Ω, where X is the tangential Cauchy-Riemann operator on S. Expressed in the coordinates of $\mathbb{C}^2$,

$$X = z_2 \frac{\partial}{\partial \bar{z}_1} - z_1 \frac{\partial}{\partial \bar{z}_2}.$$

Any function holomorphic in a neighborhood of Ω in $\mathbb{C}^2$ is a smooth CR function on Ω. Using the fact that for smooth ϕ

$$(2.1) \qquad \bar{\partial}\phi \wedge \omega = 2(X\phi)\, d\sigma$$

as measures on S, we see using Stokes' theorem that g is a smooth CR function on Ω if and only if

$$(2.2) \qquad \int_\Omega \bar{\partial}\phi \cdot g \wedge \omega = 0$$

for every $\phi \in C^\infty(\mathbb{C}^2)$ whose support meets S in a compact subset of Ω. We say that $g \in C(\Omega)$ is a *continuous CR function* on Ω, and write $g \in CR(\Omega)$, if (2.2) holds for all such ϕ. It follows easily from (2.2) that $CR(\Omega)$ is closed under uniform convergence on compact subsets of Ω. Since every $g \in A(\mathbb{B})$ is a uniform limit of polynomials on S, it follows that $g \in CR(S)$ if g is in the ball algebra and that $g \in A(\mathbb{B})$ is a smooth CR function on Ω whenever $g|_\Omega \in C^1(\Omega)$. For a compact $K \subset S$, let $CR(K)$ be the uniform closure in $C(K)$ of functions that are CR in some relatively open neighborhood of K in S.

As a replacement for the Cauchy kernel $(\zeta - z)^{-1}$ Henkin gave the kernel

$$H(\zeta, z) = \frac{\bar{\zeta}_1 \bar{z}_2 - \bar{\zeta}_2 \bar{z}_1}{|1 - \langle z, \zeta \rangle|^2}, \quad \zeta, z \in S.$$

On $S \times S, H$ is real-analytic off the diagonal $\{\zeta = z\}$, where $\langle z, \zeta \rangle = 1$. We consider as an analogue of the Cauchy transform $\hat{\mu}$ of a measure μ the transform

$$K_\mu(\zeta) = \int_S H(\zeta, z)\, d\mu(z), \quad \zeta \in S,$$

for a measure μ on S. The integral defining K_μ converges absolutely a.e-$d\sigma$ on S, and $K_\mu \in L^1(S, \sigma)$. Under the assumption that μ is a measure on S orthogonal to the restriction to S of every holomorphic polynomial on $\mathbb{C}^2$, Henkin proves

$$(2.3) \qquad \int_S \phi\, d\mu = \frac{1}{4\pi^2} \int_S \bar{\partial}\phi \wedge K_\mu \cdot \omega, \quad \phi \in C^1(S).$$

The orthogonality assumption on μ is necessary since the right-hand side of (2.3) vanishes if ϕ is the restriction to S of a polynomial.

We record below some useful properties of H and K_μ:

(i) $H(\zeta, z) = -H(z, \zeta)$;

(ii) If U is a unitary transformation of $\mathbb{C}^2$ with $\det(U) = 1$, $H(U\zeta, Uz) = H(\zeta, z)$;

(iii) $X[H(\zeta, z)] = -(1 - \langle z, \zeta \rangle)^{-2}$ (differentiation is in the ζ variable);

(iv) $K_\mu \in CR(S \setminus \text{supp}(\mu))$.

Properties (i) - (iii) are routine computations, while (iv) follows immediately from (2.3) and the definition (2.2) by taking by taking ϕ supported on an open set disjoint from $\text{supp}(\mu)$.

As an analogue of the Cauchy-Green formula (1.1), we have the following representation on S: given $\phi \in C^1(S)$, there exists a function $\Phi \in A(\mathbb{B})$ such that

$$(2.4) \qquad \phi(z) = \Phi(z) + \frac{1}{4\pi^2} \int_S \overline{\partial}\phi(\zeta) \wedge H(\zeta, z)\, \omega(\zeta), \quad z \in S.$$

This formula, which we call the *Cauchy-Green formula on S* follows directly from (2.3) as follows: set

$$K(z) = \int_S \overline{\partial}\phi(\zeta) \wedge H(\zeta, z)\, \omega(\zeta).$$

Choose a measure μ on S with μ orthogonal to polynomials. Then

$$\int_S K(z)\, d\mu(z) = \int_S \left[\int_S H(\zeta, z)\, d\mu(z) \right] \overline{\partial}\phi(\zeta) \wedge \omega(\zeta) = \int_S \overline{\partial}\phi(\zeta) \wedge K_\mu(\zeta)\, \omega(\zeta).$$

By (2.3),

$$\int_S (4\pi^2 \phi - K)\, d\mu = 0.$$

Since this holds for every μ orthogonal to polynomials, the Hahn-Banach Theorem implies that $4\pi^2 \phi - K$ belongs to the uniform closure of polynomials on S, and so is the restriction to S of a function in $A(\mathbb{B})$, giving (2.4). Other proofs of (2.4) are given in [4] and [1].

3. The Cauchy-Green formula for a domain on S

Let Ω^+ be a smoothly bounded domain on S, and fix a function $\phi \in C^1(\overline{\Omega^+})$. We wish to generalize formula (2.4) to this situation. Let Ω^- denote the complement of $\overline{\Omega^+}$ on S. Note that $\partial\Omega^- = -\partial\Omega^+$ as oriented manifolds.

THEOREM 3.1. *For $z \in \Omega^+$ we have*

$$\phi(z) = A(z) + \frac{1}{4\pi^2} \int_{\Omega^+} \overline{\partial}\phi(\zeta) \wedge H(\zeta, z)\, \omega(\zeta) - \frac{1}{4\pi^2} \int_{\partial\Omega^+} \phi(\zeta) H(\zeta, z)\, \omega(\zeta),$$

where $A \in CR(\Omega^+)$.

PROOF. We form a smooth extension $\tilde{\phi}$ of ϕ to S. By (2.4) there exists $\tilde{\Phi} \in A(\mathbb{B})$ such that for $z \in S$,

$$\tilde{\phi}(z) = \tilde{\Phi}(z) + \frac{1}{4\pi^2} \int_S \overline{\partial}\tilde{\phi}(\zeta) \wedge H(\zeta, z) \cdot \omega(\zeta)$$

and so

$$(3.1) \quad \tilde{\phi}(z) = \tilde{\Phi}(z) + \frac{1}{4\pi^2} \int_{\Omega^+} \overline{\partial}\tilde{\phi}(\zeta) \wedge H(\zeta, z)\, \omega(\zeta) + \frac{1}{4\pi^2} \int_{\Omega^-} \overline{\partial}\tilde{\phi}(\zeta) \wedge H(\zeta, z)\, \omega(\zeta).$$

We rewrite the last term of this equation as follows: for $z \in \Omega^+$ fixed, the function $\zeta \mapsto H(\zeta, z)$ is smooth on $\overline{\Omega^-}$, and so we may use Stokes' theorem to write

$$
\begin{aligned}
\int_{\partial\Omega^-} \tilde{\phi}(\zeta) H(\zeta, z)\, \omega(\zeta) &= \int_{\Omega^-} d[\tilde{\phi}(\zeta) H(\zeta, z)\, \omega(\zeta)] \\
&= \int_{\Omega^-} \overline{\partial}[\tilde{\phi}(\zeta) H(\zeta, z)\, \omega(\zeta)] \\
&= \int_{\Omega^-} \overline{\partial}\tilde{\phi}(\zeta) \wedge H(\zeta, z)\, \omega(\zeta) + \int_{\Omega^-} \tilde{\phi}(\zeta)\, \overline{\partial} H(\zeta, z) \wedge \omega(\zeta).
\end{aligned}
$$

Using (2.1) and property (iii) of the Henkin kernel the second integral on the right-hand side of the last equation can be expressed as

$$
-2 \int_{\Omega^-} \tilde{\phi}(\zeta) \frac{d\sigma(\zeta)}{(1 - \langle z, \zeta \rangle)^2}.
$$

Therefore,

$$
(3.2) \quad \int_{\Omega^-} \overline{\partial}\tilde{\phi}(\zeta) \wedge H(\zeta, z)\, \omega(\zeta) = \int_{\partial\Omega^-} \tilde{\phi}(\zeta) H(\zeta, z) \omega(\zeta) + 2 \int_{\Omega^-} \tilde{\phi}(\zeta) \frac{d\sigma(\zeta)}{(1 - \langle z, \zeta \rangle)^2},
$$

and so we may rewrite (3.1) as

$$
(3.3) \quad \tilde{\phi}(z) = A(z) + \frac{1}{4\pi^2} \int_{\Omega^+} \overline{\partial}\phi(\zeta) \wedge H(\zeta, z)\omega(\zeta) - \frac{1}{4\pi^2} \int_{\partial\Omega^+} \phi(\zeta) H(\zeta, z)\omega(\zeta)
$$

for $z \in \Omega^+$, where

$$
(3.4) \quad\quad\quad A(z) = \tilde{\Phi}(z) + \frac{1}{2\pi^2} \int_{\Omega^-} \tilde{\phi}(\zeta) \frac{d\sigma(\zeta)}{(1 - \langle z, \zeta \rangle)^2}.
$$

As noted above, since $\tilde{\Phi}(z) \in A(\mathbb{B})$, $\tilde{\Phi} \in CR(S)$, while the second term on the right-hand side of (3.4) is holomorphic in a neighborhood of Ω^+ in $\mathbb{C}^2$, and hence belongs to $CR(\Omega^+)$. This completes the proof. $\quad\square$

We refer to the formula appearing in Theorem 3.1 as the Cauchy-Green formula for Ω^+. Our goal in the remainder of this paper is to suggest how the formula of Theorem 3.1 can be used to derive approximation results for certain compact sets $K \subset S$, in much the same way that we employed the classical Cauchy-Green formula to derive the approximation results in section 1. We first note a simple corollary of Theorem 3.1; cf. the proof of the Hartogs-Rosenthal theorem in section 1.

COROLLARY 3.2. *Let K be a compact subset of S, and suppose for each $\epsilon > 0$, Ω_ϵ is a smoothly bounded domain in S containing K, with $\lim_{\epsilon \to 0^+} \sigma(\Omega_\epsilon) = 0$. For $\phi \in C^1(S)$ set*

$$
F_\epsilon(z) = \frac{1}{4\pi^2} \int_{\partial\Omega_\epsilon} \phi(\zeta) H(\zeta, z)\, \omega(\zeta), \quad z \in \Omega_\epsilon.
$$

If for each $\epsilon > 0$ there is a function $h_\epsilon \in CR(\Omega_\epsilon)$ so that $\lim_{\epsilon \to 0^+} \|F_\epsilon - h_\epsilon\|_K = 0$, then $\phi|_K \in CR(K)$.

PROOF. Apply the formula of Theorem 3.1 to the domain Ω_ϵ to obtain for $z \in \Omega_\epsilon$,

$$
(3.5) \quad \phi(z) = A(z) - h_\epsilon(z) - (F_\epsilon(z) - h_\epsilon(z)) + \frac{1}{4\pi^2} \int_{\Omega_\epsilon} \overline{\partial}\phi(\zeta) \wedge H(\zeta, z)\, \omega(\zeta)
$$

where $A - h_\epsilon \in CR(\Omega_\epsilon)$. By hypothesis, $\|F_\epsilon - h_\epsilon\|_K$ tends to zero as $\epsilon \to 0^+$, while the uniform integrability of H in z implies that the last term on the right of (3.5) tends to zero uniformly in z as $\epsilon \to 0^+$ (see the proof of Lemma 4.1 below), and so

$$\lim_{\epsilon \to 0^+} \|\phi - (A + h_\epsilon)\|_K = 0,$$

implying $\phi \in CR(K)$.   □

As Corollary 3.2 shows, the study of approximation by CR functions on compact subsets of S can be reduced to the study of CR approximation of integrals of the form

$$F_{\Omega,\phi}(z) = \int_{\partial\Omega} \phi(\zeta) H(\zeta, z) \, \omega(\zeta).$$

for smoothly bounded neighborhoods Ω of K. The remainder of this paper is devoted to an examination of such integrals. In section 4 we show that for certain Ω and ϕ, $F_{\Omega,\phi} \in CR(\Omega)$. In this case the formula of Theorem 3.1 immediately yields an estimate similar to (1.5). However, in general one cannot expect that F will be a CR function on Ω. In sections 5 and 6 we establish approximation results on certain compact subsets K of S with $\sigma(K) = 0$ by approximating $F_{\Omega_\epsilon,\phi}$ by CR functions on a sequence of domains Ω_ϵ shrinking to K, and applying Corollary 3.2.

Because the Cauchy-Green formula for domains on S lends itself to the study of approximation by CR functions, it is worthwhile to comment on the relationship between the space $CR(K)$ and the spaces $A(K), R(K)$ (consisting of uniform limits on K of functions holomorphic in neighborhood of K, and rational and holomorphic in a neighborhood of K, respectively.) Since any function holomorphic in a neighborhood U of K in $\mathbb{C}^2$ is a CR function on $U \cap S$, it follows that $A(K) \subset CR(K)$. On the other hand, if $\psi \in CR(\Omega)$ for some open neighborhood of K in S, it is well-known that there exists an open subset U of $\mathbb{B}$ containing Ω in its closure, and a function $\tilde\psi$ holomorphic in U and continuous on $\overline{U}$ with $\tilde\psi = \psi$ on Ω. There exists $t < 1$ so that $tz \in U$ for all $z \in K$, and so the function $z \mapsto \tilde\psi(tz)$ is holomorphic in a neighborhood of K in $\mathbb{C}^2$. As $t \to 1^-$, the functions $\tilde\psi(tz)$ approach ψ uniformly on K. It follows that $CR(K) \subset A(K)$, and hence $CR(K) = A(K)$. By the Stone-Weierstrass Theorem, $A(K) = C(K)$ if and only if the conjugate coordinate functions $\{\bar z_1, \bar z_2\}$ belong to $A(K)$; if (say) $z_2 \neq 0$ on $K \subset S$, then the relation $\bar z_2 = (1 - z_1 \bar z_1)/z_2$ shows that $A(K) = C(K)$ if and only if $\bar z_1 \in A(K)$.

Recall that a compact subset K of $\mathbb{C}^n$ is said to be *rationally convex* if given any point $z \in \mathbb{C}^n \setminus K$, there is a polynomial P with $P(z) = 0$ but $P \neq 0$ on K. Rational convexity is a necessary, but far from sufficient, condition for $R(K) = C(K)$ when $n > 1$. Richard Basener [2] constructed rationally convex subsets K of S for which $R(K) \neq C(K)$. Basener's sets have positive σ-measure. We have been motivated by the following analogue of the Hartogs-Rosenthal Theorem, for which we have no proof or counterexample.

CONJECTURE 3.3. *Let K be a rationally convex subset of S with $\sigma(K) = 0$. Then $R(K) = C(K)$.*

If K is rationally convex, then it can be shown that $A(K) = R(K)$, so that by the above remarks, rational approximation on K is equivalent to CR approximation.

4. A special case

LEMMA 4.1. *Given a domain $\Omega^+ \subset S$, and $\phi \in C^1(\overline{\Omega^+})$, suppose that*

$$F(z) = \int_{\partial\Omega^+} \phi(\zeta) H(\zeta, z)\, \omega(\zeta)$$

defines a CR function on Ω^+. Then for any compact subset K of Ω^+,

(4.1) $\qquad \mathrm{dist}(\phi, CR(K))) \equiv \inf\{\|\phi - \Psi\|_K : \Psi \in CR(K)\} \leq C_{\Omega^+} \cdot \|X(\phi)\|_{\overline{\Omega^+}}$

for a positive constant C_{Ω^+} independent of ϕ and K with the property that for every $\epsilon > 0$, there exists $\delta > 0$ so that $\sigma(\Omega^+) < \delta \implies C_{\Omega^+} < \epsilon$.

PROOF. By Theorem 3.1,

(4.2) $\quad \phi(z) = A(z) + \dfrac{1}{4\pi^2} \displaystyle\int_{\Omega^+} \overline{\partial}\phi(\zeta) \wedge H(\zeta, z)\omega(\zeta) - \dfrac{1}{4\pi^2} \int_{\partial\Omega^+} \phi(\zeta) H(\zeta, z)\omega(\zeta)$

with $A \in CR(\Omega^+)$. By hypothesis, the second integral on the right of (4.2) defines a function in $CR(S \setminus \partial\Omega^+)$ and so

$$\begin{aligned}
\phi(z) &= A_1(z) + \frac{1}{4\pi^2} \int_{\Omega^+} \overline{\partial}\phi(\zeta) \wedge H(\zeta, z)\omega(\zeta) \\
&= A_1(z) + \frac{1}{2\pi^2} \int_{\Omega^+} X(\phi)(\zeta) H(\zeta, z)d\sigma(\zeta)
\end{aligned}$$

with $A_1 \in CR(\Omega^+)$. Thus for any compact $K \subset \Omega^+$,

$$\mathrm{dist}(\phi, CR(K)) \leq \frac{1}{2\pi^2} \max_{\Omega^+} |X(\phi)| \int_{\Omega^+} |H(\zeta, z)|\, d\sigma(\zeta).$$

Set $C_{\Omega^+} = \sup_{z \in \Omega^+}(1/2\pi^2) \int_{\Omega^+} |H(\zeta, z)|\, d\sigma(\zeta)$. By the unitary invariance of H and $d\sigma$,

$$\int_{\Omega^+} |H(\zeta, z)|\, d\sigma(\zeta) = \int_{U_z(\Omega^+)} |H(\zeta, z^0)|\, d\sigma(\zeta) \leq \int_S |H(\zeta, z^0)| < \infty$$

where $z^0 = (1, 0)$ and U_z is a unitary transformation with $\det(U_z) = 1$ taking z to z^0. Thus C_{Ω^+} is finite and (4.1) holds. Moreover, since $H(\cdot, z^0) \in L^1(d\sigma)$, given $\epsilon > 0$ there exists $\delta > 0$ so that whenever Y is a measurable subset of S with $\sigma(Y) < \delta$, then $\int_Y |H(\zeta, z^0)|d\sigma(\zeta) < \epsilon$. The claim regarding C_{Ω^+} then follows by noting that for every z, $\sigma(\Omega^+) = \sigma(U_z\Omega^+)$. $\qquad\square$

REMARK 4.2. The corresponding estimate for $\Omega = S$ (without hypothesis on ϕ) is given in [1], Theorem 4.1.

Next we identify a special case in which the integral F appearing in Lemma 4.1 can be shown to be a CR function on Ω^+.

LEMMA 4.3. *Let Ω^+ be a smoothly bounded domain on S and let $\phi \in C^1(\overline{\Omega^+})$. Suppose there exists a smooth 3-manifold-with-boundary Σ in $\mathbb{C}^2$ with $\partial\Sigma = \partial\Omega^+$, and a function $\tilde{\phi}$ smooth on Σ and holomorphic in a neighborhood of $\Sigma \setminus \partial\Sigma$, with $\phi = \tilde{\phi}$ on $\partial\Omega^+$. Then*

$$F(z) = \int_{\partial\Omega^+} \phi(\zeta) H(\zeta, z)\, \omega(\zeta)$$

belongs to $CR(S \setminus \partial\Omega^+)$.

PROOF. Let μ be the measure $\phi\omega|_{\partial\Omega^+}$. For any polynomial P, by Stokes' theorem we have

$$\int_S P d\mu = \int_{\partial\Omega^+} P\phi\,\omega = \int_\Sigma d(P\tilde{\phi}\,\omega) = \int_\Sigma \bar{\partial}(P\tilde{\phi}\,\omega) = 0$$

by assumption on ϕ. Therefore, $K_\mu \in CR(S \setminus \partial\Omega^+)$. But

$$K_\mu(z) = \int_S H(\zeta, z)\,d\mu(\zeta) = -F(z).$$

By property (iv) of K_μ above (section 2), $K_\mu \in CR(S \setminus \mathrm{supp}(\mu)) = CR(S \setminus \partial\Omega^+)$, so $F \in CR(S \setminus \partial\Omega^+)$. $\qquad\square$

EXAMPLE 4.4. As a first example of the situation considered in Lemma 4.2, take

$$\Omega^+ = \{\zeta \in S : \mathrm{Im}(\zeta_2) > 0\}, \ \Sigma = \{z \in \bar{\mathbb{B}} : \mathrm{Im}(\zeta_2) = 0\}, \ \phi(\zeta) = \bar{\zeta}_2.$$

Then $\tilde{\phi}(\zeta) = \zeta_2$ extends ϕ from $\partial\Omega^+$ holomorphically to a neighborhood of Σ, and so

$$F(z) = \int_{\partial\Omega^+} \bar{\zeta}_2 H(\zeta, z)\,\omega(\zeta)$$

EXAMPLE 4.5. For $0 < r < 1$, let T_r be the torus

$$\{\zeta \in S : |\zeta_1| = r\}.$$

The three-manifold $\Sigma_r = \{\zeta \in \bar{\mathbb{B}} : |\zeta_1| = r\}$ has boundary T_r. Since for $\zeta \in T_r$

$$\bar{\zeta}_1 = \frac{r^2}{\zeta_1}, \ \bar{\zeta}_2 = \frac{1 - r^2}{\zeta_2},$$

while $\zeta_1, \zeta_2 \neq 0$ near Σ_r, both $\bar{\zeta}_1$ and $\bar{\zeta}_2$ extend holomorphically from T_r to a neighborhood of Σ_r. By Lemma 4.3, (taking say $\Omega^+ = \{\zeta \in S : |\zeta_1| > r\}$), $\int_{T_r} \bar{\zeta}_j H(\zeta, z)\omega(\zeta)$ is a CR function on $S \setminus T_r$, $j = 1, 2$. It follows that if $0 < a < b < 1$ and

$$\Omega = \{\zeta \in S : a < |\zeta_1| < b\},$$

then

$$\int_{\partial\Omega} \bar{\zeta}_j H(\zeta, z)\,\omega(\zeta) = \int_{T_b} \bar{\zeta}_j H(\zeta, z)\,\omega(\zeta) - \int_{T_a} \bar{\zeta}_j H(\zeta, z)\,\omega(\zeta)$$

defines a CR function on $S \setminus (T_a \cup T_b)$, $j = 1, 2$.

EXAMPLE 4.6. Let γ be a simple closed curve contained in the open unit disk of the complex plane, and let D be the region bounded by γ. Put

$$\Omega^+ = \{\zeta \in S : \zeta_1 \in D\}.$$

We will show that for any $\phi \in C(\gamma)$, $F(z) \in CR(S \setminus \partial\Omega^+)$, where

$$F(z) = \int_{\partial\Omega^+} \phi(\zeta_1)H(\zeta, z)\omega(\zeta).$$

For $\phi \in C(\gamma)$ and $\zeta \in \partial\Omega^+$, define $\tilde{\phi}(\zeta_1, \zeta_2) = \phi(\zeta_1)$. If Σ is the three-manifold

$$\Sigma = \{(\zeta_1, \zeta_2) : \zeta_1 \in \gamma, |\zeta_2| < \sqrt{1 - |\zeta_1|^2}\}$$

then $\partial\Sigma = \partial\Omega^+$. If $g \in C(\gamma)$ extends to be holomorphic in some neighborhood of γ, then $\tilde{g}$ extends to be holomorphic in a neighborhood of Σ, and so by Lemma 4.3

$$F_g(z) = \int_{\partial\Omega^+} \tilde{g}(\zeta)H(\zeta, z)\,\omega(\zeta)$$

defines a CR function on $S \setminus \partial\Omega^+$. But the functions holomorphic in a neighborhood of γ, restricted to γ, are dense in $C(\gamma)$, and so given any $\phi \in C(\gamma)$ we may choose a sequence $\{g_n\}$ of functions, each holomorphic in some neighborhood of γ, so that $\tilde{g}_n \to \tilde{\phi}$ uniformly on $\partial\Omega^+$. Then $F_{g_n} \to F$ uniformly on compact subsets of $S \setminus \partial\Omega^+$ and so $F \in CR(S \setminus \partial\Omega^+)$.

Note that by Lemma 4.1, each of the three preceding examples yields an approximation result for compact subsets of the respective domains Ω^+.

5. Certain two-spheres in S

For any $a, 0 < a < 1$, set $\lambda_a = \sqrt{1 - |a|^2}$ and let

$$S_a = \{(\zeta_1, \zeta_2) \in S : \mathrm{Im}(\zeta_2) = a\}$$

Note that S_a is a two-sphere in the hyperplane $\{\mathrm{Im}(\zeta_2) = a\}$ defined by $|\zeta_1|^2 + \mathrm{Re}(\zeta_2)^2 = \lambda_a^2$. Fix $a_0, 0 < a_0 < 1$, and fix t with $0 < t < \lambda_{a_0}$. Henceforth we assume without comment that the parameter a is sufficiently close to a_0 so that also $t < \lambda_a$. Then $\triangle_a = \{\zeta \in S_a : |\mathrm{Re}(\zeta_2)| \geq t\}$ is a nonempty subset of S_a, consisting of two disjoint sets each diffeomorphic to a closed disk, that are neighborhoods of $(0, \pm\lambda_a + ia)$ in S_a. Set $M_a = S_a \setminus \triangle_a$. Our goal in this section is to use the formula of Theorem 3.1 to establish the following result, which can also be obtained by using results on approximation on totally real manifolds (see [7], [8], or [9]).

THEOREM 5.1. *If K is any compact subset of M_{a_0}, $A(K) = C(K)$.*

PROOF. To begin, we may parameterize M_a by

$$(5.1) \qquad \zeta_1 = \sqrt{\lambda_a^2 - x^2}\, e^{i\theta}, \quad \zeta_2 = x + ia, \quad 0 \leq \theta \leq 2\pi, \ |x| < t.$$

Define

$$G_a(z) = \int_{M_a} \bar{\zeta}_1 H(\zeta, z)\omega(\zeta), \quad z \in S \setminus M_a.$$

We may use properties (i) and (iii) of the Henkin kernel from section 1 to compute $X(G_a)$, obtaining

$$(5.2) \qquad F_a(z) \equiv X(G_a)(z) = \int_{M_a} \frac{\bar{\zeta}_1 \omega(\zeta)}{(1 - \langle \zeta, z \rangle)^2}, \quad z \in S \setminus M_a.$$

Note that F_a is in fact defined for all $z \in \mathbb{B} \cup S \setminus M_a$, since for such z, $\langle z, \zeta \rangle \neq 1$ for $\zeta \in M_a$, and F_a is anti-holomorphic in $\mathbb{B}$.

LEMMA 5.2.

$$F_a(z) = 2\pi i \int_{-t}^{t} (x^2 - \lambda_a^2) \frac{dx}{(1 - (x + ia)\bar{z}_2)^2}, \quad z \in S \setminus M_a.$$

PROOF. Using the parametrization (5.1) we have

$$\omega = i\sqrt{\lambda_a^2 - x^2}\, e^{i\theta}\, d\theta \wedge dx$$

and so

$$(5.3) \qquad F_a(z) = \int_{-t}^{t} (\lambda_a^2 - x^2) \left(\int_{0}^{2\pi} \frac{id\theta}{(A - Be^{i\theta})^2} \right) dx,$$

where $A = 1 - (x + ia)\bar{z}_2$ and $B = \sqrt{\lambda_a^2 - x^2}\, \bar{z}_1$. If $|z| < 1/2$, then $|B| < 1/2$ and $|A| > 1/2$, and thus $A - B\tau \neq 0$ for $|\tau| \leq 1$. The inner integral in (5.3) can be computed by the residue theorem:

$$\int_0^{2\pi} \frac{i\,d\theta}{(A - Be^{i\theta})^2} = \int_{\tau=1} \frac{d\tau}{\tau} \frac{1}{(A - B\tau)^2} = \frac{2\pi i}{A^2}.$$

Thus

$$(5.4) \qquad F_a(z) = 2\pi i \int_{-t}^{t} (x^2 - \lambda_a^2) \frac{dx}{(1 - (x + ia)\bar{z}_2)^2}$$

for $|z| < 1/2$. Since both $F_a(z)$ and the right-hand side of (5.4) are real-analytic in $\mathbb{B}$, the equality of (5.4) in fact holds for $z \in \mathbb{B}$. Moreover, since $|x + ia|^2 = x^2 + a^2 \leq t^2 + a^2 < \lambda_a^2 + a^2 = 1$, both sides of (5.4) are continuous on $\overline{\mathbb{B}} \setminus M_a$. It follows that the equality of (5.4) obtains for $z \in \overline{\mathbb{B}} \setminus M_a$. $\qquad\square$

Define

$$I_a(z) = -\frac{2\pi i}{z_1} \int_{-t}^{t} \frac{(x^2 - \lambda_a^2)}{(x + ia)(1 - (x + ia)\bar{z}_2)}\, dx, \quad z \in S \setminus M_a,\ z_1 \neq 0.$$

Then I_a is smooth on $S \setminus M_a$ (note $|x + ia| \geq a > 0$), and a calculation gives

$$X(I_a) = F_a(z), \quad z \in S \setminus M_a$$

and therefore $X(I_a) = X(G_a)$ on $S \setminus M_a$. This yields:

LEMMA 5.3.

$$\int_{M_a} \bar{\zeta}_1 H(\zeta, z)\omega(\zeta) = I_a + \psi_a, \quad z \in S \setminus M_a$$

where $\psi_a \in CR(S \setminus M_a)$.

Now consider the domain Ω_ϵ, defined by

$$\Omega_\epsilon = \{\zeta \in S : a_0 - \epsilon < \mathrm{Im}(\zeta_2) < a_0 + \epsilon,\ |\mathrm{Re}(\zeta_2)| < t\}$$

where ϵ is chosen small enough so that $t < \lambda_a$ for all $a \in [a_0 - \epsilon, a_0 + \epsilon]$. Note

$$\partial\Omega_\epsilon = M_{a_0 - \epsilon} \cup M_{a_0 + \epsilon} \cup Y_t^\epsilon \cup Y_{-t}^\epsilon$$

with appropriate orientations, where

$$Y_{\pm t}^\epsilon = \{\zeta \in S : \mathrm{Re}(\zeta_2) = \pm t,\ a_0 - \epsilon < \mathrm{Im}(\zeta_2) < a_0 + \epsilon\},$$

and so by Lemma 5.3,

$$\int_{\partial\Omega_\epsilon} \bar{\zeta}_1 H(\zeta, z)\omega(\zeta) = h_\epsilon(z) + [I_{a_0+\epsilon}(z) - I_{a_0-\epsilon}(z)]$$

$$+ \int_{Y_t^\epsilon} \bar{\zeta}_1 H(\zeta, z)\omega(\zeta) - \int_{Y_{-t}^\epsilon} \bar{\zeta}_1 H(\zeta, z)\omega(\zeta)$$

for $z \in \Omega_\epsilon$, where $h_\epsilon \equiv \psi_{a_0-\epsilon} - \psi_{a_0+\epsilon} \in CR(\Omega_\epsilon)$. It is easy to check for $z \in M_{a_0}$ that $I_a(z)$ is a continuous function of a at a_0, so that

$$\lim_{\epsilon \to 0^+} I_{a_0+\epsilon}(z) - I_{a_0-\epsilon}(z) = 0, \quad z \in M_{a_0},$$

and in fact the limit is uniform on M_{a_0}. Moreover, restricting z to a compact subset K of M_{a_0}, the function $\zeta \mapsto \bar{\zeta}_1 H(\zeta, z)$ is bounded on $Y_{\pm t}^\epsilon$. Since the two-dimensional Hausdorff measure of Y_t approaches zero as $\epsilon \to 0$, and since ω is absolutely continuous with respect to Hausdorff measure, we see that

$$\lim_{\epsilon \to 0^+} \int_{Y_{\pm t}^\epsilon} \bar{\zeta}_1 H(\zeta, z) \omega(\zeta) = 0$$

uniformly for $z \in K$. Combining these observations, we have

$$\lim_{\epsilon \to 0^+} \int_{\partial \Omega_\epsilon} \bar{\zeta}_1 H(\zeta, z) \omega(\zeta) - h_\epsilon(z) = 0$$

uniformly on compact subsets of M_{a_0}. By Corollary 3.2, since $h_\epsilon \in CR(\Omega_\epsilon)$, $\phi(\zeta) = \bar{\zeta}_1 \in CR(K)$. Since $\zeta_2 \neq 0$ on K, we conclude that (see the remarks at the end of section 3) $A(K) = C(K)$, and the proof of Theorem 5.1 is complete. $\square$

6. Certain graphs in S

In this section we establish approximation on certain graphs in S, using the general method of section 5. Let D be a smoothly bounded plane domain with compact closure in the open disk $\mathbb{D}$, and suppose $f \in C^1(\overline{D})$ satisfies $|f(\zeta)| = \sqrt{1 - |\zeta|^2}$, so that the graph $\Gamma_f = \{(\zeta, f(\zeta)) : \zeta \in D\}$ lies in S. As in section 5, we will study the integral

$$\int_{\partial \Omega_\epsilon} \phi(\zeta) H(\zeta, z) \omega(\zeta)$$

on a sequence of domains Ω_ϵ for which $\Omega_\epsilon \downarrow \Gamma_f$ as $\epsilon \downarrow 0$. We begin with a representation for an integral of this type over Γ_f.

LEMMA 6.1. *With f, D, Γ_f as above, and $\phi \in C^1(\Gamma_f)$, set*

$$G(z) = \int_{\Gamma_f} \phi(\zeta) H(\zeta, z) \omega(\zeta)$$

for $z \in S \setminus \Gamma_f$. Then (differentiation is in the z variable)

$$\bar{z}_2 X G(z) = -\int_{\partial D} \tilde{\phi}(\zeta_1) \frac{d\zeta_1}{1 - \zeta_1 \bar{z}_1 - f(\zeta_1)\bar{z}_2} + \int_D \frac{\partial \tilde{\phi}}{\partial \bar{\zeta}_1} \frac{d\bar{\zeta}_1 \wedge d\zeta_1}{1 - \zeta_1 \bar{z}_1 - f(\zeta_1)\bar{z}_2},$$

where $\tilde{\phi}(\zeta_1) = \phi(\zeta_1, f(\zeta_1))$.

PROOF. Use properties (i) and (iii) of the Henkin kernel to write

$$\begin{aligned}
\bar{z}_2 X G(z) &= \bar{z}_2 \int_{\Gamma_f} \phi(\zeta)(1 - \langle \zeta, z \rangle)^{-2} \omega(\zeta) \\
&= \bar{z}_2 \int_D \tilde{\phi}(\zeta_1) \frac{d\zeta_1 \wedge (\partial f / \partial \bar{\zeta}_1) d\bar{\zeta}_1}{(1 - \zeta_1 \bar{z}_1 - f(\zeta_1)\bar{z}_2)^2} \\
&= \int_D \tilde{\phi}(\zeta_1) \frac{\partial}{\partial \bar{\zeta}_1} \left(\frac{1}{1 - \zeta_1 \bar{z}_1 - f(\zeta_1)\bar{z}_2} \right) d\zeta_1 \wedge d\bar{\zeta}_1
\end{aligned}$$

Rewrite the latter integral and use Stokes' Theorem to obtain

$$
\begin{aligned}
\overline{z}_2 X G(z) &= -\int_D d\left(\frac{\tilde{\phi}(\zeta_1)d\zeta_1}{1 - \zeta_1\overline{z}_1 - f(\zeta_1)\overline{z}_2}\right) + \int_D \frac{\partial\tilde{\phi}}{\partial\overline{\zeta}_1} \cdot \frac{d\overline{\zeta}_1 \wedge d\zeta_1}{1 - \zeta_1\overline{z}_1 - f(\zeta_1)\overline{z}_2} \\
&= -\int_{\partial D} \frac{\tilde{\phi}(\zeta_1)d\zeta_1}{1 - \zeta_1\overline{z}_1 - f(\zeta_1)\overline{z}_2} + \int_D \frac{\partial\tilde{\phi}}{\partial\overline{\zeta}_1} \cdot \frac{d\overline{\zeta}_1 \wedge d\zeta_1}{1 - \zeta_1\overline{z}_1 - f(\zeta_1)\overline{z}_2}.
\end{aligned}
$$

$\square$

We now restrict our attention to graphs of the following form. Fix r_0, r_1 with $0 < r_0 < r_1 < 1$ and let D be the annulus $\{\lambda \in \mathbb{C} : r_0 < |\lambda| < r_1\}$. For $\lambda \in D$ set

$$
(6.1) \qquad f(\lambda) = \sqrt{1 - |\lambda|^2}\, B\left(\frac{\lambda}{|\lambda|}\right)
$$

where B is a (fixed) finite Blaschke product:

$$
B(\lambda) = \prod_{j=1}^n \frac{\lambda - \alpha_j}{1 - \overline{\alpha}_j\lambda}
$$

with $|\alpha_j| < 1, j = 1, \ldots, n$. Denote by Γ_f the graph $\{(\lambda, f(\lambda)) : \lambda \in D\}$; note that since $|B(\tau)| = 1$ when $|\tau| = 1$, $\Gamma_f \subset S$.

THEOREM 6.2. *Assume f has the form (6.1). Then for any compact subset K of Γ_f, $A(K) = C(K)$.*

PROOF. For $t \in \mathbb{R}$ set $f_t(\lambda) = f(\lambda)e^{it}$. We consider first the integral

$$
G_t = \int_{\Gamma_{f_t}} \overline{\zeta}_1 H(\zeta, z)\, \omega(\zeta)
$$

LEMMA 6.3. *For $z \in S \setminus \Gamma_{f_t}$,*

$$
\overline{z}_2 X G_t(z) = 2\pi i\left(\frac{r_0^2}{1 - b_t(r_0)\overline{z}_2} - \frac{r_1^2}{1 - b_t(r_1)\overline{z}_2} + 2\int_{r_0}^{r_1} \frac{r\,dr}{1 - b_t(r)\overline{z}_2}\right)
$$

where

$$
b_t(r) = \sqrt{1 - r^2} \cdot e^{it}B(0).
$$

PROOF. According to Lemma 6.1,

$$
(6.2) \qquad \overline{z}_2 X G_t(z) = -\int_{\partial D} \overline{\zeta}_1 \frac{d\zeta_1}{1 - \zeta_1\overline{z}_1 - f_t(\zeta_1)\overline{z}_2} + \int_D \frac{d\overline{\zeta}_1 \wedge d\zeta_1}{1 - \zeta_1\overline{z}_1 - f_t(\zeta_1)\overline{z}_2}.
$$

Write the first integral on the right-hand side of (6.2) as $I_{r_1}(z) - I_{r_0}(z)$, where

$$
\begin{aligned}
I_r(z) &= i\int_0^{2\pi} \frac{r^2\, d\theta}{1 - re^{i\theta}\overline{z}_1 - \sqrt{1 - r^2}B(e^{i\theta})e^{it}\overline{z}_2} \\
&= \int_{|\tau|=1} \frac{r^2\, d\tau}{\tau(1 - r\tau\overline{z}_1 - \sqrt{1 - r^2}B(\tau)e^{it}\overline{z}_2)}.
\end{aligned}
$$

It is easy to check that $(1 - r\tau\overline{z}_1 - \sqrt{1 - r^2}B(\tau)\overline{z}_2)^{-1}$ is holomorphic for $|\tau| \leq 1$ and continuous for $|\tau| \leq 1$ for fixed $z \in S \setminus \Gamma_{f_t}$. Cauchy's integral formula then gives

$$
(6.3) \qquad I_r(z) = \frac{2\pi i r^2}{1 - \sqrt{1 - r^2}B(0)e^{it}\overline{z}_2} = \frac{2\pi i r^2}{1 - b_t(r)\overline{z}_2}.
$$

The second integral on the right-hand side of (6.2) can be evaluated as follows: write $\zeta_1 = re^{i\theta}$, then

$$
\int_D \frac{d\bar{\zeta}_1 \wedge d\zeta_1}{1 - \zeta_1 \bar{z}_1 - f_t(\zeta_1)\bar{z}_2} = 2 \int_{r_0}^{r_1} \left(\int_0^{2\pi} \frac{i\, d\theta}{1 - re^{i\theta}\bar{z}_1 - \sqrt{1 - r^2}B(e^{i\theta})e^{it}\bar{z}_2} \right) r\, dr
$$

$$
= 2 \int_{r_0}^{r_1} \left(\int_{|\tau|=1} \frac{d\tau}{\tau(1 - r\tau\bar{z}_1 - \sqrt{1 - r^2}B(\tau)e^{it}\bar{z}_2)} \right) r\, dr.
$$

It is easy to check that the function $(1 - r\tau\bar{z}_1 - \sqrt{1 - r^2}B(\tau)e^{it}\bar{z}_2)$ is nonvanishing for $|\tau| \leq 1$, if $z \in \mathbb{B}$ and so the Cauchy integral formula gives

$$
\int_D \frac{d\bar{\zeta}_1 \wedge d\zeta_1}{1 - \zeta_1 \bar{z}_1 - f_t(\zeta_1)\bar{z}_2} = 2 \int_{r_0}^{r_1} \frac{2\pi i}{1 - \sqrt{1 - r^2}B(0)e^{it}\bar{z}_2} r\, dr = 4\pi i \int_{r_0}^{r_1} \frac{r\, dr}{1 - b_t(r)\bar{z}_2}
$$

Again by continuity this result holds also for $z \in S \setminus \Gamma_{f_t}$. Combining (6.2), (6.3) and the last displayed equation gives the formula of Lemma 6.3. $\square$

We next proceed to construct a certain function M_t such that $XM_t = XG_t$ on $S \setminus \Gamma_t$. By Lemma 6.3,

$$
\frac{1}{2\pi i}XG_t(z) = \frac{1}{\bar{z}_2}\frac{r_0^2}{1 - b_t(r_0)\bar{z}_2} - \frac{1}{\bar{z}_2}\frac{r_1^2}{1 - b_t(r_1)\bar{z}_2} + 2 \int_{r_0}^{r_1} \frac{r\, dr}{\bar{z}_2(1 - b_t(r)\bar{z}_2)}
$$

$$
= r_0^2 \left(\frac{1}{\bar{z}_2} + \frac{b_t(r_0)}{1 - b_t(r_0)\bar{z}_2} \right) - r_1^2 \left(\frac{1}{\bar{z}_2} + \frac{b_t(r_1)}{1 - b_t(r_1)\bar{z}_2} \right)
$$

$$
- 2 \int_{r_0}^{r_1} \left(\frac{1}{\bar{z}_2} + \frac{b_t(r)}{1 - b_t(r)\bar{z}_2} \right) r\, dr.
$$

and so

$$
\frac{1}{2\pi i}XG_t(z) = \frac{r_0^2 b_t(r_0)}{1 - b_t(r_0)\bar{z}_2} - \frac{r_1^2 b_t(r_1)}{1 - b_t(r_1)\bar{z}_2} + 2 \int_{r_0}^{r_1} \frac{b_t(r)}{1 - b_t(r)\bar{z}_2} r\, dr.
$$

Note that

$$
X\left[\frac{1}{z_1} \log(1 - b_t(r)\bar{z}_2) \right] = \frac{b_t(r)}{1 - b_t(r)\bar{z}_2},
$$

and so if we set

$$
M_t(z) = \frac{2\pi i}{z_1} \left(r_0^2 \log(1 - b_t(r_0)\bar{z}_2) - r_1^2 \log(1 - b_t(r_1)\bar{z}_2) + 2 \int_{r_0}^{r_1} \log(1 - b_t(r)\bar{z}_2)\, r\, dr \right)
$$

we have

(6.4) $$XM_t = XG_t$$

on $S \setminus \Gamma_{f_t}$.

Given a compact subset K of Γ_f, for fixed $\epsilon > 0$ we define a neighborhood Ω_ϵ of K in S by

$$
\Omega_\epsilon = \{(\zeta_1, \zeta_2) : \zeta_1 \in D, \zeta_2 = f(\zeta_1)e^{it}, |t| < \epsilon\}.
$$

Note that $\{\Omega_\epsilon\}_{\epsilon>0}$ forms a decreasing family of domains with intersection Γ_f, and that

$$
\partial\Omega_\epsilon = \Gamma_{f_\epsilon} \cup \Gamma_{f_{-\epsilon}} \cup B_\epsilon
$$

where

$$
B_\epsilon = \{(\lambda, f(\lambda)e^{it}) : \lambda \in \partial D, |t| < \epsilon\}.
$$

Set

$$F_\epsilon(z) = \int_{\partial\Omega_\epsilon} \bar{\zeta}_1 H(\zeta, z)\omega(\zeta).$$

Then for $z \in \Omega_\epsilon$,

$$
\begin{aligned}
F_\epsilon(z) &= \int_{\Gamma_{f_\epsilon}} \bar{\zeta}_1 H(\zeta, z)\omega(\zeta) - \int_{\Gamma_{f_{-\epsilon}}} \bar{\zeta}_1 H(\zeta, z)\omega(\zeta) + \int_{B_\epsilon} \bar{\zeta}_1 H(\zeta, z)\omega(\zeta) \\
&= G_\epsilon(z) - G_{-\epsilon}(z) + \int_{B_\epsilon} \bar{\zeta}_1 H(\zeta, z)\omega(\zeta) \\
&= M_\epsilon(z) - M_{-\epsilon}(z) + h_\epsilon(z) + \int_{B_\epsilon} \bar{\zeta}_1 H(\zeta, z)\omega(\zeta),
\end{aligned}
$$

where $h_\epsilon = G_\epsilon - M_\epsilon - (G_{-\epsilon} - M_{-\epsilon}) \in CR(\Omega_\epsilon)$, by (6.4), and so

$$(6.5) \qquad F_\epsilon(z) - h_\epsilon(z) = M_\epsilon(z) - M_{-\epsilon}(z) + \int_{B_\epsilon} \bar{\zeta}_1 H(\zeta, z)\omega(\zeta).$$

Since the two-dimensional measure of B_ϵ tends to zero with epsilon it follows that (cf. the end of section 5)

$$\lim_{\epsilon \to 0^+} \int_{B_\epsilon} \bar{\zeta}_1 H(\zeta, z)\omega(\zeta) = 0$$

uniformly on K. Moreover, an examination of the definition of M_t shows that $\lim_{\epsilon \to 0^+} M_\epsilon - M_{-\epsilon} = 0$ uniformly on K. It follows by (6.5) that

$$\lim_{\epsilon \to 0^+} \|F_\epsilon - h_\epsilon\|_K = 0.$$

By Corollary 3.2, $\phi(\zeta) = \bar{\zeta}_1 \in CR(K)$. Since $\zeta_2 \neq 0$ on K, we conclude that (see the remarks at the end of section 3) $A(K) = C(K)$, and the proof of Theorem 6.2 is complete. $\qquad\square$

References

[1] J. T. Anderson and J. Wermer, *A Cauchy-Green Formula on the Unit Sphere in $\mathbb{C}^2$*, Proceedings of the Fourth Conference on Function Spaces, K. Jarosz, ed., A.M.S. Contemporary Mathematics Series 328, 2003, pp. 21–30.

[2] R. Basener, *On Rationally Convex Hulls*, Trans. Amer. Math. Soc. **182** (1973), pp. 353-381.

[3] S. Bell, "The Cauchy Transform, Potential Theory and Conformal Mapping," CRC Press, 1992.

[4] S.–C. Chen and M.–C. Shaw, "Partial Differential Equations in Several Complex Variables," American Mathematical Society, 2001.

[5] T. Gamelin and D. Khavinson, *The Isoperimetric Inequality and Rational Approximation*, Amer. Math. Monthly, **96** (1989), no. 1, pp. 18–30.

[6] G. M. Henkin, *The Lewy Equation and Analysis on Pseudoconvex Manifolds*, Russian Math. Surveys, **32:3** (1977); Uspehi Mat. Nauk **32:3** (1977), pp. 57–118.

[7] L. Hörmander and J. Wermer, *Uniform Approximation on Compact Subsets in $\mathbf{C}^n$*, Math. Scand **23** (1968), pp. 5-21.

[8] L. Nirenberg and R. O. Wells, *Holomorphic Approximation on Real Submanifolds of a Complex Manifold*, Bull. A.M.S. **73** (1967), pp. 378-381.

[9] A. J. O'Farrell, K.J. Preskenis, and D. Walsh, *Holomorphic Approximation in Lipschitz Norms*, in Proceedings of the Conference on Banach Algebras and Several Complex Variables, Contemporary Math. v. 32, American Mathematical Society, 1983.

DEPARTMENT OF MATHEMATICS AND COMPUTER SCIENCE, COLLEGE OF THE HOLY CROSS, WORCESTER, MA 01610-2395

E-mail address: anderson@mathcs.holycross.edu

DEPARTMENT OF MATHEMATICS, BROWN UNIVERSITY, PROVIDENCE, RI 02912

E-mail address: wermer@math.brown.edu

Contemporary Mathematics
Volume **393**, 2006

Superposition operators and the order and type of entire functions

Stephen M. Buckley and Dragan Vukotić

ABSTRACT. We distinguish between entire functions of different order or type by the behavior of their associated superposition operators on subsets of Besov spaces or the little Bloch space.

1. Introduction

Let X, Y be spaces of analytic functions on the unit disk $\mathbb{D} \subset \mathbb{C}$ which contain the polynomials. The *nonlinear superposition operator* S_φ on X (with *symbol* φ) is defined by

$$S_\varphi(f) = \varphi \circ f, \qquad f \in X.$$

We write $\varphi \in S(X, Y)$ if $S_\varphi(X) \subset Y$. It is immediate that $S(X, Y)$ is a vector space of entire functions. The following natural questions arise:

(1) Given spaces X, Y, what is $S(X, Y)$?
(2) Are there X, Y for which $S(X, Y)$ equals a given space of entire functions?

Similar problems in the context of real variables have a long history [AZ], but the above questions have only recently been studied in complex function theory. Cámera and Giménez answered (1) in the context of Bergman spaces and the Nevanlinna area class [CG] and Hardy spaces and related classes [Ca]. Together with J.L. Fernández [BFV], we answered (1) for various classes of Dirichlet-type spaces. Question (2) is most natural for spaces of entire functions φ defined in terms of a bound on the growth rate of φ, and we showed in [BFV] that many such spaces of entire functions were of the form $S(X, Y)$ for pairs of appropriate Dirichlet-type spaces X, Y. More precisely, we showed that this the case for the space of polynomials of degree at most $n \geq 0$ (see Corollary 10 and Theorem 12) and the spaces $E(\rho)$ for each $\rho > 1$ (see Corollary 25); here $E(\rho)$ is the space of functions of order less than ρ, or order ρ and finite type. Other related results on superposition operators from Bloch or Besov space to Bergman space were proved in [AMV] and [BV].

2000 *Mathematics Subject Classification.* 47H30, 31C25, 30H05.
This research of both authors was supported by MCyT grant BFM2003-07294-C02-01, Spain. The first author was also supported by Enterprise Ireland. The second author was also partially supported by MTM2004-21420-E (Acciones Especiales), MEC, Spain.

These earlier results involving $E(\rho)$ can be viewed as "tests" that distinguish between entire functions of different orders according to whether or not they map Dirichlet-type spaces to other Dirichlet-type or Bergman spaces. However none of these earlier results provide us with tests to distinguish between entire functions of the same order but different types. It is precisely this refinement that we examine in this note. In particular we show that superposition operators from Besov spaces $X = B^p$ or the little Bloch space $X = \mathcal{B}_0$ to Bergman spaces $Y = A^q$ can be used to distinguish between entire functions that have the same order $\rho \geq 1$ but different types. This distinction requires a more refined criterion than simply whether the operator S_φ maps all of one space into another. Instead the test will be whether $S_\varphi(f)$ lies in a fixed ball in Y for all f in some bounded subset of X.

Let us state here one such result which implies that superposition operators from Dirichlet space $\mathcal{D} = B^2$ to Bergman space A^q can be used in order to distinguish between entire functions of order 2 and distinct types. For a definition of the space $E(2, \tau)$, see Section 2; for now, we remark that $E(2, \tau)$ contains all functions whose order and type strictly precede $(2, \tau)$ lexicographically.

THEOREM 1.1. *Suppose* $0 < \tau, q < \infty$. *Let* $R := \sqrt{2/\tau q}$ *and* $K_R := \{f \in \mathcal{D} : f(0) = 0, \|f\|_\mathcal{D} \leq R\}$. *Then* $\varphi \in E(2, \tau)$ *if and only if* $S_\varphi(K_R)$ *is a bounded set in* A^q.

2. Background

Throughout the paper, $\mathbb{C}$ denotes the complex plane, $\mathbb{D}$ the unit disk $\{z \in \mathbb{C} : |z| < 1\}$, $\mathbb{T}$ the unit circle $\{z \in \mathbb{C} : |z| = 1\}$, and $dA(z) := \pi^{-1} \, dx dy = \pi^{-1} r \, dr d\theta$ is normalized area measure (a probability measure) on $\mathbb{D}$. A disk of radius r centered at w is denoted $D(w, r)$, while $[z, w]$ means the line segment from the point z to the point w.

Given two positive expressions A, B, we write $A \lesssim B$ or $B \gtrsim A$ to mean that $A \leq CB$ for some constant C dependent only on allowed parameters (which should be clear from the context). We write $C = C(A, B, \ldots)$ to indicate that C depends only on the parameters $A, B, \ldots$.

2.1. Order and type. Recall that the *order* of a non-constant entire function φ is

$$\rho := \limsup_{r \to \infty} \frac{\log \log M(r, \varphi)}{\log r},$$

where $M(r, \varphi) := \max\{|\varphi(z)| : |z| = r\}$. If φ has order $\rho < \infty$, then the *type* τ of φ is

$$\tau := \limsup_{r \to \infty} \frac{\log M(r, \varphi)}{r^\rho}.$$

See the first two chapters of [**Boa**] for more on entire functions. The *allowable (order-type) pairs* are (ρ, τ), $0 \leq \rho < \infty$, $0 \leq \tau \leq \infty$, together with the special pair $(\infty, *)$, where $*$ just means that type is undefined for infinite order functions. Let us define $\prec$ on this set of allowable pairs to be the *(strict) lexicographic order*. Thus $(\rho_1, \tau_1) \prec (\rho_2, \tau_2)$ if either $\rho_1 < \rho_2 \leq \infty$ or $\rho_1 = \rho_2 < \infty$ and $\tau_1 < \tau_2$.

Suppose (ρ, τ) is an allowable pair with $\rho < \infty$. We denote by $E'(\rho, \tau)$ the space of entire functions that grow more slowly than some entire function of order ρ and type τ. Equivalently, $\varphi \in E'(\rho, \tau)$ if φ is constant, has order less than ρ, or has order ρ and type at most τ. If $\tau < \infty$, we also define $E(\rho, \tau)$ to be the space of entire functions φ for which there exists a constant C such that $|\varphi(z)| \leq C \exp(\tau |z|^p)$, $z \in \mathbb{C}$. Note that $E'(\rho', \tau') \subset E(\rho, \tau) \subset E'(\rho, \tau)$ whenever $(\rho', \tau') \prec (\rho, \tau)$. Thus there is a close relationship between $E(\rho, \tau)$ and $E'(\rho, \tau)$, but for our purposes, $E(\rho, \tau)$ is much better suited to uniform norm estimates than $E'(\rho, \tau)$. We also write $E(\rho) := \bigcup_{\tau < \infty} E(\rho, \tau)$. It is convenient to define both $E(\infty, *)$ and $E'(\infty, *)$ to be the class E of all entire functions, and to define $E(t, \infty) = E(t)$. If we define the set-valued functions $f_1(\rho, \tau) = E(\rho, \tau)$ and $f_2(\rho, \tau) = E'(\rho, \tau)$ for all allowable pairs (ρ, τ), then f_1 and f_2 are increasing functions in the sense that if $(\rho', \tau') \prec (\rho, \tau)$ then $f_i(\rho', \tau') \prec f_i(\rho, \tau)$, $i = 1, 2$.

2.2. Hardy and Bergman spaces. We denote by H^p, $0 < p < \infty$, the well-known *Hardy space* of functions analytic in $\mathbb{D}$ for which

$$\|f\|_{H^p}^p := \sup_{0 < r < 1} \left(\int_0^{2\pi} |f(re^{i\theta})|^p \frac{d\theta}{2\pi} \right)^{1/p} < \infty.$$

H^p functions have radial limits $f(e^{i\theta})$ almost everywhere on $\mathbb{T}$. The *Bergman space* A^p, $0 < p < \infty$, is the space of functions analytic on $\mathbb{D}$ with

$$\|f\|_{A^p}^p := \int_{\mathbb{D}} |f(z)|^p \, dA(z) < \infty.$$

The spaces H^p and A^p are Banach spaces if $1 \leq p < \infty$.

The theory of Bergman spaces, including many recent developments, can be found in [**HKZ**] and [**DS**]. We mention two facts that we shall need. First

$$(2.1) \qquad (1 - |z|)^2 |f(z)|^p \leq \|f\|_{A^p}^p, \quad f \in A^p, \ z \in \mathbb{D},$$

as can be seen by applying the area version of the sub-mean value property to the subharmonic function $|f|^p$ on $D(z, 1 - |z|)$. Secondly, there exists $C = C(p)$ such that $\|f\|_{A^p} \leq C\|f\|_{H^{p/2}}$; this follows from [**D**, Theorem 9.4].

2.3. Besov, Dirichlet, and Bloch spaces. The *Besov space* B^p, $1 < p < \infty$, is the Banach space of functions analytic on $\mathbb{D}$ for which

$$\|f\|_{B^p} := |f(0)| + \left((p-1) \int_{\mathbb{D}} |f'(z)|^p (1 - |z|^2)^{p-2} \, dA(z) \right)^{1/p} < \infty.$$

The *Dirichlet space* is $\mathcal{D} := B^2$. The *Bloch space* $\mathcal{B}$ is the Banach space of all functions analytic in $\mathbb{D}$ for which

$$\|f\|_{\mathcal{B}} := |f(0)| + \sup_{z \in \mathbb{D}} (1 - |z|^2)|f'(z)| < \infty.$$

The *little Bloch space* $\mathcal{B}_0$ is the set of all $f \in \mathcal{B}$ such that $\lim_{|z| \to 1^-} (1 - |z|^2)|f'(z)| = 0$. It is the natural limit as $p \to \infty$ of B^p, as well as the closure of the polynomials in $\mathcal{B}$. Note that $\mathcal{D} \subset H^p$, $0 < p < \infty$ (see [**D**, Chapter 6, Exercise 7]), so functions in $\mathcal{D}$ have radial limits almost everywhere on $\mathbb{T}$. For more on these spaces, see for instance [**Z1**] and [**Z2**].

We shall need the functions

$$(2.2) \quad f_{r,p} := \begin{cases} \log(1/(1 - rz))(\log(1/(1 - r^2)))^{-1/p}, & 0 < r < 1, \ 1 < p < \infty, \\ \log((1 + rz)/(1 - rz)), & 0 < r < 1, \ p = \infty. \end{cases}$$

Then $f_{r,p} \in X^p$, if we define $X^p := B^p$ when $p < \infty$, and $X^\infty := \mathcal{B}_0$. Let $c_{r,p} := \|f_{r,p}\|_{X^p}$. Then $c_{r,\infty} \leq 2$. For fixed $p < \infty$, $c_{r,p}^p \to \Gamma(p - 1)/(\Gamma(p/2))^2$ as $r \to 1^-$. Here, Γ is the classical gamma function, and this limit can be seen by checking the proof of Theorem 1.7 in [**HKZ**] or of Lemma 4.2.2 of [**Z1**]. In particular, the class of all $f_{r,p}$, $0 < r < 1$, is a bounded subset of X^p. The functions $f_{r,2}$ are the well-known *Beurling functions*, and $\|f_{r,2}\|_{\mathcal{D}} = c_{r,2} = 1$ for all $0 < r < 1$.

3. Main results

If $1 < \rho < \infty$ and $0 < q < \infty$, then $E(\rho) = S(B^{\rho/(\rho-1)}, A^q)$; see [**BV**, Theorem 1]. The following theorem says that the action of S_φ on a certain bounded subset U_R of $B^{\rho/(\rho-1)}$ allows us to make finer distinctions between the classes $E(\rho, \tau)$.

THEOREM 3.1. *Suppose $1 < \rho < \infty$ and $0 < \tau, q < \infty$. Define $s := \rho/(\rho - 1)$, $R := (2/\tau q)^{1/\rho}$, and $U_R := \{cRf_{r,s} : 0 < r < 1,\ c \in \mathbb{T}\}$, where $f_{r,s}$ is as in (2.2). Let F be the space of entire functions φ such that $S_\varphi(U_R)$ is bounded in A^q. Then $\bigcup_{0 < \tau' < \tau} E(\rho, \tau') \subsetneq F \subset E(\rho, \tau)$.*

Theorem 1 of [**BV**] also says that $E(1) = S(\mathcal{B}_0, A^q)$ for all $0 < q < \infty$. Consequently, it is not surprising that the following analogue of Theorem 3.1 for $\rho = 1$ uses a bounded subset U_R of $\mathcal{B}_0$.

THEOREM 3.2. *Suppose $0 < \tau, q < \infty$. Define $R := 2/\tau q$, and $U_R := \{cRf_{r,\infty} : 0 < r < 1,\ c \in \mathbb{T}\}$, where $f_{r,\infty}$ is as in (2.2). Let F be the space of entire functions φ such that $S_\varphi(U_R)$ is bounded in A^q. Then $\bigcup_{0 < \tau' < \tau} E(1, \tau') \subsetneq F \subsetneq E(1, \tau)$.*

It would be nice if we could, as in Theorem 1.1, replace the set U_R in Theorem 3.1 by a ball in $B^s \cap \{f : f(0) = 0\}$. This would require a variant Trudinger inequality with sharp constants, i.e. a B^s version of the Chang-Marshall inequality. The appropriate type of variant Trudinger inequality was proven in [**BFV**, Theorem 23] (see also [**BO**]), but not with a sharp constant. We believe that both the sharp constant problem and such a variant of Theorem 3.1 are difficult open problems. Similar remarks apply to such a variant of Theorem 3.2. However the following partial results are easy corollaries of the proofs of Theorems 3.1 and 3.2.

COROLLARY 3.3. *Suppose $1 < \rho < \infty$ and $0 < \tau, q < \infty$. Define $s := \rho/(\rho-1)$, $R := \Gamma(s-1)^{1/s}\,\Gamma(s/2)^{-2/s}\,(2/\tau q)^{1/\rho}$, and $K_R := \{f \in B^s : f(0) = 0,\ \|f\|_{B^s} \leq R\}$. Let F be the space of entire functions φ such that $S_\varphi(K_R)$ is bounded in A^q. Then $F \subset E(\rho, \tau)$.*

COROLLARY 3.4. *Suppose $0 < \tau, q < \infty$. Define $R := 2/\tau q$ and $K_R := \{f \in \mathcal{B}_0 : f(0) = 0,\ \|f\|_{\mathcal{B}} \leq R\}$. Let F be the space of entire functions φ such that $S_\varphi(K_R)$ is bounded in A^q. Then $F \subset E(1, \tau)$.*

Theorem 1.1 allows us to distinguish between entire functions of order two and different types, and Theorems 3.1 and 3.2 allow us to distinguish between entire functions of different orders or types, at least when one of the orders lies in $[1, \infty)$. Perhaps these assertions deserve a little justification. Suppose that φ_i has order ρ_i and type τ_i, $i = 1, 2$, where either ρ_1 or ρ_2 lies in the interval $[1, \infty)$ and $(\rho_1, \tau_1) \prec (\rho_2, \tau_2)$. It follows that we can choose $1 \leq \rho < \infty$ and $0 < \tau' < \tau < \infty$ so that $(\rho_1, \tau_1) \prec (\rho, \tau') \prec (\rho, \tau) \prec (\rho_2, \tau_2)$. Then $\varphi_1 \in E(\rho, \tau')$ and so $\|\varphi_1 \circ f\|_{A^q}$ is uniformly bounded for all $f \in U_R$, where q, R, and U_R are as in either Theorem 3.1 or Theorem 3.2, depending on whether $1 < \rho < \infty$ or $\rho = 1$. On the other hand, $\varphi_2 \notin E(\rho, \tau)$ and so $\|\varphi_2 \circ f\|_{A^q}$ is not uniformly bounded.

Let us now prove each of our theorems.

Proof of Theorem 1.1. Suppose $\varphi \in E(2, \tau)$, and so $|\varphi(w)| \leq C \exp(\tau|w|^2)$, $w \in \mathbb{C}$. Recall that $\| \cdot \|_{A^q} \lesssim \| \cdot \|_{H^{q/2}}$. Thus if $f \in K_R$ and $g := f/R$, then

$$\left(\int_{\mathbb{D}} |\varphi \circ f|^q \, dA \right)^{1/2} \lesssim \int_{\mathbb{T}} |\varphi \circ f|^{q/2} \lesssim \int_{\mathbb{T}} \exp \left(\frac{\tau q}{2} |f|^2 \right) = \int_{\mathbb{T}} \exp(|g|^2).$$

But this last integral is uniformly bounded by the Chang-Marshall inequality [**CM,** Theorem 1] since $\|g\|_{\mathcal{D}} \leq 1$ and $g(0) = 0$.

Suppose instead that $\varphi \notin E(2, \tau)$, and so there exist a sequence of nonzero points (w_n) tending to infinity so that $|\varphi(w_n)| \geq n \exp(\tau|w_n|^2)$. Let r_n be the unique number in $(0, 1)$ for which

$$|w_n| = R f_{r_n, 2}(r_n) \equiv R \left[\log(1/(1 - r_n^2)) \right]^{1/2},$$

so that $r_n \to 1$ as $n \to \infty$. The Beurling functions $f_{r,2}$ all lie in K_1, and $g_n := R e^{i \arg w_n} f_{r_n, 2} \in K_R$ has the property that $g_n(r_n) = w_n$. Using (2.1), we see that

$$\begin{aligned}
\|\varphi \circ g_n\|_{A^q}^q &\geq (1 - |r_n|)^2 |\varphi(g_n(r_n))|^q \\
&\geq (1 - r_n)^2 \cdot n \exp \left(q\tau |w_n|^2 \right) \\
&= n(1 - r_n)^2 \exp \left(-q\tau R^2 \log(1 - r_n^2) \right) \\
&= n(1 - r_n)^2 (1 - r_n^2)^{-2}.
\end{aligned}$$

Since $r_n \to 1$, this last expression tends to infinity as $n \to \infty$. Thus $\|\varphi \circ g_n\|_{A^q}$ is not uniformly bounded. $\square$

It will actually be convenient for us to prove Theorem 3.2 first and then Theorem 3.1.

Proof of Theorem 3.2. Suppose $\varphi \in E(1, \tau')$ for some $\tau' < \tau$, and so $|\varphi(w)| \leq C \exp(\tau'|w|)$, $w \in \mathbb{C}$. Suppose also $g := c R f_{r, \infty}$ for some $c \in \mathbb{T}$, $0 < r < 1$. Then

$$|\varphi(g(z))|^q \leq C \left| \frac{1 + rz}{1 - rz} \right|^{\tau' R q}.$$

Now $\tau' R q = s$ for some $s < 2$, and it is easily verified that $|1 - rz|^{-s}$ is integrable on the unit disk, uniformly in r. It follows that $\varphi(U_R)$ is bounded is A^q.

We omit the proof that if $\varphi \notin E(1, \tau)$, then $S_\varphi(U_R)$ fails to be a bounded set, since it is very similar to the corresponding part of the proof of Theorem 1.1.

It remains to show that both containments are strict. The function $\varphi(z) := \exp(\tau z)$ lies in $E(1, \tau)$. Moreover, $\varphi(R f_{r, \infty}(z)) = [(1 + rz)/(1 - rz)]^{2/q}$, and it is readily verified that the A^q norm of this last function tends to infinity as $r \to 1^-$. To check this, just look at the Taylor series of the function inside the modulus sign and apply the standard norm formula for the Bergman space A^2 to get

$$\int_{\mathbb{D}} \left| \frac{1 + rz}{1 - rz} \right|^2 dA(z) = 1 + 4 \sum_{n=1}^{\infty} \frac{r^{2n}}{n + 1}.$$

The statement now follows from the monotone convergence theorem. Thus $\varphi \in E(1, \tau) \setminus F$.

For each $m \in \mathbb{N}$, the function

$$\varphi_m(z) := \sum_{n=m}^{\infty} \frac{\tau^n z^{n-m}}{n!}$$

is entire. Clearly, $|\varphi_m(z)| \leq |\exp(\tau|z|)|/|z|^m$ for $z \neq 0$, and $\varphi_m(z) \gtrsim \exp(\tau z)/z^m$ for all sufficiently large real z, so $\varphi_m \in E(1,\tau) \setminus \bigcup_{0<\tau'<\tau} E(1,\tau')$, $m \in \mathbb{N}$.

We claim that $|\varphi_m(cRf_{r,\infty}(z))|^q$ is integrable on $\mathbb{D}$, uniformly over all r and c, as long as $mq > 1$. Let us fix $0 < r < 1$, and partition $\mathbb{D}$ into annular sets S_j, $j \geq 0$, where $S_0 := \mathbb{D} \cap D(1, 2(1-r^2))$, and inductively,

$$S_j := \left[\mathbb{D} \cap D(1, 2^{j+1}(1-r^2))\right] \setminus S_{j-1}, \qquad j \in \mathbb{N}.$$

Let j_0 be the last integer for which $2^j(1-r^2) < 2$, and so S_j is empty for $j > j_0$. By considering separately the cases where z lies in the left and right half of $\mathbb{D}$, we see that $|1 - rz| \geq |1 - z|/2$. Since also $1 - 1/s = 1/\rho$, we get

$$|f_{r,\infty}(z)| \leq \log\left(\frac{4}{2^j(1-r^2)}\right), \qquad z \in S_j, \ 0 \leq j \leq j_0.$$

Let $g := cRf_{r,\infty}$ for some $c \in \mathbb{D}$. For $0 \leq j \leq j_0$, we have

$$I_j := \int_{S_j} |\varphi_m(g(z))|^q \, dA(z) \leq \left(\log\left(\frac{4}{2^j(1-r^2)}\right)\right)^{-mq} \cdot \left(\frac{4}{2^j(1-r^2)}\right)^{q\tau R} |S_j|.$$

where $|S_j|$ is the normalized area of S_j. Now $q\tau R = 2$ and $|S_j| \leq 2^{2(j+1)}(1-r^2)^2$, so $I_j \lesssim (j_1 + 1 - j)^{-mq}$. Thus $I := \sum_{j=0}^{j_0} I_j$ is dominated by a bounded multiple of the series $\sum_{i=n}^{\infty} n^{-mq}$, which converges for $mq > 1$, and so our claim follows. Thus $\varphi_m \in F \setminus \bigcup_{0<\tau'<\tau} E(1,\tau')$ whenever m is an integer exceeding $1/q$. $\square$

Proof of Theorem 3.1. Suppose $\varphi \in E(\rho,\tau')$ for some $\tau' < \tau$, and so $|\varphi(w)| \leq C\exp(\tau'|w|^\rho)$, $w \in \mathbb{C}$. Let us fix $0 < r < 1$, and define S_j, $j \geq 0$, and j_0 as in the last part of the proof of Theorem 3.2. Noting that $1 - 1/s = 1/\rho$, and arguing as before, we get

$$|f_{r,s}(z)| \leq \left[\log\left(\frac{4}{2^j(1-r^2)}\right)\right]^{1/\rho}, \qquad z \in S_j, \ 0 \leq j \leq j_0.$$

Let $g := cRf_{r,s}$ for some $c \in \mathbb{D}$. For $0 \leq j \leq j_0$, we have

$$I_j := \int_{S_j} |\varphi(g(z))|^q \, dA(z) \leq \left(\frac{4}{2^j(1-r^2)}\right)^{\tau'qR^\rho} |S_j|.$$

where $|S_j|$ is the normalized area of S_j. Since $|S_j| \leq 2^{2(j+1)}(1-r^2)^2$, and $\tau'qR^\rho < 2$, we see that $I_j \lesssim \left[2^j(1-r^2)\right]^\varepsilon$ for some $\varepsilon > 0$ which is independent of r and c. Thus $I := \sum_{j=0}^{j_0} I_j$ is dominated by a finite geometric series with geometric factor $2^{-\varepsilon}$ and largest term less than 2^ε, so I converges uniformly in r and c. Thus $\|\varphi \circ f\|_{A^q}$ is uniformly bounded for all $f \in U_R$, as required.

We omit the proof that if $\varphi \notin E(\rho,\tau)$, then $S_\varphi(U_R)$ fails to be a bounded set, since it is very similar to the corresponding part of the proof of Theorem 1.1.

It remains to exhibit a function $\varphi \in F \setminus \bigcup_{0<\tau'<\tau} E(\rho,\tau')$. The function $\varphi_m(z) := \sum_{n=[m/\rho]+1}^{\infty} \tau^n z^{[\rho n]-m}/n!$ is entire; here $[t]$ is the greatest integer not exceeding t. Clearly $|\varphi_m(z)| \leq |\exp(\tau|z|^\rho)|/|z|^m$ for all $z \neq 0$, and $\varphi_m(z) \gtrsim \exp(\tau z^\rho)/z^{m+1}$ for all sufficiently large real z. Thus $\varphi_m \in E(\rho,\tau) \setminus \bigcup_{0<\tau'<\tau} E(\rho,\tau')$ for all $m \in \mathbb{N}$. We leave to the reader the task of modifying the first part of this proof to verify that $z \mapsto \varphi_m(cRf_{r,s}(z))$ is uniformly in A^q as long as $qm/\tau > 1$. Thus $\phi_m \in F \setminus \bigcup_{0<\tau'<\tau} E(\rho,\tau')$ whenever m is an integer exceeding τ/q. $\square$

Corollary 3.3 follows easily from the omitted half of the proof of Theorem 3.1 and the fact that the B^s norm of $f_{r,s}$ is asymptotic to $\Gamma(s-1)^{1/s}\Gamma(s/2)^{-2/s}$. Similarly, we deduce Corollary 3.4 from the omitted half of the proof of Theorem 3.2 and the fact that $\|f_{r,\infty}\|_{\mathcal{B}} \leq 2$.

Finally, we mention two things we do not know. First, we do not know if the second containment in the conclusion of Theorem 3.1 is strict. More significantly, we know of no results similar to those in this paper that distinguish between entire functions of different *orders* $\rho_1, \rho_2 \in (0,1)$, let alone the more refined distinctions involving different *types*.

References

[AMV] V. Álvarez, M.A. Márquez, and D. Vukotić, Superposition operators between the Bloch space and Bergman spaces, *Ark. Mat.* **42** (2004), No. 2, 205–216.

[AZ] J. Appell and P.P. Zabrejko, *Nonlinear Superposition Operators*, Cambridge University Press, Cambridge, 1990.

[Be] A. Beurling, *Études sur un problème de majoration*, Thèse pour le doctorat, Almquist & Wieksell, Upsalla, 1933.

[Boa] R.P. Boas, Jr., *Entire functions*, Academic Press Inc., New York, 1954.

[BFV] S.M. Buckley, J.L. Fernández, and D. Vukotić, Superposition operators on Dirichlet type spaces, *Report Univ. Jyväskylä* **83** (2001), 41–61. (Papers on Analysis: dedicated to Olli Martio on the occasion of his 60th Birthday, editors: J. Heinonen, T. Kilpeläinen, and P. Koskela.) *Available from* http://www.math.jyu.fi/research/report83.html.

[BO] S.M. Buckley and J.O'Shea, Weighted Trudinger-type inequalities, *Indiana Univ. Math. J.* **48** (1999), 85–114.

[BV] S.M. Buckley and D. Vukotić, Univalent interpolation in Besov spaces and superposition into Bergman spaces, preprint.

[Ca] G. Cámera, Nonlinear superposition on spaces of analytic functions, in: Harmonic analysis and operator theory (Caracas, 1994), 103–116, *Contemp. Math.* **189**, Amer. Math. Soc., Providence, RI, 1995.

[CG] G. Cámera and J. Giménez, Nonlinear superposition operators acting on Bergman spaces, *Compositio Math.* **93** (1994), 23–35.

[CM] S.-Y. A. Chang and D. E. Marshall, On a sharp inequality concerning the Dirichlet integral, *Amer. J. Math.* **107** (1985), 1015–1033.

[D] P.L. Duren, *Theory of H^p Spaces*, Academic Press, New York-London 1970. Reprint: Dover, Mineola, New York, 2000.

[DS] P. L. Duren and A. P. Schuster, *Bergman Spaces*, Math. Surveys and Monographs **100**, American Mathematical Society, Providence, Rhode Island, 2004.

[HKZ] H. Hedenmalm, B. Korenblum, and K. Zhu, *Theory of Bergman Spaces*, Graduate Texts in Mathematics, Vol. **199**, Springer, New York, 2000.

[Z1] K. Zhu, *Operator Theory in Function Spaces*, Monographs and Textbooks in Pure and Applied Mathematics **139**, Marcel Dekker, New York, 1990.

[Z2] K. Zhu, Analytic Besov Spaces, *J. Math. Anal. Appl.* **157** (1991), 318–336.

[Zy] A. Zygmund, *Trigonometric series*, Vol. I, Cambridge University Press, Cambridge, 1959.

DEPARTMENT OF MATHEMATICS, NATIONAL UNIVERSITY OF IRELAND, MAYNOOTH, CO. KILDARE, IRELAND

E-mail address: sbuckley@maths.nuim.ie

DEPARTAMENTO DE MATEMÁTICAS, UNIVERSIDAD AUTÓNOMA DE MADRID, 28049 MADRID, SPAIN

E-mail address: dragan.vukotic@uam.es

Contemporary Mathematics
Volume **393**, 2006

Smooth functions in star-invariant subspaces

Konstantin Dyakonov and Dmitry Khavinson

Dedicated to Joe Cima on the occasion of his 70th birthday

ABSTRACT. In this note we summarize some necessary and sufficient conditions for subspaces invariant with respect to the backward shift to contain smooth functions. We also discuss smoothness of moduli of functions in such subspaces.

1. Introduction

For $0 < p \leq \infty$, let H^p denote the classical Hardy space of analytic functions on the disk $\mathbb{D} := \{z \in \mathbb{C} : |z| < 1\}$. As usual, we also treat H^p as a subspace of $L^p(\mathbb{T}, m)$, where $\mathbb{T} := \partial \mathbb{D}$ and m is the normalized arc length measure on $\mathbb{T}$.

Now suppose θ is an inner function on $\mathbb{D}$, that is, $\theta \in H^\infty$ and $|\theta(\zeta)| = 1$ for m-almost all $\zeta \in \mathbb{T}$. Factoring θ canonically, we get $\theta = BS$, where B is a Blaschke product and S is a singular inner function (see [**12**], Chapter II). The latter is thus of the form

$$S(z) = S_\mu(z) := \exp\left\{ -\int_{\mathbb{T}} \frac{\zeta + z}{\zeta - z} d\mu(\zeta) \right\},$$

where μ is a (positive) singular measure on $\mathbb{T}$, and we shall write $\mu = \mu_\theta (= \mu_S)$ to indicate that μ is associated with θ (or S) in this way.

We shall be concerned with the *star-invariant subspace*

$$(1.1) \qquad\qquad K_\theta := H^2 \ominus \theta H^2$$

that θ generates in H^2. Here, the term *star-invariant* stands for *invariant under the backward shift operator*

$$f \mapsto (f - f(0))/z, \qquad f \in H^2,$$

and it is well known that the general form of a (closed and proper) star-invariant subspace in H^2 is actually given by (1.1), with θ inner; see, e. g., [**5**] or [**14**].

This paper treats two questions related to the (boundary) smoothness of functions in K_θ. The first of these concerns the very existence of nontrivial functions

---

2000 *Mathematics Subject Classification.* Primary 30H05, 47B38.

Key words and phrases. Smooth functions, Lipschitz classes, star-invariant subspaces.

The first author is supported in part by grant 02-01-00267 from the Russian Foundation for Fundamental Research, DGICYT grant BFM2002-04072-C02-01, CIRIT grant 2001-SGR-00172, and by the Ramón y Cajal program (Spain).

The second author is supported in part by NSF grant DMS-0139008.

in $K_\theta \cap X$, where X is a given smoothness class. The answer should of course depend on X, but for a wide range of X's it turns out to be the same. Before we can state it, let us recall that a closed subset E of $\mathbb{T}$ is said to be a *Carleson* (or *Beurling–Carleson*) *set* if

$$\int_{\mathbb{T}} \log \operatorname{dist}(\zeta, E) \, dm(\zeta) > -\infty.$$

Originally, Carleson sets arose in [4] and the earlier work of Beurling as boundary zero-sets of analytic functions on $\mathbb{D}$ that are smooth, say of class C^1 or C^n, up to $\mathbb{T}$. Later on, they emerged in Korenblum's and Roberts' description of cyclic inner functions in Bergman spaces; see [13], [15], and [6], Chapter 8.

Our first result, Theorem 2.1, basically says that for many – or "most" – natural smoothness spaces X, one has $K_\theta \cap X = \{0\}$ if and only if θ is a singular inner function with the property that

(1.2) $\mu_\theta(E) = 0$ for every Carleson set $E \subset \mathbb{T}$.

This contrasts with the fact that the intersection $K_\theta \cap C(\mathbb{T})$ is always dense in K_θ (and hence always nontrivial), a result due to A. B. Aleksandrov; cf. [1], Theorem 6.

We admit that our Theorem 2.1 is not completely original, and the appearance of Carleson sets in this context should not be surprising. For instance, it was proved by H. S. Shapiro in [16] that if θ is a singular inner function for which (1.2) fails, then K_θ contains nonzero functions of class $C^n(\mathbb{T})$, for any fixed n. The new feature is, however, that our theorem applies to a larger scale of smoothness classes X. These range from the nicest possible space $C^\infty(\mathbb{T})$ to certain Bergman–Sobolev (or Besov) spaces that contain unbounded functions and enjoy very little smoothness indeed. In fact, those Bergman–Sobolev spaces are "almost the largest ones" for which the Korenblum–Roberts condition (1.2) is still relevant; we shall explain this in more detail below.

Our second theme is related to moduli of K_θ-functions. Roughly speaking, the question is how various smoothness properties of $f \in K_\theta$ are affected by those of $|f|$. More precisely, we seek to determine the nonnegative functions φ on $\mathbb{T}$ for which the set $\{f \in K_\theta : |f| = \varphi\}$ is nonempty and lies in a given smoothness class. This time we restrict our attention to the Lipschitz spaces Λ_ω defined in terms of a majorant ω; the solution is then given by Theorem 3.1.

2. Smooth functions in K_θ: existence

First let us fix some additional notations. We write σ for the normalized area measure on $\mathbb{D}$, and A^p for the Bergman p-space defined as the set of analytic functions in $L^p(\mathbb{D}, \sigma)$; we also need the Bergman–Sobolev spaces $A^{p,1} := \{f \in A^p : f' \in A^p\}$.

Further, we recall the definition of the Lipschitz–Zygmund spaces $\Lambda^\alpha = \Lambda^\alpha(\mathbb{T})$ with $\alpha > 0$. Given $\alpha = k + \beta$, where $k \geq 0$ is an integer and $0 < \beta \leq 1$, the space Λ^α consists of those functions $f \in C^k(\mathbb{T})$ which satisfy

$$f^{(k)}(e^{ih}\zeta) - f^{(k)}(\zeta) = O(|h|^\beta), \quad \text{if} \quad 0 < \beta < 1,$$

and

$$f^{(k)}(e^{ih}\zeta) - 2f^{(k)}(\zeta) + f^{(k)}(e^{-ih}\zeta) = O(|h|), \quad \text{if} \quad \beta = 1,$$

uniformly in $\zeta \in \mathbb{T}$ and $h \in \mathbb{R}$. Finally, we put $\Lambda_A^\alpha := H^\infty \cap \Lambda^\alpha$ and $\mathcal{A}^\infty :=$ $H^\infty \cap C^\infty(\mathbb{T})$.

THEOREM 2.1. *Let θ be an inner function on $\mathbb{D}$. The following statements are equivalent.*

(i.1) K_θ *contains a nontrivial function of class $\mathcal{A}^\infty$.*

(ii.1) K_θ *contains a nontrivial function of class $\bigcup_{p>1} A^{p,1}$.*

(iii.1) *Either θ has a zero in $\mathbb{D}$, or there is a Carleson set $E \subset \mathbb{T}$ with $\mu_\theta(E) > 0$.*

PROOF. The implication (i.1) $\Longrightarrow$ (ii.1) is obvious.

(ii.1) $\Longrightarrow$ (iii.1). Suppose (iii.1) fails, so that θ is a purely singular inner function, whose associated measure μ_θ vanishes on every Carleson set. By the Korenblum–Roberts theorem (see [6], p. 249), it follows that θ is a *cyclic* vector in each Bergman space A^q with $q \geq 1$.

Now if (ii.1) holds, then we can find a nontrivial function $F \in K_\theta \cap A^{p,1}$, with some $p > 1$. Being orthogonal to θH^2 (in H^2), this F satisfies

$$(2.1) \qquad \int_\mathbb{T} \overline{zF(z)}\, \theta(z)z^n\, dz = 0 \qquad (n = 0, 1, \dots).$$

Using Green's formula, we rewrite (2.1) as

$$(2.2) \qquad \int_\mathbb{D} \overline{f(z)}\, \theta(z)z^n\, d\sigma(z) = 0 \qquad (n = 0, 1, \dots),$$

where $f := (zF)'$. Letting $q = p/(p-1)$, we further rephrase (2.2) by saying that the family $\{\theta z^n : n \geq 0\}$ (and hence the subspace it spans in A^q) is annihilated by a nonzero functional in $(A^q)^* = A^p$, namely by the functional $g \mapsto \int_\mathbb{D} \overline{f}g\, d\sigma$. Indeed, we have $f \in A^p$ because $F \in A^{p,1}$, and $f \not\equiv 0$ because $F \not\equiv 0$. Thus θ generates a *proper* shift-invariant subspace in A^q and is, therefore, a *noncyclic* vector therein. This contradiction implies that (iii.1) holds as soon as (ii.1) does.

(iii.1) $\Longrightarrow$ (i.1). If θ has a zero z_0 in $\mathbb{D}$, then $z \mapsto (1 - \bar{z}_0z)^{-1}$ is a nontrivial function in $K_\theta \cap \mathcal{A}^\infty$.

Now assume that θ is a singular inner function and that $E \subset \mathbb{T}$ is a Carleson set with $\mu_\theta(E) > 0$. Put $\nu := \mu_\theta|E$, and let $S = S_\nu$ be the corresponding singular inner function. Since S divides θ, and hence $K_S \subset K_\theta$, it suffices to find a nontrivial $\mathcal{A}^\infty$-function in K_S. First we observe that, since ν lives on a Carleson set, S must divide the inner part of some nontrivial $\mathcal{A}^\infty$-function (see [19], Corollary 4.8). Thus, $GS \in \mathcal{A}^\infty$ for some $G \in H^\infty$, $G \not\equiv 0$, whence it actually follows (see, e. g., [19], Theorem 4.1) that $G \in \mathcal{A}^\infty$. In fact, there is no loss of generality in assuming that G is outer (once again, because division by inner factors preserves membership in $\mathcal{A}^\infty$). Next, we claim that

$$(2.3) \qquad\qquad\qquad G\bar{S} \in C^\infty(\mathbb{T}),$$

a fact we shall soon verify.

Postponing this verification for a moment, let us now use (2.3) to complete the proof. Put $\Phi := \bar{z}\bar{G}S$ and $f := P_+\Phi$, where P_+ stands for the orthogonal projection from $L^2(\mathbb{T})$ onto H^2. Our plan is to show that f is a nontrivial function in $\mathcal{A}^\infty \cap K_S$. First of all, $f \not\equiv 0$, because otherwise we would have $\Phi \in \bar{z}\overline{H^2}$, or equivalently $G\bar{S} \in H^2$, which is impossible since G is outer and S is inner. To see that f is in $C^\infty = C^\infty(\mathbb{T})$, and hence in $\mathcal{A}^\infty$, we note that $\Phi \in C^\infty$ by virtue of (2.3) and then recall that P_+ maps C^∞ into itself. Finally, since $f \in H^2$, the inclusion $f \in K_S$ will be established as soon as we check that f is orthogonal to the subspace SH^2. This we now do: if $h \in H^2$, then

$$\int_{\mathbb{T}} f \bar{S} \bar{h} \, dm = \int_{\mathbb{T}} \Phi \bar{S} \bar{h} \, dm = \int_{\mathbb{T}} \bar{z} \bar{G} \bar{h} \, dm = 0,$$

where the first equality holds because the antianalytic function $(I - P_+)\Phi$ is automatically orthogonal to Sh.

It remains to prove (2.3). Fix $\alpha > 0$ and an integer n with $n > \alpha$. This done, we invoke Proposition 1.5 of [8] which says, in particular, that given a function $F \in \Lambda_A^\alpha$ and an inner function I, the inclusions $FI^n \in \Lambda_A^\alpha$ and $F/I^n \in \Lambda^\alpha$ are equivalent. Applying this to $F = G$ and $I = S^{1/n}$, while recalling that $GS \in \mathcal{A}^\infty \subset \Lambda_A^\alpha$, we deduce that $G/S(= G\bar{S})$ is in Λ^α. And since this happens for *each* $\alpha > 0$, we finally conclude that

$$G\bar{S} \in \bigcap_{\alpha > 0} \Lambda^\alpha = C^\infty(\mathbb{T}),$$

as desired.   □

REMARK 2.2. In connection with condition (ii.1) above, we observe that K_θ always contains nontrivial H^∞-functions; one example is $1 - \overline{\theta(0)}\theta$. Now if $\mathcal{B}$ stands for the *Bloch space*, defined as the set of analytic functions f on $\mathbb{D}$ with

$$\sup_{z \in \mathbb{D}} (1 - |z|)|f'(z)| < \infty,$$

then we have

$$H^\infty \subset \mathcal{B} \subset \bigcap_{0 < p < 1} A^{p,1},$$

so the intersection $K_\theta \cap \bigcap_{0 < p < 1} A^{p,1}$ is always nontrivial. Thus, the smoothness class $\bigcup_{p > 1} A^{p,1}$ in Theorem 2.1 cannot be made "much larger" (say, by extending the union to $p > 1 - \varepsilon$) if the result is to remain true.

REMARK 2.3. We do not know, however, if the latter class can be replaced by $A^{1,1}$. The dual of A^1 being the Bloch space $\mathcal{B}$ (see [6], p. 48), the question can be rephrased in terms of weak* cyclicity of an inner function in $\mathcal{B}$. While the Korenblum–Roberts condition (1.2) on a singular inner function θ is necessary for θ to be weak* cyclic in $\mathcal{B}$, the sufficiency of that condition seems to present an open problem. This was mentioned in [3], and we are unaware of any further progress on that matter.

On the other hand, some sufficient conditions for weak* cyclicity in $\mathcal{B}$ – and a construction of an inner function satisfying them – can be found in [2]. In particular, there do exist inner functions θ with the property that $K_\theta \cap A^{1,1} = \{0\}$.

3. Smooth functions in K_θ and their moduli

This section deals with the following problem. Suppose φ is a nonnegative function on $\mathbb{T}$ that coincides a. e. with the modulus of some K_θ-function (this will be written as $\varphi \in |K_\theta|$). *When does it happen that all functions $f \in K_\theta$ with $|f| = \varphi$ are smooth, in some sense or other?*

We shall address this question when smoothness is understood as membership in $\Lambda_\omega = \Lambda_\omega(\mathbb{T})$, the Lipschitz space associated with a majorant (alias, modulus of continuity) ω. It will be assumed that $\omega : [0, 2] \to \mathbb{R}$ is a continuous increasing function with $\omega(0) = 0$ and that $\omega(t)/t$ is non-increasing. The space Λ_ω is then formed by those functions $f \in C(\mathbb{T})$ which satisfy

$$f(z_1) - f(z_2) = O\left(\omega(|z_1 - z_2|)\right), \qquad z_1, z_2 \in \mathbb{T}.$$

Thus, we want the set

$$M(\theta, \varphi) := \{f \in K_\theta : |f| = \varphi\}$$

to be contained in Λ_ω, and we shall soon characterize the pairs (θ, φ) for which this happens.

Before going any further, we recall that there is a simple characterization of the set $|K_\theta|$, as well as a parametrization of $M(\theta, \varphi)$ for $\varphi \in |K_\theta|$. These results are contained in [7] (see also Lemma 5 in [10]) and can be summarized as follows. In order that $\varphi \in |K_\theta|$, it is necessary and sufficient that $\bar{z}\varphi^2\theta \in H^1$. If that is so, we can factor the latter function as

$$(3.1) \qquad \qquad \bar{z}\varphi^2\theta = \mathcal{O}_\varphi^2 I,$$

where $\mathcal{O}_\varphi := \exp\left(\log\varphi + i\widetilde{\log\varphi}\right)$ is the outer function with modulus φ and I is an inner function. This done, it is easy to see that the functions $\mathcal{O}_\varphi$ and $\mathcal{O}_\varphi I$ are both in K_θ, and hence in $M(\theta, \varphi)$, while any other function in $M(\theta, \varphi)$ lies "in between". Precisely speaking, we have

$$(3.2) \qquad \qquad M(\theta, \varphi) = \{\mathcal{O}_\varphi J : J \in \mathcal{D}(I)\},$$

where $\mathcal{D}(I)$ stands for the set of all inner divisors of I. We also point out, for future reference, that (3.1) implies

$$(3.3) \qquad \qquad \bar{z}\bar{\mathcal{O}}_\varphi\theta = \mathcal{O}_\varphi I$$

(to see why, write $\varphi^2 = \mathcal{O}_\varphi\bar{\mathcal{O}}_\varphi$ and substitute this into (3.1)).

Finally, with an inner function θ we associate the sets

$$\Omega(\theta, \varepsilon) := \{z \in \mathbb{D} : |\theta(z)| < \varepsilon\}, \qquad 0 < \varepsilon < 1,$$

and

$$\rho(\theta) := \{z \in \mathbb{D} \cup \mathbb{T} : \liminf_{\mathbb{D}\ni w \to z} |\theta(w)| = 0\}.$$

Of course, $\rho(\theta) \cap \mathbb{D}$ is just the zero-set of θ, while $\rho(\theta) \cap \mathbb{T}$ consists of its boundary singularities.

THEOREM 3.1. *Let θ be an inner function, and let $\varphi \in |K_\theta|$. The following are equivalent:*

(i.2) $M(\theta, \varphi) \subset \Lambda_\omega$.

(ii.2) $\mathcal{O}_\varphi\theta \in \Lambda_\omega$.

(iii.2) $\mathcal{O}_\varphi \in \Lambda_\omega$ *and* $\varphi\theta \in \Lambda_\omega$.

(iv.2) $\mathcal{O}_\varphi \in \Lambda_\omega$, *and for some (or any) $\varepsilon \in (0, 1)$ one has*

$$(3.4) \qquad \qquad \mathcal{O}_\varphi(z) = O\left(\omega(1 - |z|)\right), \qquad z \in \Omega(\theta, \varepsilon).$$

PROOF. Let us begin by proving that (i.2) $\Longleftrightarrow$ (ii.2). Put $F := \mathcal{O}_\varphi$, and let I be as in (3.1). Taking (3.2) into account and recalling that division by inner factors preserves membership in $\Lambda_\omega \cap H^\infty$ (see [17]), we deduce that (i.2) holds iff the "extremal" function FI is in Λ_ω. Using the identity $FI = \bar{z}\bar{F}\theta$ (this is precisely

(3.3)), we restate the condition $FI \in \Lambda_\omega$ as $F\bar{\theta} \in \Lambda_\omega$. The latter can be further rephrased by saying that the quantity

(3.5)
$$Q(z_1, z_2) := (F\bar{\theta})(z_1) - (F\bar{\theta})(z_2)$$
$$= [F(z_1) - F(z_2)]\,\bar{\theta}(z_1) + F(z_2)\left[\bar{\theta}(z_1) - \bar{\theta}(z_2)\right]$$

is $O\left(\omega(|z_1 - z_2|)\right)$ whenever z_1, z_2 are in $\mathbb{T} \setminus \rho(\theta)$.

It should be noted that if $F\bar{\theta}$ satisfies a Λ_ω-condition over $\mathbb{T}\setminus\rho(\theta)$, then $F\bar{\theta} \in \Lambda_\omega$ (the converse being trivially true). Indeed, we can take it for granted that $F\bar{\theta}$ is at least continuous on $\mathbb{T}$, since this follows automatically from either (i.2) or (ii.2). Consequently, $F = 0$ on $\rho(\theta) \cap \mathbb{T}$. And if $F \not\equiv 0$, which we may safely assume, then we conclude that $m(\rho(\theta) \cap \mathbb{T}) = 0$ and so $\mathbb{T} \setminus \rho(\theta)$ is dense in $\mathbb{T}$.

Going back to (3.5), we observe that the first of the two terms on the right will be $O\left(\omega(|z_1 - z_2|)\right)$ as soon as

(3.6)
$$F \in \Lambda_\omega$$

(and this happens if any of the conditions (i.2)–(iv.2) holds). Therefore, the estimate

$$Q(z_1, z_2) = O\left(\omega(|z_1 - z_2|)\right), \qquad z_1, z_2 \in \mathbb{T} \setminus \rho(\theta),$$

reduces to

(3.7)
$$\varphi(z_2)|\theta(z_1) - \theta(z_2)| = O\left(\omega(|z_1 - z_2|)\right), \qquad z_1, z_2 \in \mathbb{T} \setminus \rho(\theta),$$

where we have also used the fact that $|F| = \varphi$ on $\mathbb{T}$. Thus, (i.2) is equivalent to (3.6)&(3.7) (that is, to (3.6) and (3.7) taken together).

A similar argument now enables us to rewrite the condition (3.6)&(3.7) as (ii.2). Indeed, (ii.2) says that $F\theta \in \Lambda_\omega$, which in turn means that

$$\tilde{Q}(z_1, z_2) := (F\theta)(z_1) - (F\theta)(z_2)$$

is $O\left(\omega(|z_1 - z_2|)\right)$ for $z_1, z_2 \in \mathbb{T} \setminus \rho(\theta)$. A formula similar to (3.5), but with $\tilde{Q}$ in place of Q and with no bar over θ, convinces us that the required estimate on $\tilde{Q}(z_1, z_2)$ reduces to (3.6)&(3.7), exactly as before.

We now know that (i.2) $\iff$ (ii.2). That (iii.2) is also equivalent to (3.6)&(3.7), and hence to (i.2) and (ii.2), is verified in very much the same way. Indeed, (iii.2) obviously implies (3.6) and, *a fortiori*, the weaker condition that $\varphi \in \Lambda_\omega$, while the rest follows from the formula

$$(\varphi\theta)(z_1) - (\varphi\theta)(z_2) = [\varphi(z_1) - \varphi(z_2)]\,\theta(z_1) + \varphi(z_2)\,[\theta(z_1) - \theta(z_2)].$$

Finally, the equivalence between (ii.2) and (iv.2) is contained in Theorem 5 of [11]. $\square$

REMARK 3.2. It was proved by Shirokov that an inner function θ divides (the inner part of) some nontrivial function in $\Lambda_\omega \cap H^\infty$ if and only if

(3.8)
$$\int_{\mathbb{T}} \log \omega \left(\mathrm{dist}(\zeta, \rho(\theta))\right) dm(\zeta) > -\infty;$$

see [18], Chapter III. Thus, a nontrivial function φ satisfying (i.2)–(iv.2) can only exist if the Carleson-type condition (3.8) is fulfilled.

REMARK 3.3. Under the additional assumption

$$\int_0^\delta \frac{\omega(t)}{t}\,dt \le \text{const}\cdot\omega(\delta), \qquad 0 < \delta < 1,$$

the nonnegative functions $\varphi \in \Lambda_\omega$ with the property $\mathcal{O}_\varphi \in \Lambda_\omega$ can be characterized as those satisfying $\log\varphi \in L^1(m)$ and

$$\varphi(z/|z|) - |\mathcal{O}_\varphi(z)| = O\left(\omega(1-|z|)\right);$$

see [9] and [11] for a proof. Consequently, condition (3.4) in (iv.2) is then equivalent to saying that

$$\varphi\left(\frac{z}{|z|}\right) = O(\omega(1-|z|))$$

as $|z| \to 1$, $z \in \Omega(\theta, \varepsilon)$.

References

[1] A. B. Aleksandrov, *On the existence of angular boundary values of pseudocontinuable functions*, Zap. Nauchn. Sem. S.-Peterburg. Otdel. Mat. Inst. Steklov. (POMI) **222** (1995), 5–17; English transl. in J. Math. Sci. **87** (1997), 3781–3787.

[2] J. M. Anderson, J. L. Fernández, and A. L. Shields, *Inner functions and cyclic vectors in the Bloch space*, Trans. Amer. Math. Soc. **323** (1991), 429–448.

[3] L. Brown and A. L. Shields, *Multipliers and cyclic vectors in the Bloch space*, Michigan Math. J. **38** (1991), 141–146.

[4] L. Carleson, *Sets of uniqueness for functions regular in the unit circle*, Acta Math. **87** (1952), 325–345.

[5] J. A. Cima and W. T. Ross, *The Backward Shift on the Hardy Space*, Mathematical Surveys and Monographs, 79, American Mathematical Society, Providence, RI, 2000.

[6] P. L. Duren and A. P. Schuster, *Bergman Spaces*, Mathematical Surveys and Monographs, 100, American Mathematical Society, Providence, RI, 2004.

[7] K. M. Dyakonov, *Moduli and arguments of analytic functions from subspaces in H^p that are invariant under the backward shift operator*, Sibirsk. Mat. Zh. **31** (1990), No. 6, 64–79; English transl. in Siberian Math. J. **31** (1990), 926–939.

[8] K. M. Dyakonov, *Smooth functions and coinvariant subspaces of the shift operator*, Algebra i Analiz 4 (1992), No. 5, 117–147; English transl. in St. Petersburg Math. J. **4** (1993), 933–959.

[9] K. M. Dyakonov, *Equivalent norms on Lipschitz-type spaces of holomorphic functions*, Acta Math. **178** (1997), 143–167.

[10] K. M. Dyakonov, *Zero sets and multiplier theorems for star-invariant subspaces*, J. Anal. Math. **86** (2002), 247–269.

[11] K. M. Dyakonov, *Holomorphic functions and quasiconformal mappings with smooth moduli*, Adv. Math. **187** (2004), 146–172.

[12] J. B. Garnett, *Bounded Analytic Functions*, Academic Press, New York, 1981.

[13] B. Korenblum, *Cyclic elements in some spaces of analytic functions*, Bull. Amer. Math. Soc. **5** (1981), 317–318.

[14] N. K. Nikol'skiĭ, *Treatise on the Shift Operator*, Springer-Verlag, Berlin, 1986.

[15] J. W. Roberts, *Cyclic inner functions in the Bergman spaces and weak outer functions in H^p, $0 < p < 1$*, Illinois J. Math. **29** (1985), 25–38.

[16] H. S. Shapiro, *Some remarks on weighted polynomial approximation of holomorphic functions*, Mat. Sbornik **73** (1967), No. 3, 320–330; English transl. in Math. USSR Sb. **2** (1967), 285–294.

[17] N. A. Shirokov, *Ideals and factorization in algebras of analytic functions smooth up to the boundary*, Trudy Mat. Inst. Steklov. **130** (1978), 196–222; English transl. in Proc. Steklov Inst. Math. **130** (1979).

[18] N. A. Shirokov, *Analytic Functions Smooth up to the Boundary*, Lecture Notes Math. **1312**, Springer, Berlin, 1988.

[19] B. A. Taylor and D. L. Williams, *Ideals in rings of analytic functions with smooth boundary values*, Canad. J. Math. **22** (1970), 1266–1283.

Departament de Matemàtica Aplicada i Anàlisi, Universitat de Barcelona, Gran Via 585, E-08007 Barcelona, Spain
E-mail address: dyakonov@mat.ub.es

Department of Mathematics, University of Arkansas, Fayetteville, AR 72701, U.S.A.
E-mail address: dmitry@uark.edu

Contemporary Mathematics
Volume **393**, 2006

Conjugation and Clark operators

Stephan Ramon Garcia

Dedicated to Joseph A. Cima, on the occasion of his 70th birthday.

ABSTRACT. We discuss the application of antilinear symmetries (conjugation operators) to problems connected to the compressed shift on the spaces $H^2 \ominus \varphi H^2$ where φ denotes a nonconstant inner function. For example, we discuss an explicit parameterization of the noncyclic functions for the backward shift operator and show that compressed shifts and their associated Clark operators [**10**] belong to the class of complex symmetric operators [**23, 24**]. Furthermore, we show that the partial isometry in the polar decomposition of an (invertible) compressed shift is a certain Clark unitary operator. Moreover, we are able to compute certain preferred orthonormal bases for the spaces $H^2 \ominus \varphi H^2$ (where φ denotes a nonconstant inner function) that respect the underlying symmetry. This is a self-contained exposition of several results of the author and others contained in the papers [**19, 21, 23, 24, 25**].

1. Introduction

The operators of our present interest (namely compressed shifts and their associated Clark perturbations) belong to a surprisingly large class of operators, known collectively as *complex symmetric operators*. Although there are several different characterizations of such operators, the simplest is in terms of matrix representations: A bounded operator on a Hilbert space is complex symmetric if it has a symmetric matrix representation with respect to some orthonormal basis. At the outset, it is not obvious that compressed shifts should have such matrix representations, nor is it obvious how to go about finding such a preferred orthonormal basis. These problems roughly constitute the topic of this note.

2000 *Mathematics Subject Classification.* 30D55, 47A15.

Key words and phrases. Complex symmetric operators, shift operators, compressed shift, backward shift, inner functions, Clark perturbations, Jordan operators, Volterra operator.

There are many concrete examples (besides compressed shifts) of complex symmetric operators and the classical roots of the theory are deep. This surprisingly large class includes all *normal operators*, all *Hankel operators*, all finite *Toeplitz matrices*, and the *Volterra operator* (see below). Section 2 contains some basic background material on complex symmetric operators, beginning with a brief discussion of the classical theory of complex symmetric matrices (Subsection 2.1). In particular, we highlight a few places where such matrices appear in function theory and function-related operator theory.

From an operator theoretic viewpoint, we prefer things to be coordinate-free. Indeed, we are more interested in operators themselves rather than their matrix representations and therefore a more intrinsic definition of the transpose is necessary. By exploiting the symmetry of a finite Toeplitz matrix with respect to the second diagonal (Subsection 2.2), we motivate a coordinate-free approach to the transpose symmetry $T = T^t$ enjoyed by a complex symmetric matrix. This entails the consideration of certain antilinear symmetries which are analogous to complex conjugation (Subsection 2.3). To be specific, we say that an antilinear linear operator C on a complex Hilbert space $\mathcal{H}$ is a *conjugation operator* (or simply a *conjugation*) if $C^2 = I$ and $\langle Cf, Cg \rangle = \langle g, f \rangle$ for all f, g in $\mathcal{H}$. The standard example of a conjugation operator is complex conjugation on a Lebesgue space $L^2(X, \mu)$, where no explicit reference is made to an orthonormal basis.

For a fixed conjugation operator C, we define the *transpose* of an operator by the formula $T^t = CT^*C$ (Subsection 2.4) and we say that T is *C-symmetric* if $T = CT^*C$. Furthermore, we say that T is *complex symmetric* if there exists a conjugation C such that T is C-symmetric (Subsection 2.5). A straightforward computation (see Lemma 2.7) shows that an operator T is complex symmetric if and only if it has a symmetric matrix representation with respect to some orthonormal basis.

The *model spaces* $H^2 \ominus \varphi H^2$ (where φ denotes a nonconstant inner function) possess their own natural conjugation operators. In Section 3, we study function theoretic aspects of the conjugation operator

$$Cf = \overline{f z} \varphi$$

on $H^2 \ominus \varphi H^2$. We first review several basic definitions (Subsection 3.1) before examining the connection with the theory of *pseudocontinuations*. It turns out that conjugation on $H^2 \ominus \varphi H^2$ and pseudocontinuations are closely related concepts and that they represent two different ways of looking at the same underlying phenomenon (Subsection 3.2).

Since each space $H^2 \ominus \varphi H^2$ is a reproducing kernel Hilbert space, it is natural to investigate the relationship between its reproducing kernel K_λ and C. This leads naturally to the *conjugate kernel* CK_λ which reproduces the values of $[Cf](\lambda)$ via the formula $[Cf](\lambda) = \langle CK_\lambda, f \rangle$ (Subsection 3.3).

Specifically, these functions are given by the formulas

$$K_\lambda(z) = \frac{1 - \overline{\varphi(\lambda)}\varphi(z)}{1 - \bar{\lambda}z}, \quad [CK_\lambda](z) = \frac{\varphi(z) - \varphi(\lambda)}{z - \lambda},$$

where λ and z belong to the open unit disk $\mathbb{D}$. They play a major role in our computations involving the compressed shift and its associated Clark perturbations.

Our first application of this approach is in the explicit function theoretic parameterization of the space $H^2 \ominus \varphi H^2$ (Subsection 3.4), thereby obtaining a characterization of noncyclic functions for the backward shift on H^2. In light of a famous theorem of Douglas, Shapiro, and Shields ([**13**, Thm. 2.2.1]), this also yields a description of H^2 functions which are *pseudocontinuable of bounded type*.

Unlike many standard Hilbert spaces, the model spaces do not come equipped with "convenient" orthonormal bases (Subsection 3.5). In light of the conjugation operator C, we wish to construct orthonormal bases for the spaces $H^2 \ominus \varphi H^2$ which are also fixed by C. We refer to such bases as *C-real* orthonormal bases and the remaining sections of this note are concerned with various aspects of this construction.

Section 4 is essentially an extended example in which we explicitly construct an infinite family of C-real bases for the space $H^2 \ominus \varphi H^2$ corresponding to the *atomic inner function*

$$\varphi(z) = \exp\left(\frac{z+1}{z-1}\right).$$

Our main tool is a certain integral transform $\mathcal{W} : L^2[0,1] \longrightarrow H^2 \ominus \varphi H^2$ which arose in Sarason's solution [**42**] to the *Gelfand problem* [**26**] (i.e. characterize the invariant subspace lattice for the *Volterra integration operator*

$$[Vf](x) = \int_0^x f(t)\,dt$$

on $L^2[0,1]$). Specifically, $\mathcal{W}$ provides a unitary equivalence between the Cayley transform $(I-V)(I+V)^{-1}$ of the Volterra operator and its functional model, the compressed shift on $H^2 \ominus \varphi H^2$ (Subsection 4.1).

As we mentioned earlier, the Volterra operator is a complex symmetric operator. Indeed, it is $\mathcal{C}$-symmetric with respect to the conjugation operator

$$[\mathcal{C}f](x) = \overline{f(1-x)}$$

on $L^2[0,1]$. By following this symmetry through to the compressed shift on $H^2 \ominus \varphi H^2$, we deduce that the compressed shift is C-symmetric where $C = \mathcal{W}\mathcal{C}\mathcal{W}^*$ turns out to be the conjugation operator $Cf = \overline{f}z\varphi$ that we have already mentioned.

By first solving the corresponding problem in $L^2[0,1]$, we obtain an explicit description of certain natural C-real bases for $H^2 \ominus \varphi H^2$ (Subsection 4.2). By studying this particular example, we are able to conjecture what a more general solution might look like (Subsection 4.3).

Section 5 concerns Toeplitz operators on H^2 and their compressions to the spaces $H^2 \ominus \varphi H^2$. After reviewing some basic definitions (Subsection 5.1) and discussing a simple example (Subsection 5.2), we show that the compression of any Toeplitz operator to a subspace of the form $H^2 \ominus \varphi H^2$ is C-symmetric with respect to the natural conjugation operator C on $H^2 \ominus \varphi H^2$ (Subsection 5.3).

In particular, the *compressed shift* $Sf = P_\varphi(zf)$ (where P_φ denotes the orthogonal projection from H^2 onto $H^2 \ominus \varphi H^2$) is a C-symmetric operator. This observation allows us to employ several general results on complex symmetric operators to problems concerning the compressed shift and (ultimately) its Clark perturbations.

In Section 6, we discuss a variant of the classical polar decomposition (for complex symmetric operators) which generalizes an elegant result of Godič and Lucenko [29]. Their theorem asserts that every unitary operator U decomposes as the product $U = CJ$ of two conjugations C and J (Subsection 6.1). This generalizes the simple geometric fact that any planar rotation can be written as the product of two reflections.

For a C-symmetric operator T, there is a related decomposition. For instance, if T is C-symmetric, then we can write $T = CJ|T|$ where J is an auxiliary conjugation operator which commutes with the modulus $|T| = \sqrt{T^*T}$ and with its spectral projections (Subsection 6.2). This result (from [24]) is the main tool with which we attack compressed shifts and their Clark perturbations.

In Section 7, we compute the decomposition described above for the compressed shift. More generally, we actually consider the compressions $S_\lambda f = P_\varphi(b_\lambda f)$ of Toeplitz operators whose symbols are the disk automorphisms

$$b_\lambda(z) = \frac{z - \lambda}{1 - \overline{\lambda} z}$$

vanishing at points λ in $\mathbb{D}$. The compressed shift corresponds to the special case $\lambda = 0$. After collecting some preliminary notation (Subsection 7.1), we compute the modulus $|S_\lambda|$ (Subsection 7.2), and the auxiliary conjugation J_λ (Subsection 7.3) in the decomposition

$$S_\lambda = CJ_\lambda|S_\lambda|.$$

If S_λ happens to be invertible, then the partial isometry U_λ in the polar decomposition of S_λ is actually unitary and has the Godič-Lucenko decomposition $U_\lambda = CJ_\lambda$ (Subsection 7.4). Moreover, U_λ turns out to be a Clark-type unitary operator, namely a rank-one perturbation of the compressed shift.

The appearance of Clark perturbations in the polar decomposition of compressed shifts motivates us to consider these perturbations from the complex symmetric perspective. In Section 8, we develop a slight generalization of Clark's theory to produce C-real bases for the spaces $H^2 \ominus \varphi H^2$ (subject to some function theoretic restrictions).

Using the explicit Godič-Lucenko decomposition $U_{\lambda,\alpha} = CJ_{\lambda,\alpha}$, we compute the eigenvalues and eigenvectors (if any exist) of $U_{\lambda,\alpha}$ (Subsection 8.1). These computations proceed along entirely different lines than Clark's original computations and provide an alternate approach to the calculation of the eigenstructure of Clark operators. Under certain circumstances, it is possible to obtain a C-real orthonormal basis for $H^2 \ominus \varphi H^2$ by this method (Subsection 8.2).

Acknowledgments. The author wishes to thank D. Sarason (co-author of [25] and advisor) for his numerous comments and suggestions regarding this note. The author is also grateful to M. Putinar, coauthor of [22, 23, 24] and general mentor. Moreover, we are indebted to D.Z. Arov, J.A. Cima, J.W. Helton, D. Khavinson, S. Richter, H.S. Shapiro, and C. Sundberg for their constructive comments during the writing of [23] and/or [24]. The author would especially like to thank G. Karaali for reading a preliminary version of this note.

2. Complex Symmetric Operators

As we have mentioned in the introduction, we will make extensive use of complex symmetric operators in our discussion of problems related to compressed shifts and Clark operators. In this section, we discuss the basic properties of this class of operators and provide a few simple examples. Much of the material in this section comes from [23].

2.1. Complex symmetric matrices. Before proceeding to discuss complex symmetric operators in full generality, it is useful to reflect for a moment on the classical theory of complex symmetric matrices (see the texts [18, 32]). In this context, we can motivate our results and see the connection between symmetric matrices and Toeplitz operators.

We say that a square matrix T (with complex entries) is *complex symmetric* if T coincides with its transpose T^t. Obvious examples of such matrices include real symmetric matrices and complex Hankel matrices.

A notable early contribution to the subject was made by Takagi [47, Thm. II], who established a useful decomposition (reproved by Hua [33], Jacobson [36], Schur [43], and Siegel [46], among others) for complex symmetric matrices and used it to provide an elegant proof of a holomorphic interpolation theorem of Carathéodory and Fejér [7].

Specifically, this theorem states that if $a_0, a_1, \ldots, a_{n-1}$ are complex numbers, then there exists an analytic function f on the open unit disk $\mathbb{D}$ satisfying $\| f \|_\infty \leq 1$ and whose first n nonnegative Fourier coefficients are precisely $a_0, a_1, \ldots, a_{n-1}$ if and only if the associated $n \times n$ Hankel matrix

$$
\begin{pmatrix}
a_{n-1} & a_{n-2} & \cdots & a_0 \\
a_{n-2} & a_{n-3} & \cdots & 0 \\
\vdots & \vdots & & \vdots \\
a_0 & 0 & \cdots & 0
\end{pmatrix}
$$

has operator norm less than or equal to one. In the course of his proof of the Carathéodory and Fejér theorem, Takagi showed that any complex symmetric matrix can be written in the form UDU^t where D is a nonnegative diagonal matrix and U is unitary. This elegant decomposition has many applications in matrix theory and linear algebra, some of which are detailed in [**32**].

In the theory of univalent functions, the Goluzin inequality (see [**15**, Cor. 9, p.128]) can be phrased in terms of complex symmetric matrices. Specifically, recall that a holomorphic function f on $\mathbb{D}$ is *univalent* if it is injective and *normalized* if it satisfies $f(0) = 0$ and $f'(0) = 1$. A normalized univalent function is called *schlicht*.

It turns out that f is schlicht if and only if for any n distinct complex numbers $z_1, z_2, \ldots, z_n$, the inequality

$$
\left| \sum_{j,k=1}^{n} w_j w_k \log \left(\frac{z_j z_k}{f(z_j) f(z_k)} \cdot \frac{f(z_j) - f(z_k)}{z_j - z_k} \right) \right| \leq \sum_{j,k=1}^{n} w_j \overline{w_k} \log \frac{1}{1 - z_j \overline{z_k}}
$$

holds for all choices of $w_1, w_2, \ldots, w_n$. In terms of matrices, this represents the majorization

$$
|\langle Aw, w \rangle| \leq \langle Bw, w \rangle,
$$

for any $w = (w_1, w_2, \ldots, w_n)$ in $\mathbb{C}^n$, of the complex symmetric matrix

$$
[A]_{jk} = \log \left(\frac{z_j z_k}{f(z_j) f(z_k)} \cdot \frac{f(z_j) - f(z_k)}{z_j - z_k} \right)
$$

by the positive matrix

$$
[B]_{jk} = \log \frac{1}{1 - z_j \overline{z_k}}.
$$

More information on such hermitian-symmetric inequalities can be found in [**16, 17**].

In physical applications, complex symmetric matrices also appear in the theory of wave propagation in continuous media [**45**] and in the study of certain chemical reaction problems [**5**]. Perhaps the ubiquity of complex symmetric matrices can best be seen from the fact that *any* square complex matrix is similar to a complex symmetric one. We will sketch a proof of this result in our discussion below.

Since our interest in this note lies closer to operator theory than linear algebra, we will develop a coordinate free approach. If C denotes the operation of complex conjugation of a vector in $\mathbb{C}^n$:

$$
C(z_1, z_2, \ldots, z_n) = (\overline{z_1}, \overline{z_2}, \ldots, \overline{z_n}), \tag{2.1}
$$

then the transpose symmetry $T = T^t$ of a complex symmetric matrix T is equivalent to the algebraic condition $T = CT^*C$, where T^* denotes the adjoint matrix of T. We will take this equation as our starting point.

2.2. Toeplitz matrices. The algebraic condition $T = CT^*C$ satisfied by a complex symmetric matrix is readily extended to the more general setting of bounded operators on a Hilbert space. One advantage to generalizing the transpose symmetry $T = T^t$ in this way is that the adjoint operation $T \mapsto T^*$ is intrinsic to a given Hilbert space and does not require a preferred orthonormal basis to begin with.

The simple example of a finite Toeplitz matrix described below illustrates that even in finite dimensions, it may not be immediately evident that an operator has a symmetric matrix representation with respect to some orthonormal basis.

Let T denote a *Toeplitz matrix*

$$T = \begin{pmatrix} a_0 & a_1 & a_2 & \cdots & a_{n-1} \\ a_{-1} & a_0 & a_1 & \cdots & a_{n-2} \\ a_{-2} & a_{-1} & a_0 & \cdots & a_{n-3} \\ \vdots & \vdots & \vdots & & \vdots \\ a_{-(n-1)} & a_{-(n-2)} & a_{-(n-3)} & \cdots & a_0 \end{pmatrix} \tag{2.2}$$

and observe the symmetry of T with respect to the *second diagonal*. Although the Toeplitz matrix T is not in general symmetric, it does satisfy the algebraic condition $T = CT^*C$ where C denotes the antilinear operator

$$C(z_0, z_1, \ldots, z_{n-1}) = (\overline{z_{n-1}}, \overline{z_{n-2}}, \ldots, \overline{z_0}) \tag{2.3}$$

on $\mathbb{C}^n$. Clearly C shares many of the fundamental properties of complex conjugation. Indeed, we easily see that C is *involutive*, meaning that

$$C^2 = I$$

and that C is *isometric* in the sense that

$$\langle Cf, Cg \rangle = \langle g, f \rangle$$

for any vectors f, g in $\mathbb{C}^n$. Moreover, it is not hard to find orthonormal bases of $\mathbb{C}^n$ with respect to which C can be represented simply as complex conjugation.

EXAMPLE 2.1. In the case $n = 4$, the vectors

$$e_1 = \frac{1}{\sqrt{2}}(1, 0, 0, 1)$$
$$e_2 = \frac{1}{\sqrt{2}}(i, 0, 0, -i)$$
$$e_3 = \frac{1}{\sqrt{2}}(0, 1, 1, 0)$$
$$e_4 = \frac{1}{\sqrt{2}}(0, i, -i, 0)$$

form an orthonormal basis of $\mathbb{C}^4$ which is fixed by the antilinear involution $C(z_0, z_1, z_2, z_3) = (\overline{z_3}, \overline{z_2}, \overline{z_1}, \overline{z_0})$. Thus

$$C(a_1 e_1 + a_2 e_2 + a_3 e_3 + a_4 e_4) = \overline{a_1} e_1 + \overline{a_2} e_2 + \overline{a_3} e_3 + \overline{a_4} e_4$$

for any complex constants a_1, a_2, a_3, a_4 and hence C is unitarily equivalent to complex conjugation with respect to the standard basis of $\mathbb{C}^4$. The reader will not have difficulty generalizing this procedure for other values of n.

With respect to such a basis, one can show that the matrix representation of any Toeplitz matrix T is symmetric (see Lemma 2.7 below). In particular, this implies that any (finite) Toeplitz matrix is *unitarily equivalent* to a complex symmetric matrix. While this result is classical (see [18, 32] for background), it is not necessarily well-known outside of linear algebra and matrix analysis circles.

A simple, yet important, example of a finite Toeplitz matrix is an $n \times n$ Jordan block (with eigenvalue 0):

EXAMPLE 2.2. Let T denote the matrix

$$T = \begin{pmatrix} 0 & 1 & & & & \\ & 0 & 1 & & & \\ & & 0 & & & \\ & & & \ddots & & \\ & & & & 0 & 1 \\ & & & & & 0 \end{pmatrix}.$$

Since T is a finite Toeplitz matrix, it satisfies $T = CT^*C$ where C is the antilinear operator defined by (2.3).

In particular, the preceding example indicates that every Jordan block is *unitarily equivalent* to a symmetric matrix. This runs somewhat counter to our intuition since Jordan blocks were specifically designed to cope with situations in linear algebra where symmetry (in a somewhat vague sense) is lacking. Moreover, by considering direct sums of Jordan blocks, one easily obtains the following theorem:

THEOREM 2.3. *Every square matrix is similar to a complex symmetric matrix.*

Along these lines, symmetric canonical forms for arbitrary square matrices have been investigated by several authors, the earliest (1930) perhaps being Wellstein [49] (see also [11, 44] and the texts [18, 32]). We also remark that, since a *real* symmetric matrix must be self-adjoint, the preceding theorem generally requires the use of complex similarity transformations and complex symmetric matrices. Moreover, Theorem 2.3 implies that there are no restrictions on the Jordan canonical form of a complex symmetric matrix.

The connection between complex symmetric matrices and operator related function theory begins to reveal itself when one observes that

$$\begin{pmatrix} 0 & 1 & & & \\ & 0 & 1 & & \\ & & 0 & & \\ & & & \ddots & \\ & & & 0 & 1 \\ & & & & 0 \end{pmatrix} \begin{pmatrix} a_0 \\ a_1 \\ a_2 \\ \vdots \\ a_{n-2} \\ a_{n-1} \end{pmatrix} = \begin{pmatrix} a_1 \\ a_2 \\ a_3 \\ \vdots \\ a_{n-1} \\ 0 \end{pmatrix}$$

and hence an $n \times n$ Jordan block (with eigenvalue 0) represents a "part" of the backward shift operator. We will discuss the connections between complex symmetric matrices, shift operators, and function theory in a later section. First, we must generalize some of these simple observations to the setting of operators on an arbitrary Hilbert space.

2.3. Conjugation operators. In our consideration of finite Toeplitz matrices, the antilinear operator (2.3) arose. With this C, we recognized that the symmetry of a Toeplitz matrix with respect to the second diagonal can be represented by the simple algebraic formula $T = CT^*C$. We now wish to consider such antilinear symmetries in a more general setting.

If $\mathcal{H}$ is a Hilbert space, then we say that C is a *conjugation operator* on $\mathcal{H}$ if the following conditions hold:

(a) C is *antilinear*:

$$C(a_1 f_1 + a_2 f_2) = \overline{a_1} C f_1 + \overline{a_2} C f_2$$

for all $a_1, a_2 \in \mathbb{C}$ and $f_1, f_2 \in \mathcal{H}$.

(b) C is *isometric*:

$$\langle Cf, Cg \rangle = \langle g, f \rangle$$

for all $f, g \in \mathcal{H}$.

(c) C is *involutive*: $C^2 = I$.

The astute reader will note that these are not the "minimal" axioms which will yield the same effect. For instance, condition (a) follows easily from (b) and (c). Nevertheless, we prefer to be more explicit than necessary for the sake of clarity.

EXAMPLE 2.4. The most obvious (and trivial) example of a conjugation operator is simply complex conjugation on the one-dimensional complex Hilbert space $\mathbb{C}$. Indeed, this is exactly the notion that we are attempting to generalize. Other obvious examples are complex conjugation with respect to the standard basis in $\mathbb{C}^n$ (2.1) and the conjugation operator (2.3) that arose in our consideration of finite Toeplitz matrices.

EXAMPLE 2.5. Another simple example is given by pointwise conjugation $[Cf](x) = \overline{f(x)}$ on a Lebesgue space $L^2(X, \mu)$. If our measure space (X, μ) has some symmetry, then we might take that into account as well.

For example, we will see that the conjugation operator $[Cf](x) = \overline{f(1-x)}$ on $L^2[0,1]$ arises in the consideration of the Volterra integration operator (see Lemma 4.1).

Despite a multitude of different manifestations, conjugation operators are actually quite simple objects. As the following easy lemma shows, any conjugation operator can be represented as complex conjugation with respect to a certain orthonormal basis:

LEMMA 2.6. *If C is a conjugation operator on $\mathcal{H}$, then there exists an orthonormal basis $(e_n)_{n=1}^{\dim \mathcal{H}}$ of $\mathcal{H}$ such that $Ce_n = e_n$ for all n.*

We refer to such a basis as a *C-real orthonormal basis* for $\mathcal{H}$. The importance of such a basis lies in the fact that the action of a possibly abstract conjugation operator C can be represented simply as complex conjugation with respect to the basis $(e_n)_{n=1}^{\dim \mathcal{H}}$. In other words, any conjugation operator is unitarily equivalent to complex conjugation on an l^2 space of the appropriate dimension.

2.4. The transpose of an operator. For a fixed conjugation operator C on a Hilbert space $\mathcal{H}$, we *define* the transpose T^t of a bounded linear operator $T : \mathcal{H} \longrightarrow \mathcal{H}$ to be the linear operator

$$T^t = CT^*C.$$

It is important to note that the definition is *coordinate free* since it depends only upon the conjugation operator C and not on a particular choice of basis for $\mathcal{H}$. Indeed, one frequently considers function theoretic conjugation operators which are not a priori defined in terms of complex conjugation with respect to an orthonormal basis.

In terms of matrix representations, the following lemma shows that the definition $T^t = CT^*C$ for the transpose of an operator is justified:

LEMMA 2.7. *If $(e_n)_{n=1}^{\dim \mathcal{H}}$ is a C-real orthonormal basis for $\mathcal{H}$ and $[T]_{jk}$ denotes the jk-th entry of the matrix representation for T with respect to the basis $(e_n)_{n=1}^{\dim \mathcal{H}}$, then $[T^t]_{jk} = [T]_{kj}$ for all $j, k = 1, 2, \ldots, \dim \mathcal{H}$.*

PROOF. The proof is a straightforward computation based on the definition $T^t = CT^*C$ and the isometric property (b) of conjugation:

$$
\begin{aligned}
[T^t]_{jk} &= \langle T^t e_j, e_k \rangle \\
&= \langle CT^*C e_j, e_k \rangle \\
&= \langle C e_k, T^*C e_j \rangle \\
&= \langle e_k, T^* e_j \rangle \\
&= \langle T e_k, e_j \rangle \\
&= [T]_{kj}.
\end{aligned}
$$

Thus the matrix for T^t with respect to the basis $(e_n)_{n=1}^{\infty}$ is simply the matrix transpose of the matrix for T with respect to the same basis. $\qquad \square$

2.5. Complex symmetric operators. Suppose now that $\mathcal{H}$ is a Hilbert space equipped with a conjugation operator C. We say that a bounded operator T is *C-symmetric* if $T^t = T$. In other words, T is a C-symmetric operator if and only if $T = CT^*C$. More generally, we will say that T is *complex symmetric* if there exists a conjugation operator C such that T is C-symmetric.

There are many examples of complex symmetric operators. In fact, they are more common than one might initially imagine.

EXAMPLE 2.8. All normal operators are complex symmetric. Indeed, one need only produce a conjugation C commuting with the spectral measure of the given normal operator. In finite dimensions, this amounts to nothing more than the observation that a normal operator can be represented by a diagonal (and hence symmetric) matrix with respect to a certain orthonormal basis. In this case, C can simply be taken to be complex conjugation with respect to that basis.

In general, it suffices to show that the operator M_z of multiplication by the independent variable on a Lebesgue space $L^2(\mu)$ (where μ is a compactly supported Borel measure on $\mathbb{C}$) is complex symmetric. In this case, one can immediately verify the equation $M_z = CM_z^*C$ where C denotes complex conjugation on $L^2(\mu)$.

EXAMPLE 2.9. Since a Hankel matrix is symmetric, any (finite or infinite) Hankel matrix defines a complex symmetric operator on its associated l^2 space.

EXAMPLE 2.10. Another family of examples is furnished by considering integral operators whose kernels possess certain functional symmetries. For example, (bounded) integral operators of the form

$$[Tf](x) = \int_X K(x,y)f(y)d\mu(y)$$

whose kernels $K(x,y)$ are symmetric (in the sense that $K(x,y) = K(y,x)$ for all x, y in X) are C-symmetric with respect to complex conjugation $[Cf](x) = \overline{f(x)}$ on the corresponding Lebesgue space $L^2(X, \mu)$.

EXAMPLE 2.11. One can also combine complex conjugation on a Lebesgue space $L^2(X, \mu)$ with a measure-preserving geometric symmetry of the underlying measure space (X, μ). For example, the *Volterra integration operator*

$$[Vf](x) = \int_0^x f(t)\,dt,$$

defined on $L^2[0,1]$, is C-symmetric with respect to the conjugation $[Cf](x) = \overline{f(1-x)}$ (see Lemma 4.1). We will discuss this example and its relationship to compressed shifts in greater detail in Section 4.

EXAMPLE 2.12. All *compressed Toeplitz operators* (with which the remainder of this note is concerned) are complex symmetric operators. The details and definitions are postponed until Section 5.

Although we will not do so in this note, one can also consider unbounded complex symmetric operators as well. More information and references on the unbounded theory can be found in [**22, 24**].

It is natural to ask, for a fixed C, what operations preserve the class of C-symmetric operators. Multiplication, for instance, does not. Indeed, it is not hard to find examples of 2×2 symmetric matrices whose product is not symmetric. The following lemma (whose proof is omitted) provides one important method for constructing new complex symmetric operators from old ones:

LEMMA 2.13. *If T is a C-symmetric operator, then $p(T)$ is C-symmetric for any polynomial $p(z)$.*

By applying the lemma to the polynomial $p(z) = z - \lambda$, we see that if T is a C-symmetric operator, then λ is an eigenvalue of T if and only if $\overline{\lambda}$ is an eigenvalue for T^*. Moreover, it is not hard to see that C provides an antilinear isometric bijection between $\ker(T - \lambda I)^n$ and $\ker(T^* - \overline{\lambda} I)^n$ for all λ in $\mathbb{C}$ and all n. This is an example of the "spectral symmetry principle", satisfied by any complex symmetric operator. Loosely put, if T is a complex symmetric operator, then T and T^* are mirror images in every way and their spectral structures correspond under complex conjugation.

In the finite dimensional setting, the preceding comments do not amount to much. In infinite dimensions, however, we can easily exclude a wide variety of operators from the complex symmetric class.

EXAMPLE 2.14. The unilateral shift M_z on H^2 is not complex symmetric. Indeed, M_z has no eigenvalues while M_z^* (the backward shift) has many [**8, 30**]. Here is another straightforward proof. If there were a conjugation operator C on H^2 such that M_z was C-symmetric, then the equation $CM_z = M_z^*C$ would imply that

$$I = M_z^*M_z = (M_z^*C)(CM_z) = CM_zM_z^*C$$

and hence $M_zM_z^* = I$. This is absurd since $M_zM_z^*$ is the orthogonal projection onto the subspace zH^2.

More generally, one can show that T^*T and TT^* are unitarily equivalent whenever T is complex symmetric [**24**].

3. Conjugation on Model Spaces

The conjugation operator that we are most concerned with in this note resides on the so-called *model spaces*, which we briefly describe.

3.1. Model spaces. A famous theorem of Beurling characterizes the invariant subspaces of the unilateral shift operator M_z on H^2. It asserts that the proper, nontrivial invariant subspaces for M_z are precisely the subspaces

$$\varphi H^2 = \{\varphi f : f \in H^2\}$$

where φ is a nonconstant *inner function*, a bounded analytic function on the unit disk $\mathbb{D}$ with nontangential limiting values of unit modulus a.e. on $\partial\mathbb{D}$. Since inner functions and Beurling's theorem are well-understood, we refer the reader to the standard texts [**14, 35**].

It follows from Beurling's theorem that the proper, nontrivial invariant subspaces for the backward shift operator M_z^* are precisely the subspaces $H^2 \ominus \varphi H^2$, where φ is a nonconstant inner function. We refer to these subspaces as *model spaces* in light of an important theorem of Sz.-Nagy and Foaiş which roughly states that *any* Hilbert space contraction T such that T^n tends strongly to 0 is unitarily equivalent to a suitable vector-valued analogue of the *compressed shift* (also called a *model* or *Jordan* operator) $S : H^2 \ominus \varphi H^2 \longrightarrow H^2 \ominus \varphi H^2$ defined by

$$Sf = P_\varphi(zf) \tag{3.1}$$

where P_φ denotes the orthogonal projection from H^2 onto $H^2 \ominus \varphi H^2$. We remark also that the hypothesis that T^n tends strongly to zero is not restrictive and is meant to exclude the possibility of a unitary direct summand. For further details, the reader may consult the recent text [**38**].

The following lemma shows that each model space carries a natural conjugation operator:

LEMMA 3.1. *If φ is a nonconstant inner function, then*

$$Cf = \overline{fz}\varphi \tag{3.2}$$

defines a conjugation operator on $H^2 \ominus \varphi H^2$ which preserves outer factors.

PROOF. Let f be an arbitrary function in $H^2 \ominus \varphi H^2$ and consider the function $\overline{fz}\varphi$ in $L^2(\partial\mathbb{D})$. Although it appears at first that this is not the boundary function of an analytic function on $\mathbb{D}$, the short computation

$$\langle \overline{fz}\varphi, \overline{zh} \rangle = \langle \varphi h, f \rangle = 0$$

shows that $\overline{fz}\varphi$ is orthogonal to every anti-analytic function which vanishes at the origin. This implies that $\overline{fz}\varphi$, despite its appearance, belongs to H^2. The similar computation

$$\langle \overline{fz}\varphi, \varphi h \rangle = \langle \overline{zh}, f \rangle = 0$$

shows that $\overline{fz}\varphi$ is also orthogonal to φH^2. In other words, the antilinear operator C defined by (3.2) maps $H^2 \ominus \varphi H^2$ to itself. Since $|\varphi| = 1$ a.e. on $\partial\mathbb{D}$, it is not hard to verify conditions (*b*) and (*c*) in the definition of a conjugation operator. To see that C also preserves outer factors, observe that $|Cf| = |f|$ a.e. on $\partial\mathbb{D}$. □

Although the conjugation (3.2) is defined in terms of boundary functions, in the case where φ is a finite Blaschke product, the conjugation can be quite explicitly realized.

EXAMPLE 3.2. Let φ denote a finite Blaschke product

$$\varphi(z) = \prod_{k=1}^{n} \frac{z - \lambda_k}{1 - \overline{\lambda_k} z}$$

with n (not necessarily distinct) zeros λ_k. It is not hard to show that each function f in $H^2 \ominus \varphi H^2$ is of the form

$$f(z) = \frac{a_0 + a_1 z + \cdots + a_{n-1} z^{n-1}}{(1 - \overline{\lambda_1} z) \cdots (1 - \overline{\lambda_n} z)}$$

and that the conjugate function is given simply by

$$[Cf](z) = \frac{\overline{a_{n-1}} + \overline{a_{n-2}} z + \cdots + \overline{a_0} z^{n-1}}{(1 - \overline{\lambda_1} z) \cdots (1 - \overline{\lambda_n} z)}.$$

In particular, the conjugation operator (2.3) we considered in our discussion of Toeplitz matrices corresponds (under a suitable interpretation) to the special case $\varphi = z^n$. Further details about the conjugation operator on finite dimensional model spaces can be found in [**20, 23**].

In the case of a general inner function, the computations are not quite as simple. However, we will see in Subsection 3.4 that the conjugation operator (3.2) can be used to obtain an explicit function theoretic representation for the space $H^2 \ominus \varphi H^2$.

3.2. Conjugation and pseudocontinuation. In this subsection, we discuss the relationship between the concept of *pseudocontinuation*, non-cyclicity for the backward shift M_z^*, and our conjugation operator (3.2) on the model space $H^2 \ominus \varphi H^2$.

Due to the fact that each space $H^2 \ominus \varphi H^2$ is invariant under the backward shift M_z^*, one says that a function f belonging to one of the spaces $H^2 \ominus \varphi H^2$ is *noncyclic* since the closed linear span of its iterates under M_z^* fails to be all of H^2. On the other hand, one says that a function f is *cyclic* if it does not belong to a subspace of the form $H^2 \ominus \varphi H^2$. Equivalently, f is cyclic if and only if the closed linear span of its iterates under M_z^* is H^2.

A remarkable characterization of noncyclic functions is provided by a theorem of Douglas, Shapiro, and Shields. Their theorem requires a bit of explanation. Let $\mathbb{D}_e$ denote the complement of the closed unit disk in the extended complex plane. If f and $\tilde{f}$ are meromorphic functions on $\mathbb{D}$ and $\mathbb{D}_e$, respectively, with nontangential limiting values that agree almost everywhere on $\partial \mathbb{D}$, then f and $\tilde{f}$ are called *pseudocontinuations* of one another. A meromorphic function on $\mathbb{D}_e$ is of *bounded type* if it is the quotient of bounded analytic functions on $\mathbb{D}_e$. The following theorem from [**13**, Thm. 2.2.1] relates pseudocontinuations to the backward shift operator:

THEOREM 3.3 (Douglas, Shapiro, Shields). *A function $f \in H^2$ is non-cyclic for the backward shift operator if and only if f has a pseudocontinuation to $\mathbb{D}_e$ which is of bounded type.*

Letting $PCBT$ denote the class of H^2 functions which have pseudo-continuations to $\mathbb{D}_e$ which are of bounded type, we see that there exists a nonconstant inner function φ such that f belongs to $H^2 \ominus \varphi H^2$ if and only if f belongs to $PCBT$. The reader is invited to consult the recent books [8, 40] for a more detailed discussion of the preceding theorem and its generalizations to other function spaces. In particular, [40, Sect. 6.2, 6.3] contains numerous instructive examples.

The simplest nontrivial examples of pseudocontinuable functions are inner functions:

EXAMPLE 3.4. If φ is an inner function, then

$$\tilde{\varphi}(z) = 1/\overline{\varphi(1/\overline{z})}$$

is a pseudocontinuation of φ to $\mathbb{D}_e$ which is of bounded type.

EXAMPLE 3.5. Functions with isolated branch points on $\partial \mathbb{D}$, such as $\sqrt{1+z}$, do not possess pseudocontinuations at all and are therefore cyclic vectors for the backward shift (see [40, Ex. 6.2.3] for a thorough explanation).

To see how Theorem 3.3 fits in the context of conjugation operators, note that a function f belongs to $H^2 \ominus \varphi H^2$ if and only if there exists a function g in H^2 such that $f = \overline{gz}\varphi$ holds almost everywhere on $\partial \mathbb{D}$. Indeed, the proof of this fact is essentially contained in our proof of Lemma 3.1. If f is orthogonal to φH^2, then

$$0 = \langle f, \varphi h \rangle = \langle \overline{\varphi} f, h \rangle$$

for all h in H^2. This implies that $\overline{\varphi} f$ is orthogonal to H^2 and hence $\overline{\varphi} f = \overline{gz}$ for some g in H^2.

Letting g^* denote the function $g^*(z) = \overline{g(\overline{z})}$ obtained by conjugating the Taylor coefficients of g, we can interpret the equation $f = \overline{gz}\varphi$ as saying that the functions $f(z)/\varphi(z)$ on $\mathbb{D}$ and $\frac{1}{z}g^*(\frac{1}{z})$ on $\mathbb{D}_e$ have matching nontangential limits almost everywhere on $\partial \mathbb{D}$.

The approach that we take here is to think of the complementary function g as Cf, a function belonging to $H^2 \ominus \varphi H^2$ and hence with domain $\mathbb{D}$, as opposed to a function on $\mathbb{D}_e$. Indeed, the definition (3.2) of C indicates that

$$f = \overline{(Cf)z}\varphi.$$

In some sense, we see that Lemma 3.1 (on the existence of C) is a reinterpretation of the Douglas, Shapiro, and Shields result (Theorem 3.3). We will say considerably more about noncyclic functions in Subsection 3.4.

3.3. Conjugation and the reproducing kernel. In this subsection, we investigate the relationship between the conjugation operator (3.2) on a model space $H^2 \ominus \varphi H^2$ and the reproducing kernel Hilbert space structure of $H^2 \ominus \varphi H^2$.

Recall that the reproducing kernel for H^2 is the *Szegö kernel*

$$e_\lambda(z) = \frac{1}{1 - \overline{\lambda}z}, \tag{3.3}$$

where λ, z belong to $\mathbb{D}$. These functions have the property that

$$f(\lambda) = \langle f, e_\lambda \rangle \tag{3.4}$$

for every f in H^2 and all λ in $\mathbb{D}$. In fact, the preceding equation follows directly from the Cauchy integral formula or by considering the Taylor expansions of f and e_λ. It is not hard to derive from (3.4) and the definition of $H^2 \ominus \varphi H^2$ that the equation

$$f(\lambda) = \langle f, K_\lambda \rangle \tag{3.5}$$

holds for every f in $H^2 \ominus \varphi H^2$. Here K_λ denotes the *reproducing kernel*

$$K_\lambda(z) = \frac{1 - \overline{\varphi(\lambda)}\varphi(z)}{1 - \overline{\lambda}z} \tag{3.6}$$

for $H^2 \ominus \varphi H^2$.

Since the model space $H^2 \ominus \varphi H^2$ is a reproducing kernel Hilbert space, it is natural to investigate the relationship between the reproducing kernel (3.6) and the conjugation operator (3.2). The reproducing property (3.5) of K_λ and the isometric property of C imply that

$$[Cf](\lambda) = \langle Cf, K_\lambda \rangle = \langle CK_\lambda, f \rangle$$

for all f in $H^2 \ominus \varphi H^2$ and λ in $\mathbb{D}$. In other words, the *conjugate kernel* function CK_λ reproduces the values of Cf via the formula $[Cf](\lambda) = \langle CK_\lambda, f \rangle$. Indeed, the existence of such a conjugate kernel could have been deduced from the Riesz representation theorem since the map $f \mapsto \overline{[Cf](\lambda)}$ is a bounded linear functional for every $|\lambda| < 1$.

Using the definition (3.2) of C, we can give an explicit formula for the conjugate kernel function CK_λ. We need only keep in mind that our computations take place on $\partial\mathbb{D}$, where φ is unimodular almost everywhere:

$$
\begin{aligned}
[CK_\lambda](z) &= \overline{\left(\frac{1 - \overline{\varphi(\lambda)}\varphi(z)}{1 - \overline{\lambda}z}\right)} \overline{z}\varphi(z) \\
&= \frac{1 - \varphi(\lambda)\overline{\varphi(z)}}{1 - \lambda\overline{z}} \cdot \frac{\varphi(z)}{z} \\
&= \frac{\varphi(z) - \varphi(\lambda)}{z - \lambda}. \tag{3.7}
\end{aligned}
$$

Thus the conjugation kernel function CK_λ for $H^2 \ominus \varphi H^2$ is none other than the difference quotient (3.7).

The last statement of Lemma 3.1 asserts that C preserves outer factors. In particular, this applies to the functions K_w and CK_w. This is an important fact which we verify directly. For each λ, the reproducing kernel K_λ is an outer function, being the quotient of the two outer functions $1 - \overline{\varphi(\lambda)}\varphi(z)$

and $1 - \bar{\lambda}z$. Since C preserves outer factors, it follows that the corresponding conjugate function CK_λ is the product of K_λ with an inner function. This is indeed the case:

$$[CK_\lambda](z) \;=\; \frac{\varphi(z) - \varphi(\lambda)}{z - \lambda}$$

$$=\; \frac{\varphi(z) - \varphi(\lambda)}{1 - \overline{\varphi(\lambda)}\varphi(z)} \cdot \frac{1 - \bar{\lambda}z}{z - \lambda} \cdot \frac{1 - \overline{\varphi(\lambda)}\varphi(z)}{1 - \bar{\lambda}z}$$

$$=\; \frac{b_{\varphi(\lambda)}(\varphi(z))}{b_\lambda(z)} K_\lambda(z).$$

Here b_a denotes the disk automorphism

$$b_a(z) = \frac{z - a}{1 - \bar{a}z}$$

where a belongs to $\mathbb{D}$. In particular, we observe that the inner factor of CK_λ is the inner function $b_{\varphi(\lambda)}(\varphi(z))/b_\lambda(z)$.

A remarkable thing happens as $|\lambda| \longrightarrow 1$. Let ζ be a point on $\partial\mathbb{D}$ such that φ has a nontangential limiting value at ζ of unit modulus and suppose that the limit function K_ζ formally obtained by substituting ζ for λ in (3.6) belongs to H^2 (and hence $H^2 \ominus \varphi H^2$). Since ζ and $\varphi(\zeta)$ are unimodular, we see that

$$[CK_\zeta](z) \;=\; \frac{\varphi(z) - \varphi(\zeta)}{z - \zeta}$$

$$=\; \frac{\varphi(\zeta)}{\zeta} \cdot \frac{1 - \overline{\varphi(\zeta)}\varphi(z)}{1 - \bar{\zeta}z}$$

$$=\; \bar{\zeta}\varphi(\zeta)K_\zeta(z).$$

Thus, the functions K_ζ and CK_ζ corresponding to a point ζ on $\partial\mathbb{D}$ differ only by the unimodular constant $\bar{\zeta}\varphi(\zeta)$. Selecting any branch of the square root function, it follows from the antilinearity of C that the functions

$$(\bar{\zeta}\varphi(\zeta))^{\frac{1}{2}}K_\zeta$$

are fixed by C (whenever they belong to H^2). This is an idea that we will return to several times in this note.

3.4. Description of noncyclic functions. Using the conjugation operator (3.2), we can obtain an explicit function theoretic characterization of the functions in $H^2 \ominus \varphi H^2$. The construction that we pursue in this subsection can be found in [19].

Since $C^2 = I$, we can split a given function f in $H^2 \ominus \varphi H^2$ into C-*real* and C-*imaginary* parts. Each f in $H^2 \ominus \varphi H^2$ decomposes as

$$f = \tfrac{1}{2}(f + Cf) + i\tfrac{1}{2i}(f - Cf)$$

where the terms $\tfrac{1}{2}(f + Cf)$ and $\tfrac{1}{2i}(f - Cf)$ are both fixed by C. This motivates us to find the fixed points for C.

Suppose that f is fixed by C. In other words, f satisfies

$$f = \overline{f} z \varphi \tag{3.8}$$

almost everywhere on $\partial \mathbb{D}$. Now select a point ζ on $\partial \mathbb{D}$ where φ has a nontangential limiting value of unit modulus. Since φ is an inner function, almost every ζ will have this property. We may, without loss of generality, assume that $\varphi(\zeta) = \zeta$ since this may be obtained by multiplying φ by a suitable unimodular constant.

The boundary kernel function

$$K_\zeta(z) = \frac{1 - \overline{\varphi(\zeta)} \varphi(z)}{1 - \overline{\zeta} z}$$

obtained from (3.6) by letting $\lambda \to \zeta$ nontangentially yields a function in the Smirnov class N^+ (see [8, 14] for more information). Although K_ζ might not belong to H^2, it is the quotient of two bounded analytic functions and hence belongs to N^+.

The appearance of the term $\overline{z} \varphi$ in (3.8) is fortuitous as the following calculation will justify. Using the fact that $\varphi(\zeta) = \zeta$ we see that

$$\begin{aligned}
K_\zeta(z)/\overline{K_\zeta(z)} &= \frac{1 - \overline{\varphi(\zeta)} \varphi(z)}{1 - \overline{\zeta} z} \cdot \frac{1 - \zeta \overline{z}}{1 - \varphi(\zeta) \overline{\varphi(z)}} \\
&= \frac{1 - \overline{\varphi(\zeta)} \varphi(z)}{1 - \varphi(\zeta) \overline{\varphi(z)}} \cdot \frac{1 - \zeta \overline{z}}{1 - \overline{\zeta} z} \\
&= \overline{\varphi(\zeta)} \varphi(z) \left(\frac{\varphi(\zeta) \overline{\varphi(z)} - 1}{1 - \varphi(\zeta) \overline{\varphi(z)}} \right) \cdot \zeta \overline{z} \left(\frac{\overline{\zeta} z - 1}{1 - \overline{\zeta} z} \right) \\
&= \zeta \overline{\varphi(\zeta)} \overline{z} \varphi(z) \\
&= \overline{z} \varphi(z)
\end{aligned}$$

and thus (3.8) can be written in the symmetric form

$$f/K_\zeta = \overline{f/K_\zeta}.$$

This shows that the function f/K_ζ belongs to N^+ and is *real* almost everywhere on $\partial \mathbb{D}$. Thus each f in $H^2 \ominus \varphi H^2$ can be written in the form

$$f(z) = [a(z) + ib(z)] K_\zeta(z)$$

where a and b are functions in N^+ which are real a.e. on $\partial \mathbb{D}$.

A function f belonging to the Smirnov class N^+ is called a *real Smirnov function* if its boundary function is real valued a.e. on $\partial \mathbb{D}$. The set R^+ of all real Smirnov functions is a real subalgebra of N^+ that was explicitly described by Helson [31]. He showed that if ψ_1 and ψ_2 are relatively prime inner functions such that $\psi_1 - \psi_2$ is outer, then the function

$$f(z) = i \frac{\psi_1 + \psi_2}{\psi_1 - \psi_2} \tag{3.9}$$

is a real Smirnov function and every real Smirnov function arises this way. Although elegant, the representation (3.9) has its limitations. For example, the inner functions ψ_1 and ψ_2 are often difficult to identify and there are no general criteria describing when the difference of inner functions is outer. A simple proof of Helson's formula (3.9) can be based on linear fractional transformations and the factorization theory for N^+ [**25**, Sec. 3].

In [**25**], Sarason and the author proved that any outer function F belonging to R^+ can be represented as a locally uniformly convergent product

$$F(z) = |F(0)| \prod_{n=1}^{\infty} \frac{T(\varphi_n^+)}{T(\varphi_n^-)},$$

where T denotes the linear fractional transformation

$$T(z) = i\frac{1 - iz}{1 + iz},$$

and the inner functions φ_n^+ and φ_n^- are naturally associated with the boundary values of $\arg F$ on $\partial\mathbb{D}$. The proof of this fact is somewhat technical and relies heavily on the Cauchy A-integral (see [**4**] and the recent book [**9**]), due to the fact that $\arg F$, unlike $\log|F|$, might not be integrable on $\partial\mathbb{D}$.

By using the Koebe function $k(z) = z/(1-z)^2$ (see [**15**]), we can reduce the description of functions in R^+ to the preceding case of so-called *real outer functions*. If ψ is a nonconstant inner function, then $k(\psi)$ belongs to R^+. Since the inner factor of $k(\psi)$ is precisely ψ, the construction of the general function in R^+ can be reduced to the product (3.4). A more detailed description of this reduction can be found in [**21**].

In short, real Smirnov functions can be described explicitly and reasonably constructively in terms of inner functions. In light of Example 3.4, these computations explain the theorem of Douglas, Shapiro, and Shields (Theorem 3.3). Specifically, pseudocontinuations of bounded type arise from inner functions.

Similar considerations lead to the following parameterization of the invariant subspaces for the backward shift on H^p (for $1 < p < \infty$):

$$H^p \cap \varphi \overline{zH^p} = \{\, [a(z) + ib(z)]K_\zeta(z) : a, b \in R^+\} \cap H^p.$$

Even further, one can use the same method to parametrize the kernels of Toeplitz operators on H^p for $1 < p < \infty$ [**19**].

3.5. C-real bases for model spaces. One of our main concerns in this note is to produce C-real (fixed by C) orthonormal bases for the model spaces $H^2 \ominus \varphi H^2$. Our interest in this problem stems not only from the desire to understand the function theoretic aspects of the conjugation operator (3.2), but also to understand its *operator theoretic* implications.

Unfortunately, an explicit description of C-real bases for model spaces is not always readily apparent. Indeed, finding explicit orthonormal bases for model spaces is nontrivial. The Malmquist-Walsh lemma [**48**] (see also [**37**, Lec. V.]) for instance, applies only in the case where φ is a Blaschke product.

Moreover, as the reader may check, this procedure does not produce a C-real basis for $H^2 \ominus \varphi H^2$.

Instead, we parallel and generalize the method of Clark [10] which, it turns out, often produces C-real bases for model spaces. Those familiar with this method will recognize that there is no mention of the conjugation operator (3.2) in [10]. Indeed, that article was concerned with producing only orthonormal bases for the model spaces. A more detailed study of complex symmetric operators and their properties will ultimately explain exactly why Clark's method often succeeds in producing C-real bases.

4. Conjugation and the Volterra Operator

In this section, we consider the specific problem of constructing C-real bases for the model space $H^2 \ominus \varphi H^2$ where φ denotes the singular inner function

$$\varphi(z) = \exp\left(\frac{z+1}{z-1}\right)$$

arising from a unit point mass at $z = 1$. In other words, we wish to find an orthonormal basis for $H^2 \ominus \varphi H^2$ that is fixed by the conjugation operator $Cf = \overline{f}z\varphi$. Finding such a basis is natural from the viewpoint of complex symmetric operators, for the matrix representation of the compression of a Toeplitz operator to $H^2 \ominus \varphi H^2$ will be symmetric with respect to such a basis.

Although we will shortly develop a much more general approach to this problem, a careful analysis of this special case will be instructive. Indeed, this simple case will illustrate some of our main points and highlight the relationship between our approach and several well-known and classical results.

We now consider the relationship between our basis problem and Sarason's solution to the Gelfand problem.

4.1. The Gelfand problem. In 1938, I.M. Gelfand raised the question of characterizing the invariant subspace lattice of the Volterra integration operator

$$[Vf](x) = \int_0^x f(t)\,dt$$

on $L^2[0,1]$ [26]. It is clear that for any a in $[0,1]$, the subspace $\chi_{[0,a]} L^2[0,1]$ is an invariant subspace for V. Here $\chi_{[0,a]}$ denotes the characteristic function of the closed interval $[0,a]$.

It turns out that these are the only invariant subspaces for the Volterra operator and hence Lat V is linearly ordered, or *unicellular*. This result was proved first in 1949 by Agmon [1] and later by Sakhnovich [41], Brodskii [6], Donoghue [12], Kalisch [34], and Sarason [42]. We are interested here not in the Gelfand problem itself, but rather in a clever technique used by Sarason, which was later generalized by Ahern and Clark [3] (see also [37, Lec. 5]).

What Sarason noted was that the lattice of invariant subspaces for the Volterra operator corresponds, under a certain integral transform, to the lattice of invariant subspaces for the compressed shift on the model space $H^2 \ominus \varphi H^2$ corresponding to the atomic inner function

$$\varphi(z) = \exp\left(\frac{z+1}{z-1}\right).$$

Once this is proved, the unicellularity of Lat V follows immediately from Beurling's theorem. Indeed, the inner functions dividing φ are precisely the functions φ^t for $0 \leq t \leq 1$ where φ^t is unambiguously defined by the formula

$$[\varphi(z)]^t = \exp\left(t\frac{z+1}{z-1}\right).$$

The integral transform

$$[\mathcal{W}g](z) = \frac{\sqrt{2}i}{z-1} \int_0^1 g(t)[\varphi(z)]^t \, dt \qquad (4.1)$$

is a unitary map from $L^2[0,1]$ to $H^2 \ominus \varphi H^2$. Indeed, this is simply a unimodular constant multiple of the transformation appearing in the introduction to Clark's original paper [10] on compressed shifts. Although at first the formula (4.1) might appear a little mysterious, it is easy to motivate once one observes that the integrand $g(t)\varphi^t(z)$ (as a function of z) belongs to $H^2 \ominus \varphi H^2$ for all t in $[0,1]$. In fact, the operator $\mathcal{W}$ yields the unitary equivalence of the characteristic function of the Cayley transform

$$(I-V)(I+V)^{-1}$$

of the Volterra operator V with its functional model, the compressed shift S on $H^2 \ominus \varphi H^2$ (for full details, see [38] and [37, Lec. 5]). In other words, the operators S and V are related by the formula

$$S = \mathcal{W}[(I-V)(I+V)^{-1}]\mathcal{W}^*.$$

The desired solution to the Gelfand problem then follows from a simple resolvent lemma (see [38, Lem 2.1.10]) which shows that

$$\text{Lat } V = \text{Lat}(I-V)(I+V)^{-1}$$

and hence Lat V is isomorphic to the linearly ordered lattice for the compressed shift.

The unitary operator $\mathcal{W} : L^2[0,1] \longrightarrow H^2 \ominus \varphi H^2$ is therefore useful in passing between properties of the Volterra operator V and the compressed shift S. We will use this unitary transformation to investigate the complex symmetric properties of the compressed shift using known properties of the Volterra operator. For instance, a straightforward computation based on the fact that

$$[V^*f](x) = \int_x^1 f(t) \, dt$$

shows that the Volterra operator is complex symmetric:

LEMMA 4.1. *The Volterra integration operator* $V : L^2[0,1] \longrightarrow L^2[0,1]$ *is C-symmetric with respect to the conjugation operator* $[\mathcal{C}f](x) = \overline{f(1-x)}$ *on* $L^2[0,1]$.

As noted in [**24**], the equation $\mathcal{C}V = V^*\mathcal{C}$ ultimately reflects the functional symmetry $K(x,y) = K(1-y, 1-x)$ of the *Volterra kernel* $K(x,y)$, the characteristic function of the triangle

$$\{(x,y) : 0 \leq y \leq x \leq 1\}.$$

Equivalently, one might say that the $\mathcal{C}$-symmetry of V arises from the geometric symmetry of the triangle.

Since V is a $\mathcal{C}$-symmetric operator, it follows (by passing to the limit in Lemma 2.13) that the Cayley transform $(I - V)(I + V)^{-1}$ of V is also $\mathcal{C}$-symmetric. Since this operator is unitarily equivalent (via $\mathcal{W}$) to the compressed shift S, it follows that S is a complex symmetric operator with respect to the conjugation operator $\mathcal{W}\mathcal{C}\mathcal{W}^*$ on $H^2 \ominus \varphi H^2$. Remarkably, the corresponding conjugation on $H^2 \ominus \varphi H^2$ is precisely the conjugation $Cf = \overline{f}z\varphi$ discussed in the preceding pages.

LEMMA 4.2. *The conjugation operator*

$$[\mathcal{C}g](x) = \overline{g(1-x)}$$

on $L^2[0,1]$ *and the conjugation operator*

$$Cf = \overline{f}z\varphi$$

on $H^2 \ominus \varphi H^2$ *are related by the unitary operator* $\mathcal{W}$:

$$C = \mathcal{W}\mathcal{C}\mathcal{W}^*.$$

PROOF. We prove the lemma by establishing that $\mathcal{W}\mathcal{C} = C\mathcal{W}$. For any g in $L^2[0,1]$, the integrands of the following integrals are all dominated by $\max\{|g(t)|, |g(1-t)|\}$ and it follows that

$$
\begin{aligned}
[\mathcal{W}\mathcal{C}g](z) &= \frac{\sqrt{2}i}{z-1} \int_0^1 [\mathcal{C}g](t)[\varphi(z)]^t dt \\
&= \frac{\sqrt{2}i}{z-1} \int_0^1 \overline{g(1-t)}[\varphi(z)]^t dt \\
&= \frac{\sqrt{2}i}{z-1} \int_0^1 \overline{g(s)}[\varphi(z)]^{1-s} ds \\
&= \overline{z}\varphi(z)\overline{\frac{\sqrt{2}i}{1-z} \int_0^1 g(s)[\varphi(z)]^s ds} \\
&= \overline{[\mathcal{W}g](z)}z\varphi(z)
\end{aligned}
$$

for almost every z on $\partial\mathbb{D}$.   □

Using the unitary transformation $\mathcal{W} : L^2[0,1] \longrightarrow H^2 \ominus \varphi H^2$, we can now transform $\mathcal{C}$-real bases in $L^2[0,1]$ into C-real bases in the model space $H^2 \ominus \varphi H^2$.

4.2. Constructing C-real bases. As we have noted before, one of the primary difficulties in working with model spaces is the lack of convenient orthonormal bases. On the other hand, the space $L^2[0,1]$ has many natural orthonormal bases, the most obvious being the exponential basis $(e^{2\pi inx})_{n\in\mathbb{Z}}$. Since the Sarason transform (4.1) is a unitary operator from $L^2[0,1]$ onto $H^2\ominus\varphi H^2$, it follows that the image of any orthonormal basis in $L^2[0,1]$ will be an orthonormal basis for $H^2\ominus\varphi H^2$. Moreover, Lemma 4.2 guarantees that the image under W of a C-real basis of $L^2[0,1]$ will be C-real in $H^2\ominus\varphi H^2$. In particular, this will provide "natural" C-real bases for the model space $H^2\ominus\varphi H^2$ with respect to which the compressed shift can be represented as a complex symmetric matrix.

We will now construct a continuous family (indexed by a parameter α) of C-real bases for $H^2\ominus\varphi H^2$ by transforming the C-real bases of $L^2[0,1]$ provided by the following lemma:

LEMMA 4.3. *For each fixed α in $[0,2\pi)$, the vectors $(e_n)_{n\in\mathbb{Z}}$ defined by*

$$e_n(x) = \exp[i(\alpha + 2\pi n)(x - \tfrac{1}{2})] \qquad (4.2)$$

form a C-real basis of $L^2[0,1]$.

PROOF. A direct computation shows that each e_n is fixed by C.

$$
\begin{aligned}
[Ce_n](x) &= \overline{e^{i(\alpha+2\pi n)((1-x)-\frac{1}{2})}} \\
&= e^{-i(\alpha+2\pi n)(\frac{1}{2}-x)} \\
&= e^{i(\alpha+2\pi n)(x-\frac{1}{2})} \\
&= e_n(x).
\end{aligned}
$$

To see that $(e_n)_{n\in\mathbb{Z}}$ is an orthonormal basis for $L^2[0,1]$, simply observe the identity

$$
\begin{aligned}
e^{i(\alpha+2\pi n)(x-\frac{1}{2})} &= e^{i(\alpha x+2\pi nx-\frac{\alpha}{2}-\pi n)} \\
&= (-1)^n e^{-\frac{i}{2}\alpha} e^{i\alpha x} e^{i2\pi nx}
\end{aligned}
$$

and note that multiplication by $e^{i\alpha x}$ is a unitary operator on $L^2[0,1]$. $\qquad\square$

For each α in $[0,2\pi)$, we obtain an orthonormal basis $(e_n)_{n\in\mathbb{Z}}$ of $L^2[0,1]$ that is fixed by C. In particular, for each α we obtain a natural basis of $L^2[0,1]$ with respect to which C-symmetric operators, such as the Volterra operator and its Cayley transform, have symmetric matrix representations. We now wish to compute the image of this basis in $H^2\ominus\varphi H^2$ under the unitary transformation W.

Writing each basis vector e_n in the form

$$e_n(x) = e^{-i(\alpha+2\pi n)/2} e^{i(\alpha+2\pi n)x}, \qquad (4.3)$$

we see that it suffices to compute $e^{-i\gamma/2}We^{i\gamma t}$ for real γ. Once this is done, we need only substitute $\gamma = \alpha + 2\pi n$ to find the corresponding basis for our model space. Since this computation is somewhat lengthy, we will present it through a sequence of lemmas.

The following lemma already shows that for a particular α, the image of the basis $(e_n)_{n\in\mathbb{Z}}$ in $H^2 \ominus \varphi H^2$ will be a family of reproducing kernels corresponding to a sequence of points on the unit circle. Since φ is analytically continuable across any arc of the unit circle $\partial\mathbb{D}$ not containing $z = 1$, the reproducing kernel corresponding to any boundary point $\zeta \neq 1$ in $\partial\mathbb{D}$ is well-defined and belongs to $H^2 \ominus \varphi H^2$.

LEMMA 4.4. *If γ is real, then*

$$[We^{i\gamma x}](z) = \frac{(1-\zeta)}{\sqrt{2i}\zeta} K_\zeta(z),$$

where $K_\zeta(z)$ is the reproducing kernel (3.6) for $H^2 \ominus \varphi H^2$ corresponding to the point ζ on $\partial\mathbb{D}$ defined by

$$\zeta = \frac{\gamma+i}{\gamma-i}.$$

In particular, we have $\varphi(\zeta) = e^{-i\gamma}$.

PROOF. Fixing γ in $\mathbb{R}$ and using the definition (4.1) of W we find that

$$\begin{aligned}
[We^{i\gamma x}](z) &= \frac{\sqrt{2i}}{z-1} \int_0^1 e^{i\gamma t} \exp\left(t\frac{z+1}{z-1}\right) dt \\
&= \frac{\sqrt{2i}}{z-1} \int_0^1 \exp\left[t\left(i\gamma + \frac{z+1}{z-1}\right)\right] dt \\
&= \frac{\sqrt{2i}}{z-1} \cdot \frac{1}{i\gamma + \frac{z+1}{z-1}} \left(\exp\left[\left(i\gamma + \frac{z+1}{z-1}\right)\right] - 1\right) \\
&= \frac{\sqrt{2i}}{i\gamma(z-1)+(z+1)} \left(e^{i\gamma}\varphi(z) - 1\right) \\
&= \frac{-\sqrt{2i}}{(1-i\gamma)+(1+i\gamma)z} \left(1 - \overline{e^{-i\gamma}}\varphi(z)\right) \\
&= \frac{-\sqrt{2i}}{1-i\gamma} \cdot \frac{1}{1 + \frac{1+i\gamma}{1-i\gamma}z} \left(1 - \overline{e^{-i\gamma}}\varphi(z)\right). \quad (4.4)
\end{aligned}$$

Seeking to construct a scalar multiple of a reproducing kernel from (4.4), we wish to write $e^{-i\gamma} = \varphi(\zeta)$ for some ζ. Defining ζ by

$$\gamma = i\frac{\zeta+1}{\zeta-1} \quad \Longleftrightarrow \quad \zeta = \frac{\gamma+i}{\gamma-i}, \quad (4.5)$$

we see that ζ lies on the unit circle $\partial\mathbb{D}$, $\varphi(\zeta) = e^{-i\gamma}$, and that

$$-\bar{\zeta} = \frac{1+i\gamma}{1-i\gamma}.$$

Substituting these values into (4.4) shows that

$$[We^{i\gamma x}](z) = \frac{-\sqrt{2i}}{1-i\gamma} \cdot \frac{1 - \overline{\varphi(\zeta)}\varphi(z)}{1-\bar{\zeta}z}$$

$$= \frac{-\sqrt{2}i}{1 + \frac{\zeta+1}{\zeta-1}} K_\zeta(z)$$

$$= \frac{-\sqrt{2}i(\zeta - 1)}{(\zeta - 1) + (\zeta + 1)} K_\zeta(z)$$

$$= \frac{-\sqrt{2}i(\zeta - 1)}{2\zeta} K_\zeta(z)$$

$$= \frac{(1 - \zeta)}{\sqrt{2}i\zeta} K_\zeta(z), \tag{4.6}$$

where $K_\zeta(z)$ denotes the reproducing kernel (3.6) evaluated at the boundary point ζ. This proves the desired formula. □

It is important to note that in Clark's paper [10], it is simply noted that $\mathcal{W}$ maps the exponential functions $e^{i\gamma t}$ onto *constant multiples* of reproducing kernels corresponding to certain points on $\partial\mathbb{D}$. We emphasize here that the exact constants that appear are the key to the complex symmetry. By keeping careful track of the constants that appear in the following computations, we will ultimately be able to follow the symmetry through from $L^2[0,1]$ to $H^2 \ominus \varphi H^2$.

Since $\mathcal{W}$ is a unitary transformation, the image of $\mathcal{W}e^{i\gamma x}$ in $H^2 \ominus \varphi H^2$ must be a unit vector. This suggests that we must simplify the expression (4.6) further and examine the constant appearing there more carefully.

LEMMA 4.5. *If γ is real, then*

$$[\mathcal{W}e^{i\gamma x}](z) = \frac{(1 - \zeta)}{|1 - \zeta|\zeta i} k_\zeta(z),$$

where $k_\zeta(z)$ is the normalized reproducing kernel for $H^2 \ominus \varphi H^2$ corresponding to ζ.

PROOF. Since $\gamma \neq \infty$, we have $\zeta \neq 1$ and the norm of the function K_ζ can be explicitly computed in terms of φ. In fact, if z approaches ζ nontangentially we have

$$\|K_\zeta\|^2 = \lim_{z \to \zeta} \langle K_z, K_z \rangle$$

$$= \lim_{z \to \zeta} \frac{1 - |\varphi(z)|^2}{1 - |z|^2}$$

$$= \lim_{z \to \zeta} \frac{1 - |\varphi(z)|}{1 - |z|} \cdot \frac{1 + |\varphi(z)|}{1 + |z|}$$

$$= \lim_{z \to \zeta} \frac{1 - |\varphi(z)|}{1 - |z|}$$

$$= |\varphi'(\zeta)|.$$

The interested reader may consult [39] for further details. A straightforward computation using the definition of φ shows that

$$\| K_\zeta \| = \frac{\sqrt{2}}{|1 - \zeta|}.$$

With this computation in mind, we continue from (4.6) and find that

$$
\begin{aligned}
[\mathcal{W} e^{i\gamma x}](z) &= \frac{(1 - \zeta)}{|1 - \zeta|\zeta i} \frac{|1 - \zeta|}{\sqrt{2}} K_\zeta(z) \\
&= \frac{(1 - \zeta)}{|1 - \zeta|\zeta i} k_\zeta(z),
\end{aligned}
$$

where k_ζ denotes the *normalized* kernel function corresponding to ζ. This proves the desired formula. $\square$

Although it is evident from the preceding lemma that $\mathcal{W} e^{i\gamma x}$ is a unit vector in $H^2 \ominus \varphi H^2$, this was not our primary objective. Indeed, recall that we wanted to show that for any real γ, the function $e^{-i\gamma/2} \mathcal{W} e^{i\gamma x}$ is fixed by C. It turns out that the apparently unwieldy constant

$$\frac{(1 - \zeta)}{|1 - \zeta|\zeta i} \tag{4.7}$$

in preceding formulas appears there for exactly this reason.

The square of (4.7) equals

$$\frac{-(1 - \zeta)^2}{(1 - \zeta)(1 - \overline{\zeta})\zeta^2} = \overline{\zeta}$$

and hence the constant (4.7) is simply one of the square roots of $\overline{\zeta}$. The particular branch of the square root of $\overline{\zeta}$ represented by (4.7) is not particularly important for our purposes (although it is easily computable for a given γ) and we henceforth denote the constant (4.7) by $\overline{\zeta}^{1/2}$.

What we have shown so far is that for any real γ,

$$[\mathcal{W} e^{i\gamma x}](z) = \overline{\zeta}^{1/2} k_\zeta(z),$$

where the unimodular constant ζ satisfies $\varphi(\zeta) = e^{-i\gamma}$. To complete the evaluation of $e^{-i\gamma/2} \mathcal{W} e^{i\gamma t}$, we simply use (4.5) to describe the value of the constant $e^{-i\gamma/2}$ in terms of the inner function φ:

$$e^{-i\gamma/2} = e^{\frac{1}{2} \frac{\zeta + 1}{\zeta - 1}} = [\varphi(\zeta)]^{\frac{1}{2}}.$$

As before, we are not concerned with the particular branch of the square root that appears in the preceding formula. We summarize our computations in the following lemma:

LEMMA 4.6. *If γ is real, then*

$$e^{-i\gamma/2}[\mathcal{W} e^{i\gamma x}](z) = [\overline{\zeta}\varphi(\zeta)]^{\frac{1}{2}} k_\zeta(z) \tag{4.8}$$

*where $k_\zeta(z)$ is the normalized reproducing kernel for $H^2 \ominus \varphi H^2$ correspond-
ing to the point ζ. Each function (4.8) is fixed by the conjugation operator
$Cf = \overline{f z \varphi}$ on $H^2 \ominus \varphi H^2$.*

PROOF. The first portion of the lemma is simply a summary of the
preceding computations. Only the last statement requires proof. If $|z| = 1$
and $z \neq 1$, then

$$
\begin{aligned}
k_\zeta(z)/\overline{k_\zeta(z)} &= \frac{1 - \overline{\varphi(\zeta)}\varphi(z)}{1 - \overline{\zeta}z} \cdot \frac{1 - \zeta \overline{z}}{1 - \varphi(\zeta)\overline{\varphi(z)}} \\
&= \frac{1 - \overline{\varphi(\zeta)}\varphi(z)}{1 - \varphi(\zeta)\overline{\varphi(z)}} \cdot \frac{1 - \zeta \overline{z}}{1 - \overline{\zeta}z} \\
&= \overline{\varphi(\zeta)}\varphi(z) \left(\frac{\varphi(\zeta)\overline{\varphi(z)} - 1}{1 - \varphi(\zeta)\overline{\varphi(z)}} \right) \cdot \zeta \overline{z} \left(\frac{\overline{\zeta}z - 1}{1 - \overline{\zeta}z} \right) \\
&= \zeta \overline{\varphi(\zeta)} \overline{z} \varphi(z)
\end{aligned}
$$

since $\varphi(z)$ is unimodular. This identity, along with a short calculation, shows
that the function $[\overline{\zeta}\varphi(\zeta)]^{\frac{1}{2}} k_\zeta(z)$ is fixed by C. □

4.3. Summary. Putting this all together, for each fixed α in $[0, 2\pi)$ we
obtain an orthonormal basis $(e_n)_{n \in \mathbb{Z}}$ of $L^2[0, 1]$ defined by (4.2):

$$e_n(x) = \exp[i(\alpha + 2\pi n)(x - \tfrac{1}{2})],$$

which is fixed by the conjugation operator

$$[\mathcal{C}f](x) = \overline{f(1 - x)}$$

on $L^2[0, 1]$. In light of (4.8), the image of this basis in $H^2 \ominus \varphi H^2$ under the
Sarason transform

$$[\mathcal{W}g](z) = \frac{\sqrt{2}i}{z - 1} \int_0^1 g(t)[\varphi(z)]^t \, dt$$

is given by

$$
\begin{aligned}
[\mathcal{W}e_n](z) &= [\overline{\zeta_n}\varphi(\zeta_n)]^{1/2} k_{\zeta_n}(z) \\
&= \overline{\zeta_n}^{1/2} e^{-\frac{i}{2}\gamma} k_{\zeta_n}(z),
\end{aligned}
$$

where k_{ζ_n} denotes the *normalized* reproducing kernel corresponding (via
(4.5)) to the point

$$\zeta_n = \frac{(\alpha + 2\pi n) + i}{(\alpha + 2\pi n) - i}$$

on the unit circle. The functions $\mathcal{W}e_n$ are fixed by the conjugation operator

$$Cf = \overline{f z \varphi}$$

on $H^2 \ominus \varphi H^2$ and hence the matrix representation of any compressed Toeplitz
operator (including the compressed shift) on $H^2 \ominus \varphi H^2$ with respect to the

basis $(\mathcal{W}e_n)_{n\in\mathbb{Z}}$ will be complex symmetric. Finally, the points ζ_n can be characterized as the inverse image of $e^{-i\alpha}$ under φ:

$$\varphi(\zeta_n) = e^{-i\alpha}$$

for all n in $\mathbb{Z}$.

In short, we have constructed a continuous (indexed by the parameter α) family of C-real bases for the model space $H^2 \ominus \varphi H^2$. Each such basis naturally corresponds to a "level set" $(\zeta_n)_{n\in\mathbb{Z}}$ of the original inner function. The members of our bases are unimodular scalar multiples of the corresponding normalized reproducing kernels k_{ζ_n}.

Our computations in this special case illustrate a more general phenomenon, namely that one can often obtain a continuous family of C-real bases for a model space by looking at appropriate level sets of the associated inner function. The key to the present construction was the integral transform $\mathcal{W}$, which mapped $L^2[0, 1]$ unitarily onto the model space $H^2 \ominus \varphi H^2$. Although generalizations of this integral transform exist for general inner functions (see [**3**]), the details of our construction quickly become too cumbersome. Indeed, the lengthy algebraic manipulations in the case of our simple atomic inner function should be enough to convince the reader that a different approach is necessary.

To develop a more general procedure to construct C-real bases for model spaces, we will need to appeal to several structure theorems for complex symmetric operators.

5. Compressed Toeplitz Operators

In this section, we recall some of the elementary properties of Toeplitz operators and their compressions to model spaces. The most important result of this section is Theorem 5.1, which shows that compressed Toeplitz operators are complex symmetric.

5.1. Toeplitz operators. For each u in $L^\infty(\partial\mathbb{D})$, the *Toeplitz operator* with *symbol* u is the operator $T_u : H^2 \to H^2$ defined by $T_u f = P(uf)$ where P denotes the orthogonal projection from L^2 onto H^2. Note that if u belongs to H^∞, then T_u is simply the operator M_u of multiplication by u. It is not hard to show that the adjoint of a Toeplitz operator is given by the simple formula $T_u^* = T_{\bar{u}}$.

The most important examples of Toeplitz operators arise from the function $u(z) = z$. In this case, T_z and T_z^* are simply the unilateral shift operator M_z and the backward shift operator M_z^*, respectively. Indeed, a straightforward computation shows that $T_z f = zf$ shifts the Taylor coefficients (at the origin) of f to the right, while

$$[T_z^* f](z) = \frac{f(z) - f(0)}{z}$$

shifts them to the left.

A *compressed Toeplitz operator* is an operator of the form $P_\varphi T_u P_\varphi$ where T_u is a standard Toeplitz operator and P_φ denotes the orthogonal projection from H^2 onto a model space $H^2 \ominus \varphi H^2$. With a slight abuse of notation, we regard compressed Toeplitz operators as operators acting on the space $H^2 \ominus \varphi H^2$, rather than on H^2 itself.

5.2. The Backward Shift.

Recall (from Example 2.14) that the backward shift M_z^* is not a complex symmetric operator. That is, there does not exist a conjugation C with respect to which M_z^* is C-symmetric. Indeed, this operator is about as far from complex symmetric as possible, for it violates almost any necessary criterion for being complex symmetric that one can devise.

With respect to the orthonormal basis $(z^n)_{n=0}^\infty$ for H^2, we can represent M_z^* as an infinite matrix:

$$
\left(\begin{array}{ccc|cccc}
0 & 1 & 0 & 0 & 0 & \cdots \\
0 & 0 & 1 & 0 & 0 & \cdots \\
0 & 0 & 0 & 1 & 0 & \cdots \\
\hline
0 & 0 & 0 & 0 & 1 & \cdots \\
0 & 0 & 0 & 0 & 0 & \cdots \\
\vdots & \vdots & \vdots & \vdots & \vdots &
\end{array}\right)
\begin{pmatrix} a_0 \\ a_1 \\ a_2 \\ a_3 \\ a_4 \\ \vdots \end{pmatrix}
=
\begin{pmatrix} a_1 \\ a_2 \\ a_3 \\ a_4 \\ a_5 \\ \vdots \end{pmatrix}.
$$

For the sake of illustration, we have isolated the upper-left 3×3 block of the matrix above. We immediately recognize this corner as the *Jordan block* (with $\lambda = 0$)

$$
\begin{pmatrix}
0 & 1 & 0 \\
0 & 0 & 1 \\
0 & 0 & 0
\end{pmatrix}
$$

which *is* a complex symmetric operator (see Example 2.2). Indeed, this is a finite *Toeplitz matrix* and we have already noted that all finite Toeplitz matrices are complex symmetric operators.

In terms of function theory, this Jordan block represents the compression of M_z^* to the M_z^*-invariant subspace $H^2 \ominus z^3 H^2$. The symmetry of this Toeplitz matrix with respect to the second diagonal corresponds to the map

$$
C(a_0 + a_1 z + a_2 z^2) = \overline{a_2} + \overline{a_1} z + \overline{a_0} z^2
$$

on $H^2 \ominus z^3 H^2$. In terms of boundary functions, this is simply

$$
Cf = \overline{f} z^2 = \overline{f} z \varphi
$$

where $\varphi = z^3$ is our original inner function.

Similar computations, focused on the upper left $n \times n$ block of our infinite Toeplitz matrix, suggest that the compression of the backward shift M_z^* to the M_z^*-invariant subspaces $H^2 \ominus \varphi H^2$ are complex symmetric operators. The remarkable thing is that this is true in complete generality.

5.3. C-symmetry of compressed Toeplitz operators. It turns out that compressed Toeplitz operators are complex symmetric operators with respect to the conjugation operator (3.2) on $H^2 \ominus \varphi H^2$. To be explicit, we recall the following theorem from [**23**]:

THEOREM 5.1. *If φ is a nonconstant inner function, u belongs to $L^\infty(\partial\mathbb{D})$, and P_φ denotes the orthogonal projection from H^2 onto $H^2 \ominus \varphi H^2$, then the compression $P_\varphi T_u P_\varphi$ of the Toeplitz operator T_u to $H^2 \ominus \varphi H^2$ is C-symmetric with respect to the conjugation operator* (3.2).

PROOF. Let $T = P_\varphi T_u P_\varphi$ denote the compressed Toeplitz operator and let P denote the orthogonal projection from L^2 onto H^2. If f and g belong to $H^2 \ominus \varphi H^2$, then

$$
\begin{aligned}
\langle CTf, g \rangle &= \langle Cg, Tf \rangle = \langle Cg, P_\varphi T_u P_\varphi f \rangle \\
&= \langle P_\varphi Cg, T_u f \rangle = \langle Cg, P(uf) \rangle \\
&= \langle PCg, uf \rangle = \langle Cg, uf \rangle \\
&= \langle \overline{g}\overline{z}\varphi, uf \rangle = \langle \overline{f}z\varphi, ug \rangle \\
&= \langle Cf, ug \rangle = \langle PP_\varphi Cf, ug \rangle \\
&= \langle P_\varphi Cf, T_u g \rangle = \langle Cf, P_\varphi T_u P_\varphi g \rangle \\
&= \langle Cf, Tg \rangle = \langle T^*Cf, g \rangle.
\end{aligned}
$$

Thus $CT = T^*C$ and T is C-symmetric. □

In light of Theorem 5.1, every compressed Toeplitz operator (including the compression of the unilateral shift) is a complex symmetric operator. In other words:

COROLLARY. *If φ is a nonconstant inner function, then the model operator $Sf = P_\varphi(zf)$ is C-symmetric with respect to the conjugation operator* (3.2).

Moreover, we have:

COROLLARY. *If φ is a nonconstant inner function, then there exists an orthonormal basis of $H^2 \ominus \varphi H^2$ with respect to which every compressed Toeplitz operator $P_\varphi T_u P_\varphi$ has a symmetric matrix representation.*

PROOF. The corollary follows from Theorem 5.1 along with Lemmas 2.6 and 2.7. □

This again suggests the problem of finding and computing such bases. The remainder of this note concerns the solution to this problem. In the next section we examine a nontrivial example in detail.

6. Polar Decomposition of C-symmetric Operators

In order to generalize our construction of C-real bases for the model spaces $H^2 \ominus \varphi H^2$ (where φ denotes a nonconstant inner function), we first require some background material on the polar decomposition of C-symmetric operators. It turns out that the polar decompositions of complex symmetric operators have many special properties.

Our interest in this subject stems from the fact that compressed shifts are complex symmetric operators (Theorem 5.1). Much of the material in section appears in [**24**].

6.1. A theorem of Godič and Lucenko. The first result we require is not a statement about polar decompositions, but rather a statement about the structure of unitary operators themselves. The following old theorem of Godič and Lucenko [**29**] generalizes the simple geometric notion that a planar rotation can be expressed as the product of two reflections:

THEOREM 6.1 (Godič-Lucenko). *If U is a unitary operator on a Hilbert space $\mathcal{H}$, then there exist conjugation operators C and J on $\mathcal{H}$ such that $U = CJ$.*

This theorem is remarkable for, among other things, stating that all unitary operators (on a fixed Hilbert space $\mathcal{H}$) can be constructed using essentially *a single antilinear operator*. Indeed, we know from Lemma 2.6 that any conjugation on $\mathcal{H}$ can be represented as complex conjugation with respect to a certain orthonormal basis. In this sense, C and J are structurally identical objects and the fine structure of the unitary U arises entirely in how two copies of the same object are put together. Of course, one can also say that since C and J are unitarily equivalent objects, the theorem of Godič and Lucenko is quite natural.

The following easy example of such a factorization is somewhat instructive:

EXAMPLE 6.2. Let $U : \mathbb{C}^n \longrightarrow \mathbb{C}^n$ be a unitary operator. Suppose that U has n (necessarily unimodular) eigenvalues $\lambda_1, \lambda_2, \ldots, \lambda_n$ with corresponding orthonormal eigenvectors $e_1, e_2, \ldots, e_n$. One may define conjugation operators C and J on $\mathbb{C}^n$ by setting

$$Ce_k = \lambda_k e_k, \quad Je_k = e_k$$

for $k = 1, 2, \ldots, n$ and extending this definition antilinearly to all of $\mathbb{C}^n$. It is clear from this construction that $U = CJ$. By introducing offsetting unimodular parameters in the definitions of C and J, one can see that the decomposition of U as the product of conjugation operators is not uniquely determined. For instance, if we desired more symmetry we could select a branch of the square root and use

$$Ce_k = \lambda_k^{1/2} e_k, \quad Je_k = \overline{\lambda_k}^{-1/2} e_k$$

instead.

This simple finite dimensional example suggests the method of proof for Theorem 6.1. The conjugation J was chosen to commute with the spectral projections of U and hence one suspects that the spectral theorem is ultimately involved. Indeed, by the spectral theorem it suffices to prove

Theorem 6.1 in the special case where U is the operator of multiplication by $e^{i\theta}$ on a Lebesgue space $L^2(\partial\mathbb{D}, \mu)$ where μ is a finitely supported Borel measure on $\partial\mathbb{D}$.

EXAMPLE 6.3. If U denotes the unitary operator $[Uf](e^{i\theta}) = e^{i\theta}f(e^{i\theta})$ on $L^2(\partial\mathbb{D}, \mu)$, then $U = CJ$ where

$$[Cf](e^{i\theta}) = e^{\frac{i}{2}\theta}\overline{f(e^{i\theta})}, \quad [Jf](e^{i\theta}) = e^{-\frac{i}{2}\theta}\overline{f(e^{i\theta})}$$

for all f in $L^2(\partial\mathbb{D}, \mu)$.

The more determined reader may wish to consider the Godič-Lucenko decomposition of the Fourier-Plancherel transform on $L^2(-\infty, \infty)$ as well as those of other well-known unitary operators (see [24] and the original paper [29]).

The converse of Theorem 6.1 is also true:

LEMMA 6.4. If C and J are conjugation operators on a Hilbert space $\mathcal{H}$, then $U = CJ$ is a unitary operator. Moreover, U is both C-symmetric and J-symmetric.

PROOF. If $U = CJ$, then (by the isometric property of C and J) it follows that

$$\langle f, U^*g \rangle = \langle Uf, g \rangle = \langle CJf, g \rangle = \langle Cg, Jf \rangle = \langle f, JCg \rangle$$

for all f, g in $\mathcal{H}$. Thus $U^* = JC$ from which $CU = U^*C$ and $JU = U^*J$ both follow.   □

Although we will frequently refer to Godič-Lucenko theorem, we do not actually require it. In fact, we will explicitly construct the Godič-Lucenko decomposition for Clark operators and then use the decomposition to obtain information about their spectral decompositions.

Before proceeding to discuss the structure of Clark operators, we first need to investigate the *complex symmetric* structure of compressed shifts. To do this, we will need a generalization of Theorem 6.1 to the class of all C-symmetric operators.

6.2. The partial isometry is C-symmetric. Recall that the polar decomposition $T = U|T|$ of an operator T expresses T uniquely as the product of a positive operator $|T| = \sqrt{T^*T}$ (the *modulus*) and a partial isometry U (the *argument*) which satisfies $\ker U = \ker T$ and maps the *initial space* $(\ker T)^\perp$ onto the *final space* $\mathrm{cl}(\mathrm{ran}\, T)$ (the closure of the range of T). Moreover, recall that the modulus satisfies $\ker |T| = \ker T$ and $\mathrm{cl}(\mathrm{ran}\, |T|) = \mathrm{cl}(\mathrm{ran}\, T^*)$.

The following theorem states that the partial isometry appearing in the polar decomposition of a C-symmetric operator is also C-symmetric (with respect to the same C):

THEOREM 6.5. If $T = U|T|$ is the polar decomposition of a C-symmetric operator T, then the partial isometry U is also C-symmetric: $CU = U^*C$.

The proof of the general assertion can be found in [**24**] and the reader will have no difficulty proving it in the case that T is invertible. Indeed, if T is invertible then we have the obvious explicit formula $U = T(T^*T)^{-\frac{1}{2}}$ for the partial isometry. A standard argument shows that U is the limit of polynomials of the form $Tp(T^*T)$ where the polynomial $p(x)$ has *real* coefficients. It is not hard to see (by first considering terms of the form $T(T^*T)^n$) that any operator of the form $Tp(T^*T)$ is also C-symmetric.

The preceding theorem places severe restrictions on the class of partial isometries that can occur in the polar decomposition of a C-symmetric operator. For instance, Example 2.14 shows that the unilateral shift cannot appear in this context.

6.3. Refined polar decomposition. Theorem 6.5 is the key observation that allows us to relate the partial isometry U in the polar decomposition of a C-symmetric operator T to the modulus $|T|$ and the conjugation C itself.

We first consider the special case of invertible C-symmetric operators before considering the general situation.

LEMMA 6.6. *If T is an invertible C-symmetric operator with polar decomposition $T = U|T|$, then the antilinear operator $J = CU$ is a conjugation operator which commutes with the spectral projections of $|T|$.*

PROOF. Since T is invertible, the partial isometry U is unitary. By Theorem 6.5, U is C-symmetric and it follows that

$$J = CU = U^*C$$

which shows that $U^* = JC$. Since U is unitary, it follows that

$$J^2 = (JC)(CJ) = U^*U = I$$

and

$$\langle Jf, Jg \rangle = \langle CUf, CUg \rangle = \langle Ug, Uf \rangle = \langle g, f \rangle$$

for all f, g in $\mathcal{H}$ and thus J is a conjugation operator. Moreover, the equation $CT = T^*C$ is equivalent to

$$CU|T| = |T|U^*C,$$

which is equivalent to $J|T| = |T|J$. This implies that J also commutes with $p(|T|)$ for any polynomial $p(x)$ with real coefficients and hence with every spectral projection of $|T|$. $\qquad\square$

For invertible C-symmetric operators, the factorization described above leads to a substantial refinement of the polar decomposition. Specifically, we may write $T = CJ|T|$ where J is a conjugation operator which commutes with $|T|$. This can be viewed as a generalization of Theorem 6.1 since the *unitary* operator U in the polar decomposition $T = U|T|$ is the product of two conjugation operators.

A similar construction holds for general C-symmetric operators, although some minor difficulties arise. The fundamental problem is that the partial isometry U in the polar decomposition $T = U|T|$ of an arbitrary C-symmetric operator need not be unitary and it becomes necessary to consider so-called *partial conjugations*. The details, which essentially reduce to glorified bookkeeping of initial spaces and final spaces, can be found in [**24**]. Fortunately, these issues do not present a major obstacle and one can obtain the following broad generalization of the theorem (Theorem 6.1) of Godič and Lucenko:

THEOREM 6.7. *If T is a C-symmetric operator, then $T = CJ|T|$ where J is a conjugation operator which commutes with the spectral projections of $|T|$.*

By insisting that J is supported on all of $\mathcal{H}$ (rather than $\mathrm{cl}(\mathrm{ran}\,|T|)$, where the partial isometry U is supported), one in general can no longer assert that U is given by the formula $U = CJ$. Indeed, if C and J were conjugation operators, then U would be unitary, which is not always the case. For our purposes, however, these sacrifices are not significant and Theorem 6.7 will be sufficient.

7. Polar decomposition of compressed shifts

In this section we compute the polar decomposition of the compression of the unilateral shift to the subspace $H^2 \ominus \varphi H^2$ corresponding to a nonconstant inner function φ. Furthermore, we also compute the decomposition guaranteed by Theorem 6.7. In fact, we also are able to consider a slight generalization of the compressed shift with little additional effort. Namely, we will examine the compression of the operator of multiplication by a disk automorphism to $H^2 \ominus \varphi H^2$.

The most interesting aspect of these computations is the involvement of Clark-type unitary operators, for which we obtain an explicit Godič-Lucenko decomposition (see Theorem 6.1). In Section 8, we will study *generalized Clark operators* and the following computations will be necessary to consider the eigenstructures of these operators.

7.1. Preliminaries. Since the computations in this section and the next will be quite involved, let us collect some of the notational conventions we have adopted earlier in this note.

For each λ in the open unit disk $\mathbb{D}$, we let b_λ denote the disk automorphism vanishing at λ:

$$b_\lambda(z) = \frac{z - \lambda}{1 - \overline{\lambda}z}. \tag{7.1}$$

We also require the reproducing kernels K_λ and the corresponding conjugate kernels CK_λ for $H^2 \ominus \varphi H^2$:

$$K_\lambda(z) = \frac{1 - \overline{\varphi(\lambda)}\varphi(z)}{1 - \overline{\lambda}z}, \quad [CK_\lambda](z) = \frac{\varphi(z) - \varphi(\lambda)}{z - \lambda}. \tag{7.2}$$

Furthermore, we will frequently refer to the normalized kernel functions

$$k_\lambda = K_\lambda/\|\,K_\lambda\,\|, \quad Ck_\lambda = CK_\lambda/\|\,K_\lambda\,\|.$$

For each λ in $\mathbb{D}$, we consider the compression

$$S_\lambda f = P_\varphi(b_\lambda f) \tag{7.3}$$

of the operator of multiplication by b_λ to $H^2 \ominus \varphi H^2$. Here, as before, P_φ denotes the orthogonal projection from H^2 onto $H^2 \ominus \varphi H^2$. The operators S_λ are therefore straightforward generalizations of the compressed shift operator

$$S_0 f = P_\varphi(zf),$$

otherwise known as the *model* or *Jordan* operator.

We can also say that S_λ is the compression of the analytic Toeplitz operator T_{b_λ} to the M_z^*-invariant subspace $H^2 \ominus \varphi H^2$. Since $T_{b_\lambda}^* = T_{\overline{b_\lambda}}$, the adjoint of S_λ is given by

$$S_\lambda^* f = P_\varphi(\overline{b_\lambda} f).$$

As a special case of Theorem 5.1, it follows that the operators S_λ are all C-symmetric with respect to the conjugation operator (3.2) on $H^2 \ominus \varphi H^2$. In other words, we have

$$CS_\lambda = S_\lambda^* C$$

for every λ in $\mathbb{D}$.

Our aim in this section is to explicitly compute the factors in the refined polar decomposition (guaranteed by Theorem 6.7)

$$S_\lambda = CJ_\lambda |S_\lambda|$$

for each of these operators. Throughout our computations, we will require a simple computational lemma, which generalizes [**10**, Lem. 2.1].

LEMMA 7.1. *For each λ in $\mathbb{D}$, the following statements hold:*

(1) $S_\lambda f = b_\lambda f$ *if and only if f is orthogonal to Ck_λ.*
(2) $S_\lambda^* f = f/b_\lambda$ *if and only if f is orthogonal to k_λ.*

PROOF. Clearly $S_\lambda^* f = f/b_\lambda$ if and only if f/b_λ belongs to H^2. This happens if and only if $f(\lambda) = 0$, or equivalently, if and only if $\langle f, k_\lambda \rangle = 0$. By the preceding, $S_\lambda^* Cf = (Cf)/b_\lambda$ if and only if $\langle Cf, k_\lambda \rangle = 0$, or equivalently, if and only if f is orthogonal to Ck_λ. Since $CS_\lambda = S_\lambda^* C$, this implies that $S_\lambda f = C[Cf/b_\lambda] = b_\lambda f$ if and only if f is orthogonal to Ck_λ. $\square$

The preceding lemma stresses the importance of the normalized conjugate kernel Ck_λ in our discussion of the operator S_λ. Lemma 7.1 essentially states that on the orthocomplement of the one-dimensional subspace spanned by Ck_λ, the operator S_λ acts *isometrically* (since $\|\,b_\lambda f\,\| = \|\,f\,\|$ for all f).

7.2. The modulus $|S_\lambda|$. To find the modulus

$$|S_\lambda| = \sqrt{S_\lambda^* S_\lambda}$$

of S_λ, we need only compute the positive operator $S_\lambda^* S_\lambda$. By Lemma 7.1, it follows that if f is orthogonal to Ck_λ (in other words if $Cf(\lambda) = 0$), then

$$\begin{aligned} S_\lambda^* S_\lambda f &= S_\lambda^*(b_\lambda f) \\ &= f. \end{aligned}$$

Hence $|S_\lambda|$ restricts to the identity operator on the orthocomplement of the one-dimensional subspace spanned by the function Ck_λ. This tells us, among other things, that $|S_\lambda|$ must map the function Ck_λ onto a nonnegative constant multiple of itself. Using the following lemma, we will ultimately be able to compute this constant and hence obtain an explicit formula for S_λ.

LEMMA 7.2. $S_\lambda C k_\lambda = -\varphi(\lambda) k_\lambda$.

PROOF. Since multiplication by a positive real constant will not affect the desired formulas, it suffices to prove them with K_λ in place of k_λ. Using (7.2) and letting $e_\lambda = (1 - \bar\lambda z)^{-1}$ denote the Szegő kernel we find that

$$\begin{aligned} S_\lambda C K_\lambda &= S_\lambda \left(\frac{\varphi - \varphi(\lambda)}{z - \lambda} \right) \\ &= P_\varphi \left(b_\lambda \frac{\varphi - \varphi(\lambda)}{z - \lambda} \right) \\ &= P_\varphi \left(\frac{\varphi - \varphi(\lambda)}{1 - \bar\lambda z} \right) \\ &= P_\varphi(\varphi e_\lambda) - \varphi(\lambda) P_\varphi(e_\lambda) \\ &= 0 - \varphi(\lambda)\langle e_\lambda, K_z \rangle \\ &= -\varphi(\lambda)\overline{K_z(\lambda)} \\ &= -\varphi(\lambda) K_\lambda. \end{aligned}$$

Hence $S_\lambda C k_\lambda = -\varphi(\lambda) k_\lambda$ as claimed. □

For vectors u, v in $H^2 \ominus \varphi H^2$, we will let $u \otimes v$ denote the rank-one operator

$$(u \otimes v)f = \langle f, v \rangle u.$$

In particular, note that such an operator is an orthogonal projection if and only if $u = v$ and $\| u \| = 1$. Using Lemma 7.2 and the C-symmetry of the operator S_λ, we are now in a position to compute its modulus:

THEOREM 7.3. *The modulus $|S_\lambda|$ of S_λ is given by the formula*

$$|S_\lambda| = (I - Ck_\lambda \otimes Ck_\lambda) + |\varphi(\lambda)|(Ck_\lambda \otimes Ck_\lambda). \tag{7.4}$$

The first term represents the orthogonal projection onto the orthocomplement of the one-dimensional subspace spanned by Ck_λ.

PROOF. We already noted (in the discussion prior to Lemma 7.2) that $|S_\lambda|$ restricts to the identity operator on the orthocomplement of the one-dimensional subspace spanned by Ck_λ and therefore it suffices to consider the action of $|S_\lambda|$ on Ck_λ. Applying Lemma 7.2 twice and using the anti-linearity of C we see that

$$(S_\lambda C)k_\lambda = |\varphi(\lambda)|k_\lambda.$$

However, since the operator S_λ is C-symmetric, it satisfies the equation $CS_\lambda^* = S_\lambda C$ and hence

$$\begin{aligned} CS_\lambda^* S_\lambda Ck_\lambda &= S_\lambda C S_\lambda Ck_\lambda \\ &= (S_\lambda C)^2 k_\lambda \\ &= |\varphi(\lambda)|k_\lambda. \end{aligned}$$

Applying C again we obtain $S_\lambda^* S_\lambda Ck_\lambda = |\varphi(\lambda)|Ck_\lambda$ which implies the desired result. $\qquad\square$

In light of (7.4) and Lemma 7.1, we will henceforth assume that $\varphi(\lambda) \neq 0$, since otherwise the polar decomposition of S_λ is already evident. Indeed, if $\varphi(\lambda) = 0$, then $\ker S_\lambda$ equals the one-dimensional subspace spanned by Ck_λ and S_λ acts isometrically (multiplication by b_λ) on the orthocomplement of this subspace. In this case, one sees from (7.4) that $|S_\lambda|$ is not invertible. Rather, $|S_\lambda|$ reduces to the orthogonal projection onto the orthocomplement of the one-dimensional space spanned by Ck_λ. The difficulties incurred by this situation are not significant and involve only keeping track of initial and final spaces. Moreover, the partial isometry in the polar decomposition of S_λ then decomposes as the product CJ_λ where J_λ is a so-called *partial conjugation* (see [24]).

The condition $\varphi(\lambda) \neq 0$ actually makes our job slightly more difficult, but more interesting. By insisting that $\varphi(\lambda) \neq 0$, we ensure that S_λ is invertible. The partial isometry U_λ appearing in the polar decomposition $S_\lambda = U_\lambda|S_\lambda|$ is therefore unitary and hence decomposes as the product CJ_λ of two conjugation operators (by Theorem 6.7).

7.3. The conjugation J_λ. By Theorem 6.7, we may write

$$S_\lambda = CJ_\lambda|S_\lambda|$$

where J_λ is a conjugation operator on $H^2 \ominus \varphi H^2$ which commutes with the spectral projections of $|S_\lambda|$. Thus J_λ preserves the spectral subspaces of $|S_\lambda|$, namely the one-dimensional span of Ck_λ and its orthocomplement. We can verify this assertions rather explicitly.

THEOREM 7.4. *The conjugation operator J_λ in the refined polar decomposition $S_\lambda = CJ_\lambda|S_\lambda|$ is given by the explicit formula*

$$J_\lambda f = \begin{cases} C(b_\lambda f) & f \perp Ck_\lambda \\ \overline{\alpha}Ck_\lambda & f = Ck_\lambda \end{cases} \tag{7.5}$$

where the unimodular constant α is given by $\alpha = -\varphi(\lambda)/|\varphi(\lambda)|$.

PROOF. To find J_λ, we simply write

$$J_\lambda|S_\lambda| = CS_\lambda \tag{7.6}$$

and compute the action of J_λ on the two spectral subspaces of $|S_\lambda|$. If $f \perp Ck_\lambda$, then $|S_\lambda|f = f$ by (7.4) and hence

$$J_\lambda f = J_\lambda|S_\lambda|f = CS_\lambda f = C(b_\lambda f)$$

by (7.6) and Lemma 7.1. Since $\varphi(\lambda) \neq 0$ (by our nontriviality assumption) we have

$$\begin{aligned}
|\varphi(\lambda)|J(Ck_\lambda) &= J|S_\lambda|(Ck_\lambda) \\
&= C(S_\lambda Ck_\lambda) \\
&= -\overline{\varphi(\lambda)}Ck_\lambda,
\end{aligned}$$

the two equalities following from (7.6) and Lemma 7.2, respectively. Putting these calculations together yields the desired explicit formula for J_λ. □

7.4. The unitary U_λ. We can now compute the partial isometry

$$U_\lambda = CJ_\lambda$$

in the polar decomposition $S_\lambda = U_\lambda|S_\lambda|$ of S_λ using (7.5). We note that U_λ is actually *unitary* since both C and J_λ are conjugation operators on all of $H^2 \ominus \varphi H^2$ (recall that we are assuming that $\varphi(\lambda) \neq 0$ so that (7.4) does not reduce to a projection).

Applying C to (7.5) yields the piecewise formula

$$U_\lambda f = \begin{cases} b_\lambda f & f \perp Ck_\lambda \\ \alpha k_\lambda & f = Ck_\lambda \end{cases}$$

where α once again denotes the unimodular constant $\alpha = -\varphi(\lambda)/|\varphi(\lambda)|$. It turns out that U_λ is actually a rank-one unitary perturbation of the original operator S_λ, although this is not immediately obvious. The following lemma is the key:

LEMMA 7.5. *U_λ is given by the formula*

$$U_\lambda = S_\lambda(I - Ck_\lambda \otimes Ck_\lambda) + \alpha(k_\lambda \otimes Ck_\lambda). \tag{7.7}$$

PROOF. It suffices to verify the formula in the two cases $f \perp Ck_\lambda$ and $f = Ck_\lambda$. If $f \perp Ck_\lambda$, then the operator defined by the right hand side of (7.7) applied to f clearly yields $S_\lambda f = b_\lambda f$ (using Lemma 7.1). This agrees with U_λ for such f. On the other hand, the operator defined by the right hand side of (7.7) applied to Ck_λ obviously yields αk_λ, which again agrees with U_λ. This establishes the desired formula for U_λ. □

Another formula for U_λ is provided by the following lemma:

LEMMA 7.6. *U_λ is given by the formula*

$$U_\lambda = S_\lambda + (\alpha + \varphi(\lambda))(k_\lambda \otimes Ck_\lambda). \tag{7.8}$$

PROOF. Using (7.7) and the fact that $S_\lambda C k_\lambda = -\varphi(\lambda) C k_\lambda$ (Lemma 7.2), it follows that

$$\begin{aligned}
U_\lambda &= S_\lambda - (S_\lambda C k_\lambda \otimes C k_\lambda) + \alpha(k_\lambda \otimes C k_\lambda \\
&= S_\lambda + (\alpha + \varphi(\lambda))(k_\lambda \otimes C k_\lambda)
\end{aligned}$$

which is the desired formula.   □

We remark at this point that U_λ is a rank-one, unitary perturbation of S_λ. Moreover, it is C-symmetric. Although Theorem 6.5 and Lemma 6.4 assert that U_λ is automatically C-symmetric, this can be seen directly. Indeed, we already know that S_λ is C-symmetric (by Theorem 5.1) and it is easy to verify that the rank-one perturbing operator $\alpha(k_\lambda \otimes C k_\lambda)$ is also C-symmetric. In fact, it is not hard to show that a rank one operator $u \otimes v$ is C-symmetric if and only if it is a constant multiple of $u \otimes C u$.

We summarize our results in the following theorem:

THEOREM 7.7. *Let φ denote a nonconstant inner function, let P_φ denote the orthogonal projection from H^2 onto $H^2 \ominus \varphi H^2$, and let λ be a point in $\mathbb{D}$ such that $\varphi(\lambda) \neq 0$. The polar decomposition of the compressed Toeplitz operator*

$$S_\lambda f = P_\varphi(\tfrac{z-\lambda}{1-\bar{\lambda}z} f)$$

is given by $S_\lambda = U_\lambda |S_\lambda|$ where U_λ is the rank-one, unitary, C-symmetric perturbation of S_λ given by (7.8) and $|S_\lambda|$ is given by (7.4). Moreover, $U_\lambda = C J_\lambda$ where the auxiliary conjugation operator J_λ is given by (7.5).

As we mentioned before, one can treat the case $\varphi(\lambda) = 0$ similarly, although U_λ will not longer be unitary and J_λ will then be only a *partial conjugation* [24].

8. Generalized Clark Operators

The operator U_λ defined by (7.8) is *not* the only rank-one C-symmetric unitary perturbation of S_λ. Indeed, for *any* unimodular constant α, the operator

$$U_{\lambda,\alpha} = S_\lambda + (\alpha + \varphi(\lambda))(k_\lambda \otimes C k_\lambda) \tag{8.1}$$

is C-symmetric (since S_λ and $k_\lambda \otimes C k_\lambda$ are both C-symmetric) and unitary, regardless of whether the inner function φ vanishes at λ. That $U_{\lambda,\alpha}$ is unitary can be seen by writing $U_{\lambda,\alpha}$ in a form analogous to (7.7) and using the lemmas of the preceding section.

We refer to operators of the form (8.1) as *generalized Clark operators* due to their similarity to the operators considered by Clark in [10]. The most significant departure from the approach taken there, however, comes in our frequent use of antilinear operators. For instance:

THEOREM 8.1. *Each generalized Clark operator $U_{\lambda,\alpha}$ factors as the product of two conjugation operators:*

$$U_{\lambda,\alpha} = C J_{\lambda,\alpha} \tag{8.2}$$

where the conjugation operator $J_{\lambda,\alpha}$ is given by

$$J_{\lambda,\alpha}f = \begin{cases} C(b_\lambda f) & f \perp Ck_\lambda \\ \overline{\alpha}Ck_\lambda & f = Ck_\lambda. \end{cases} \tag{8.3}$$

PROOF. The proof is almost identical to the proof of Theorem 7.4. □

8.1. Eigenstructure of generalized Clark operators. In the traditional approach to the Godič-Lucenko decomposition (Theorem 6.1), one first applies the spectral theorem to a given unitary operator to obtain a representation of that unitary as the product of two conjugations. In the present situation, things are reversed. Theorem 8.1 already provides us with the decomposition $U_{\lambda,\alpha} = CJ_{\lambda,\alpha}$ of any generalized Clark operator. We can actually use this representation to explore the eigenstructure of the operators $U_{\lambda,\alpha}$.

THEOREM 8.2. *If a function f in H^2 is a constant multiple of the function*

$$f_\xi(z) = \frac{1 - \overline{\left(\frac{\alpha + \varphi(\lambda)}{1 + \varphi(\lambda)\alpha}\right)}\varphi(z)}{1 - \overline{\left(\frac{\xi + \lambda}{1 + \overline{\lambda}\xi}\right)}z}, \tag{8.4}$$

then f belongs to $H^2 \ominus \varphi H^2$ and is an eigenvector of the generalized Clark operator $U_{\lambda,\alpha}$ corresponding to the eigenvalue ξ. Moreover, all eigenvectors are simple and are of the above form. A necessary (but not sufficient) condition for the function (8.4) to belong to H^2 is that

$$\varphi\left(\frac{\xi + \lambda}{1 + \overline{\lambda}\xi}\right) = \frac{\alpha + \varphi(\lambda)}{1 + \overline{\varphi(\lambda)}\alpha}$$

holds in the sense of nontangential limiting values.

PROOF. A function f is an eigenvector of the generalized Clark operator $U_{\lambda,\alpha} = CJ_{\lambda,\alpha}$ corresponding to the (necessarily unimodular) eigenvalue ξ if and only if

$$J_{\lambda,\alpha}f = \overline{\xi}Cf. \tag{8.5}$$

In light of the explicit formula (8.3) for $J_{\lambda,\alpha}$, we take the orthogonal decomposition of f with respect to the one-dimensional subspace spanned by Ck_λ. After possibly multiplying through by a constant, we may assume that f is of the form

$$f = g + CK_\lambda$$

where $g \perp CK_\lambda$. Substituting this into (8.5) we deduce that

$$J_{\lambda,\alpha}(g + CK_\lambda) = \overline{\xi}(Cg + K_\lambda).$$

By (8.3), this can be rewritten

$$C(b_\lambda g) + \overline{\alpha}CK_\lambda = \overline{\xi}Cg + \overline{\xi}K_\lambda.$$

Applying C to the equation above gives us

$$b_\lambda g + \alpha K_\lambda = \xi g + \xi C K_\lambda.$$

Solving for g we find that

$$g = \frac{\xi C K_\lambda - \alpha K_\lambda}{b_\lambda - \xi}.$$

We can now solve for the eigenvector f:

$$
\begin{aligned}
f &= g + C K_\lambda \\
&= \frac{\xi C K_\lambda - \alpha K_\lambda}{b_\lambda - \xi} + C K_\lambda \\
&= \frac{\xi C K_\lambda - \alpha K_\lambda + b_\lambda C K_\lambda - \xi C K_\lambda}{b_\lambda - \xi} \\
&= \frac{b_\lambda C K_\lambda - \alpha K_\lambda}{b_\lambda - \xi}.
\end{aligned}
$$

Using the formulas (7.2) for K_λ and $C K_\lambda$ we find that

$$
\begin{aligned}
f &= \frac{\frac{z-\lambda}{1-\bar\lambda z} \cdot \frac{\varphi - \varphi(\lambda)}{z-\lambda} - \alpha \frac{1-\overline{\varphi(\lambda)}\varphi}{1-\bar\lambda z}}{\frac{z-\lambda}{1-\bar\lambda z} - \xi} \\
&= \frac{\frac{\varphi - \varphi(\lambda)}{1-\bar\lambda z} - \alpha \frac{1-\overline{\varphi(\lambda)}\varphi}{1-\bar\lambda z}}{\frac{z-\lambda-\xi+\xi\bar\lambda z}{1-\bar\lambda z}} \\
&= \frac{\varphi - \varphi(\lambda) - \alpha + \alpha\overline{\varphi(\lambda)}\varphi}{z - \lambda - \xi + \xi\bar\lambda z} \\
&= \frac{\varphi(1 + \overline{\varphi(\lambda)}\alpha) - (\alpha + \varphi(\lambda))}{z(1 + \bar\lambda\xi) - (\lambda + \xi)} \\
&= \left(\frac{\alpha + \varphi(\lambda)}{\lambda + \xi}\right) \frac{1 - \left(\frac{1+\overline{\varphi(\lambda)}\alpha}{\alpha+\varphi(\lambda)}\right)\varphi}{1 - \left(\frac{1+\bar\lambda\xi}{\lambda+\xi}\right)z}
\end{aligned}
$$

and hence f is a constant multiple of the function f_ξ defined by (8.4). Conversely, we see that if ξ is a unimodular constant such that f_ξ belongs to H^2, then f_ξ is an eigenvector of $U_{\lambda,\alpha}$ corresponding to the eigenvalue ξ. Moreover, the computation above shows that the eigenspaces of $U_{\lambda,\alpha}$ are one-dimensional. $\qquad\square$

A necessary condition for a function of the form (8.4) to belong to H^2 is that φ have the nontangential limiting value $b_{-\varphi(\lambda)}(\alpha)$ at the point $b_{-\lambda}(\xi)$. In other words, the condition

$$\varphi\left(\frac{\xi + \lambda}{1 + \bar\lambda\xi}\right) = \frac{\alpha + \varphi(\lambda)}{1 + \overline{\varphi(\lambda)}\alpha} \tag{8.6}$$

is necessary for f_ξ to be an eigenvector of $U_{\lambda,\alpha}$ corresponding to the eigenvalue ξ. In general, this condition is not sufficient and we must examine the angular derivative (either directly or via the local Dirichlet integral [**39**]) of φ at the point $b_{-\lambda}(\xi)$. We do not wish to pursue the function theoretic details here and simply remark that (8.6) completely generalizes the original result of Clark [**10**, Thm. 3.2].

8.2. C-real bases for model spaces. We have seen that many natural operators associated with the model spaces are C-symmetric. Indeed, we have noted that all compressed Toeplitz operators (including the compressed shift) and all generalized Clark operators (defined by equation (8.1)) are C-symmetric.

In general, the spaces $H^2 \ominus \varphi H^2$ do not come equipped with "natural" orthonormal bases and hence we must construct them. Recall that we wish to find C-real orthonormal bases (those whose vectors are fixed by C) since, among other things, the matrix representation of a C-symmetric operator with respect to a C-real basis is symmetric (Lemma 2.7).

From this point of view, our interest in the generalized Clark operators $U_{\lambda,\alpha}$ lies in the fact that they are unitary. Under certain circumstances, the eigenvectors of $U_{\lambda,\alpha}$ might furnish a C-real orthonormal basis for $H^2 \ominus \varphi H^2$.

We have already computed the Godič-Lucenko decomposition $U_{\lambda,\alpha} = CJ_{\lambda,\alpha}$ (Theorem 8.1) for these operators and used it to compute their eigenvalues and eigenvectors (Theorem 8.2). Since the eigenvectors of any generalized Clark operator are simple and pairwise mutually orthogonal, we need only address the question of whether a given $U_{\lambda,\alpha}$ has a complete set of eigenvectors and whether the associated one-dimensional eigenspaces are fixed by C.

The first question is essentially a function theoretic one and can be discussed in terms of the analytic properties of the inner function φ. Several practical conditions guaranteeing that a standard Clark operator has pure point spectrum can be found in [**10**]. We do not wish to go into further detail here, although we do point out that we are now free to vary the parameter λ throughout $\mathbb{D}$ (as well as α on $\partial\mathbb{D}$) and we only need a single generalized Clark operator $U_{\lambda,\alpha}$ to have pure point spectrum in order to produce a complete set of eigenvectors.

The following general lemma addresses the second question. Namely, it shows that we may select a unit vector, fixed by C, from each of the (necessarily one-dimensional) eigenspaces of $U_{\lambda,\alpha}$:

LEMMA 8.3. *If T is a normal C-symmetric operator, then the eigenspaces of T are fixed by C.*

PROOF. Since T is normal, the equation $Tf = \lambda f$ implies that $T^*f = \overline{\lambda}f$. Applying C to the preceding equation implies that $T(Cf) = \lambda(Cf)$ and hence the eigenspaces of T are invariant under C.   □

Thus, if λ and α are values (in $\mathbb{D}$ and on $\partial\mathbb{D}$, respectively) such that the generalized Clark operator $U_{\lambda,\alpha}$ has pure point spectrum, then we can construct a C-real orthonormal basis of $H^2 \ominus \varphi H^2$. Indeed, let $(\zeta_n)_{n=1}^{\dim \mathcal{H}}$ be an enumeration of the points on $\partial\mathbb{D}$ such that

$$\varphi(\zeta_n) = \frac{\alpha + \varphi(\lambda)}{1 + \overline{\varphi(\lambda)}\alpha}$$

(as nontangential limiting values) and such that the normalized reproducing kernels k_{ζ_n} belong to H^2 (and hence $H^2 \ominus \varphi H^2$).

In this case, the numbers

$$\xi_n = \frac{\zeta_n - \lambda}{1 - \overline{\lambda}\zeta_n}$$

will be simple eigenvalues for $U_{\lambda,\alpha}$ and corresponding unit eigenvectors are given by the formula

$$e_n(z) = \left(\overline{\zeta_n \frac{\alpha + \varphi(\lambda)}{1 + \overline{\varphi(\lambda)}\alpha}} \right)^{\frac{1}{2}} k_{\zeta_n}.$$

The particular branch of the square root used does not matter and the fact that $Ce_n = e_n$ for all n follows from a computation similar to that in the proof of Lemma 4.6.

References

[1] Agmon, S., *Sur un problème de translations*, C.R. Acad. Sci. Paris, **229** (1949), no.11, 540–542.

[2] Ahern, P., Clark, D., *On inner functions with H^p derivative*, Michigan Math. J. **21** (1974), 115–127.

[3] Ahern, P., Clark, D., *On functions orthogonal to invariant subspaces*, Acta Math. **124** (1970), 191–204.

[4] Aleksandrov, A.B., *On A-integrability of boundary values of harmonic functions*. Mat. Zametki **30** (1981), 59–72.

[5] Bar-On, I., Ryaboy, V., *Fast diagonalization of large and dense complex symmetric matrices, with applications to quantum reaction dynamics*, SIAM J. Sci. Comput. **18** (1997), no. 5, 1412–1435.

[6] Brodskii, M.S., *On a problem of I.M. Gelfand*, Uspehi Matem. Nauk, **12** (1957), no. 2, 129–132.

[7] Carathéodory, C., Fejér, L., *Über den Zusammenhang der Extremen von harmonischen Funktionen mit ihren Koeffizienten und über den Picard-Landau'schen Satz*, Rend. Circ. Mat. Palermo **31** (1911), 218–239.

[8] Cima, J.A., Ross, W.T., *The Backward Shift on the Hardy Space*, American Mathematical Society, Providence R.I., 2000.

[9] Cima, J.A., Ross, W.T., Matheson, A., *The Cauchy Transform*, (preprint).

[10] Clark, D., *One dimensional perturbations of restricted shifts*, J. Anal. Math. **25** (1972), 169–191.

[11] Craven, B.D., *Complex symmetric matrices*, J. Austral. Math. Soc. **10** (1969), 341–354.

[12] Donoghue, W.F., *The lattice of invariant subspaces of a completely continuous quasinilpotent transformation*, Pac. J. Math. **7** (1957), 1031–1035.

[13] Douglas, R.G., Shapiro, H.S., Shields, A.L., *Cyclic vectors and invariant subspaces for the backward shift operator*, Ann. Inst. Fourier (Grenoble) **20** (1970), no.1, 37–76.

[14] Duren, P.L., *Theory of H^p Spaces*, Pure and Appl. Math., Vol **38**, Academic Press, New York, 1970.

[15] Duren, P.L.. *Univalent Functions*, Springer-Verlag, New York, 1983.

[16] FitzGerald, C.H., Horn, R.A., *On the structure of hermitian-symmetric inequalities*, J. London Math. Soc. (2) **15** (1977), no. 3, 419–430.

[17] FitzGerald, C.H., Horn, R.A., *On quadratic and bilinear forms in function theory*, Proc. London Math. Soc. (3) **44** (1982), no. 3, 554–576.

[18] Gantmacher, F.R., *The Theory of Matrices* (Vol. 2), Chelsea, New York, 1989.

[19] Garcia, S.R., *Conjugation, the backward shift, and Toeplitz kernels*, (To appear: J. Operator Theory).

[20] Garcia, S.R., *Inner matrices and Darlington synthesis*, Methods. Funct. Anal. Topology. **11**, no. 1, (2005), 37-47.

[21] Garcia, S.R., *A ∗-closed subalgebra of the Smirnov class*, Proc. Amer. Math. Soc. **133**, no. 7, (2005), 2051-2059.

[22] Garcia, S.R., Prodan, E., Putinar, M., *Norm estimates of complex symmetric operators applied to quantum systems*, (submitted)

[23] Garcia, S.R., Putinar, M., *Complex symmetric operators and applications*, (To appear: Trans. Amer. Math. Soc.).

[24] Garcia, S.R., Putinar, M., *Complex symmetric operators and applications II*, (submitted).

[25] Garcia, S.R., Sarason, D., *Real outer functions*, Indiana Univ. Math. J., **52** (2003), 1397–1412.

[26] Gelfand, I.M., *A problem* (Russian), Uspehi Matem. Nauk, **5** (1938), 233.

[27] Glazman, I.M., *Direct methods of the qualitative spectral theory of singular differential operators* (Russian), Gos. Iz. Fiz.-Mat. Lit., Moscow, 1963.

[28] Glazman, I.M., *An analogue of the extension theory of hermitian operators and a non-symmetric one-dimensional boundary-value problem on a half-axis* (Russian), Dokl. Akad. Nauk SSSR **115** (1957), 214–216.

[29] Godič, V.I., Lucenko, I.E., *On the representation of a unitary operator as a product of two involutions* (Russian), Uspehi Mat. Nauk **20** (1965), 64–65.

[30] Halmos, P.R., *A Hilbert Space Problem Book* (Second Edition), Springer-Verlag, 1982.

[31] Helson, H., *Large analytic functions, II*, Analysis and Partial Differential Equations, Lecture Notes in Pure and Appl. Math., **122**, Dekker, New York, 1990, 217–220.

[32] Horn, R.A., Johnson, C.R., *Matrix Analysis*, Cambridge Univ. Press, Cambridge, 1985.

[33] Hua, L.-K., *On the theory of automorphic functions of a matrix variable I: Geometrical basis*, Amer. J. Math. **66** (1944), 470–488.

[34] Kalisch, G.K., *A functional anaysis proof of Titchmarch's theorem on convolution*, J. Math. Anal. Appl., **5** (1962), no.2, 176–183.

[35] Koosis, P., *Introduction to H_p Spaces*. Cambridge University Press, Cambridge, 1998.

[36] Jacobson, N., *Normal semi-linear transformations*, Amer. J. Math. **61** (1939), 45–58.

[37] Nikolski, N.K., *Treatise on the Shift Operator*, Springer-Verlag, New York, 1986.

[38] Nikolski, N.K., *Operators, Functions, and Systems: An Easy Reading*, Volume **2**: Model Operators and Systems, American Mathematical Society, Providence R.I., 2002.

[39] Richter, S., Sundberg, C., *A formula for the local Dirichlet integral*, Michigan Math. J. **38** (1991), 355–379.

[40] Ross, W.T., Shapiro, H.S., *Generalized Analytic Continuation*, University Lecture Series, Volume **25**, American Mathematical Society, Providence R.I., 2002.

[41] Sakhnovich, L.A., *Spectral analysis of Volterra operators and inverse problems* (Russian), Doklady Akad. Nauk SSSR, **115** (1957), no.4, 666–669.

[42] Sarason, D., *A remark on the Volterra operator*, J. Math. Anal. Appl., **12** (1965), 244–246.

[43] Schur, I., *Ein Satz über Quadratische Formen mit Komplexen Koeffizienten*, Amer. J. Math. **67** (1945), 472–480.

[44] Scott, N.H., *A new canonical form for complex symmetric matrices*, Proc. Roy. Soc. London Ser. A **441** (1993), no. 1913, 625–640.

[45] Scott, N.H., *A theorem on isotropic null vectors and its application to thermoelasticity*, Proc. Roy. Soc. London Ser. A **440** (1993), no. 1909, 431–442.

[46] Siegel, C.L., *Symplectic geometry*, Amer. J. Math. **65** (1943), 1–86.

[47] Takagi, T., *On an algebraic problem related to an analytic theorem of Caratheodory and Fejer and on an allied theorem of Landau*, Japan J. Math. **1** (1925), 83–93.

[48] Walsh, J.L., *Interpolation and approximations by rational functions in the complex domain*, Amer. Math. Soc. Coll. Publ. 20., 1935.

[49] Wellstein, J., *Über symmetrische, alternierende und orthogonale Normalformen von Matrizen*, J. Reine Angew. Math. **163** (1930), 166–182.

DEPARTMENT OF MATHEMATICS, UNIVERSITY OF CALIFORNIA AT SANTA BARBARA, SANTA BARBARA, CALIFORNIA, 93106-3080

E-mail address: garcias@math.ucsb.edu

URL: http://math.ucsb.edu/~garcias

Contemporary Mathematics
Volume **393**, 2006

A class of conformal mappings with applications to function spaces

Daniel Girela

Dedicated to Professor Joseph Cima on the occasion of his 70th birthday

ABSTRACT. We discuss some of the properties of a class of conformal mappings introduced in a joint work with O. Blasco and M. A. Márquez and show that they can be used to prove the sharpness of a number of embedding theorems between certain spaces of analytic functions and, also, to prove that, for $1 \leq p < \infty$, the space $\mathcal{B} \cap H^p$ (where $\mathcal{B}$ is the Bloch space) does not have the so called f-property of Havin.

1. Some results on conformal mappings

1.1. Basic notation. Let $\mathbb{D}$ denote the open unit disk of the complex plane $\mathbb{C}$. If $0 < r < 1$ and f is an analytic function in $\mathbb{D}$ (abbreviated $f \in H(\mathbb{D})$) we set

$$M_p(r, f) = \left(\frac{1}{2\pi} \int_0^{2\pi} |f(re^{it})|^p \, dt \right)^{1/p}, \quad I_p(r, f) = M_p^p(r, f), \ (0 < p < \infty),$$

$$M_\infty(r, f) = \sup_{0 \leq t \leq 2\pi} |f(re^{it})|.$$

For $0 < p \leq \infty$ the Hardy space H^p consists of those functions $f \in H(\mathbb{D})$ for which $\|f\|_{H^p} \overset{\text{def}}{=} \sup_{0 < r < 1} M_p(r, f) < \infty$. We refer to [9] for the theory of Hardy spaces. The space $BMOA$ consists of those functions $f \in H^1$ whose boundary values have bounded mean oscillation on $\mathbb{T} = \partial\mathbb{D}$ (cf. [4], [11] and [14]).

2000 *Mathematics Subject Classification.* Primary 30C20, 30D45, 30D55; Secondary 30C35, 30D50.

Key words and phrases. Conformal mapping, Starlike domain, Circularly symmetric domain, Bloch functions, Normal functions, Mean Lipschitz spaces, Hardy spaces, Division by inner functions.

This research has been supported in part by grants from el Ministerio de Educación y Ciencia (Spain) and FEDER (MTM2004-00078 and MTM2004-21420-E) and by a grant from "La Junta de Andalucía" (FQM-210).

1.2. On the relation between $\mathbf{M_\infty(r, F)}$ and $\mathbf{M_1(r, F')}$ when $\mathbf{F}$ is a conformal mapping from the unit disk onto a certain domain. Let Ω be a simply connected domain contained in $\mathbb{C}$ with $0 \in \Omega \neq \mathbb{C}$ and let F be the conformal mapping from $\mathbb{D}$ onto Ω with $F(0) = 0$ and $F'(0) > 0$. We note that, for $0 < r < 1$, the quantity $2\pi r M_1(r, F')$ represents the length of the closed curve C_r which is the image under F of the circle $\{|z| = r\}$. Clearly, we have that $M_\infty(r, F) \leq \operatorname{length}(C_r)$, $0 < r < 1$, and, hence,

$$(1.1) \qquad M_\infty(r, F) = O\left(M_1(r, F')\right), \quad \text{as } r \to 1.$$

It is natural to look for a geometric condition on the domain Ω which is sufficient to imply that the quantities $M_1(r, F')$ and $M_\infty(r, F)$ are comparable. Keogh [24] and Hayman [22] proved that being starlike with respect to the origin is not such a sufficient condition.

1.3. Circularly symmetric domains. A domain Ω in $\mathbb{C}$ is said to be circularly symmetric if, for every r with $0 < r < \infty$, $\Omega \cap \{|z| = r\}$ is either empty, is the whole circle $|z| = r$, or is a single arc on $|z| = r$ which contains $z = r$ and is symmetric with respect to the real axis.

Let Ω be a simply connected and circularly symmetric domain in $\mathbb{C}$ with $\Omega \neq \mathbb{C}$ and $0 \in \Omega$. Let F be the conformal mapping from $\mathbb{D}$ onto Ω with $F(0) = 0$ and $F'(0) > 0$. Then (see [23] or the Corollary in p. 154 of [3]) it is known that:
   (a) $M_\infty(r, F) = F(r)$, $0 < r < 1$.
   (b) For every r with $0 < r < 1$, $|F(re^{i\theta})|$ is a decreasing function of θ in $[0, \pi]$.

1.4. The geometric condition. In the following theorem, which was obtained in a joint work with O. Blasco and M. A. Márquez [5], we give a geometric condition on Ω which implies that $M_\infty(r, F) \asymp M_1(r, F')$, as desired.

THEOREM 1.1. ([5, Theorem 3.1]). Let Ω be a domain in $\mathbb{C}$ with $0 \in \Omega$ and $\Omega \neq \mathbb{C}$. Suppose that Ω is both circularly symmetric and starlike with respect to 0. Let F be the conformal mapping from $\mathbb{D}$ onto Ω with $F(0) = 0$ and $F'(0) > 0$. Then there exists a positive constant C such that

$$(1.2) \qquad 2\pi r M_1(r, F') \leq C M_\infty(r, F), \quad 0 < r < 1.$$

Consequently,

$$(1.3) \qquad M_\infty(r, F) \asymp M_1(r, F'), \quad \text{as } r \to 1.$$

To prove Theorem 1.1 we use (b) together with the well known fact that (see p. 43 of [29] or p. 41 of [10]), since Ω is starlike with respect to the origin, for every r with $0 < r < 1$, $\arg F(re^{i\theta})$ is an increasing function of θ in $[0, 2\pi]$.

1.5. A special type of domain. In this section we consider a certain family of domains which satisfy the conditions of Theorem 1.1 and which have been useful to solve a number of problems in the theory of spaces of analytic functions.

Let $\Gamma = \{\gamma_n\}_{n=1}^\infty$ be a decreasing sequence of positive numbers with $\gamma_1 < 1$. We set

$$(1.4) \qquad \Omega_\Gamma = \Omega_{\{\gamma_n\}} = \mathbb{D} \cup \left(\bigcup_{n=1}^\infty \{z : |\operatorname{Im} z| < \gamma_n, \operatorname{Re} z > 0, n \leq |z| < n+1\} \right).$$

It is clear that the domain Ω_Γ is circularly symmetric and starlike with respect to the origin. Consequently, if F_Γ is the conformal mapping from $\mathbb{D}$ onto Ω_Γ with $F_\Gamma(0) = 0$ and $F_\Gamma'(0) > 0$, we have that:

(i) $M_\infty(r, F_\Gamma) = F_\Gamma(r)$, $0 < r < 1$.

(ii) $M_\infty(r, F_\Gamma) \asymp M_1(r, F_\Gamma')$, as $r \to 1$.

Furthermore, $|\operatorname{Im} F_\Gamma|$ is bounded and, hence

(iii) $F_\Gamma \in BMOA$.

Using the Carathéodory kernel convergence theorem (cf. [**29**, pp. 28-31]), we can prove the following result (see the proof of Theorem 3.2 of [**5**] for the details).

THEOREM 1.2. *Let $\varphi : [0, 1) \to [0, \infty)$ be a function with $\varphi(r) \to \infty$ as $r \to 1$. Then there exists a decreasing sequence of positive numbers $\Gamma = \{\gamma_n\}_{n=1}^\infty$ with $\gamma_1 < 1$ such that, if F_Γ is the conformal mapping from $\mathbb{D}$ onto Ω_Γ with $F_\Gamma(0) = 0$ and $F_\Gamma'(0) > 0$, then we have that $F_\Gamma(r) \to \infty$, as $r \to 1$ and*

$$(1.5) \qquad M_\infty(r, F_\Gamma) = F_\Gamma(r) \le \varphi(r), \quad 0 < r < 1.$$

2. On the sharpness of certain embedding theorems

2.1. Bloch functions and normal functions. We recall that a function f analytic in $\mathbb{D}$ is a Bloch function if

$$\sup_{z \in \mathbb{D}} (1 - |z|^2)|f'(z)| < \infty.$$

The space of all Bloch functions is denoted by $\mathcal{B}$. It is well known that

$$H^\infty \subset BMOA \subset \mathcal{B}.$$

The little Bloch space $\mathcal{B}_0$ consists of those functions f which are analytic in $\mathbb{D}$ and satisfy

$$\lim_{|z| \to 1} (1 - |z|^2)|f'(z)| = 0.$$

The space $\mathcal{B}_0$ is a closed subspace of the Bloch space. In fact, $\mathcal{B}_0$ is the closure of the polynomials in the Bloch norm. We mention [**2**] as a general reference for the theory of Bloch functions

A function f which is meromorphic in $\mathbb{D}$ is said to be a normal function if

$$\sup_{z \in \mathbb{D}} (1 - |z|^2) \frac{|f'(z)|}{1 + |f(z)|^2} < \infty.$$

We refer to [**2**] and [**29**] for the theory of normal functions. For simplicity, let $\mathcal{N}$ denote the set of all holomorphic normal functions in $\mathbb{D}$. Any Bloch function is a normal function, that is, $\mathcal{B} \subset \mathcal{N}$.

2.2. Mean Lipschitz spaces of analytic functions. If f is a function which is analytic in $\mathbb{D}$ and has a non-tangential limit $f(e^{i\theta})$ at almost every $e^{i\theta} \in \mathbb{T}$, we define

(2.1)

$$\omega_p(\delta, f) = \sup_{0 < |t| \le \delta} \left(\frac{1}{2\pi} \int_{-\pi}^{\pi} \left| f(e^{i(\theta + t)}) - f(e^{i\theta}) \right|^p d\theta \right)^{1/p}, \quad \delta > 0, \quad \text{if } 1 \le p < \infty,$$

$$\omega_\infty(\delta, f) = \sup_{0 < |t| \le \delta} \left(\operatorname*{ess\,sup}_{\theta \in [-\pi, \pi]} |f(e^{i(\theta + t)}) - f(e^{i\theta})| \right), \quad \delta > 0.$$

Then $\omega_p(., f)$ is the integral modulus of continuity of order p of the boundary values $f(e^{i\theta})$ of f.

Throughout the paper $\omega : [0, 1] \to [0, \infty)$ will be a continuous and increasing function with $\omega(0) = 0$. Then, for $1 \leq p \leq \infty$, the mean Lipschitz space $\Lambda(p, \omega)$ consists of those functions $f \in H^p$ which satisfy

$$\omega_p(\delta, f) = O(\omega(\delta)), \quad \text{as } \delta \to 0.$$

If $0 < \alpha \leq 1$ and $\omega(\delta) = \delta^\alpha$, we shall write Λ_α^p instead of $\Lambda(p, \omega)$, that is, we set

$$\Lambda_\alpha^p = \Lambda(p, \delta^\alpha), \quad 0 < \alpha \leq 1, \quad 1 \leq p \leq \infty.$$

A classical result of Hardy and Littlewood [20] (see also Chapter 5 of [9]) asserts that for $1 \leq p \leq \infty$ and $0 < \alpha \leq 1$, we have that

$$(2.2) \qquad \Lambda_\alpha^p = \left\{ f \text{ analytic in } \mathbb{D} \colon M_p(r, f') = O\left(\frac{1}{(1-r)^{1-\alpha}}\right), \quad \text{as } r \to 1 \right\}.$$

Blasco and de Souza found in [6] conditions on ω so that the result of Hardy and Littlewood can be extended to the spaces $\Lambda(p, \omega)$. Namely, they proved that if ω satisfies the so called Dini condition and the condition b_1 (see [6], [5] or [13] for the precise definitions) then, for $1 \leq p \leq \infty$,

$$(2.3) \qquad \Lambda(p, \omega) = \left\{ f \text{ analytic in } \mathbb{D} \colon M_p(r, f') = O\left(\frac{\omega(1-r)}{1-r}\right), \quad \text{as } r \to 1 \right\}.$$

Let $1 \leq p \leq \infty$ and let ϕ be a non-negative function defined in $[0, 1)$. We define $\mathcal{L}(p, \phi)$ as the space of all functions f analytic in $\mathbb{D}$ for which

$$(2.4) \qquad M_p(r, f') = O(\phi(r)), \quad \text{as } r \to 1.$$

We remark that $\mathcal{L}(p, \phi) \subset \mathcal{L}(q, \phi)$, if $1 \leq q < p$. Notice that (2.2) can be written as

$$(2.5) \qquad \Lambda_\alpha^p = \mathcal{L}\left(p, \frac{1}{(1-r)^{1-\alpha}}\right), \quad 1 \leq p \leq \infty, \quad 0 < \alpha \leq 1,$$

and (2.3) is equivalent to

$$(2.6) \qquad \Lambda(p, \omega) = \mathcal{L}\left(p, \frac{\omega(1-r)}{1-r}\right).$$

Cima and Petersen proved in [8] that $\Lambda_{1/2}^2 \subset BMOA$. This result was extended by Bourdon, Shapiro and Sledd who proved the following result in [7]:

$$(2.7) \qquad \Lambda_{1/p}^p \subset BMOA, \quad 1 < p < \infty.$$

For $p = 1$ we have a stronger result. Indeed, a classical result of Privalov [9, Th. 3.11] asserts that a function f which is analytic in $\mathbb{D}$ has a continuous extension to the closed unit disk $\overline{\mathbb{D}}$ whose boundary values are absolutely continuous on $\partial\mathbb{D}$ if and only if $f' \in H^1$. In particular, we have

$$(2.8) \qquad f' \in H^1 \Rightarrow f \in \mathcal{A} \subset H^\infty,$$

where, $\mathcal{A}$ denotes the disk algebra. Using (2.2), we see that the the condition $f' \in H^1$ is equivalent to saying that $f \in \Lambda_1^1$. Hence, (2.8) can be written as

$$(2.9) \qquad \Lambda_1^1 \subset \mathcal{A} \subset H^\infty.$$

The author proved in [12] that (2.8) (or, equivalently, (2.9)) is sharp in a very strong sense:

THEOREM 2.1. ([**12**, Theorem 1]). Let ϕ be any positive continuous function defined in $[0, 1)$ with $\phi(r) \to \infty$, as $r \to 1$. Then, there exists a function $f \in \mathcal{L}(1, \phi)$ which is not a normal function.

Hence, no condition on the growth of $M_1(r, f')$ other that its boundedness is enough to conclude that $f \in \mathcal{N}$. We recall here that $\mathcal{A} \subset H^\infty \subset BMOA \subset \mathcal{B} \subset \mathcal{N}$.

The original proof of Theorem 2.1 was constructive. The constructed function f was of the form $f = BF$ where B is a Blaschke product and F is a function given by a series of analytic functions in $\mathbb{D}$ which converges uniformly on every compact subset of $\mathbb{D}$. The constructions of B and F were very involved, and made use in an essential way of certain sequences introduced by K. I. Oskolkov in several contexts (see, e.g., [**26, 27, 28**]). But now is possible to give a much simpler proof of Theorem 2.1, independent of the Oskolkov's sequences, using Theorem 1.2 and the following result about Blaschke products proved in collaboration with C. González.

THEOREM 2.2. ([**15**, Theorem 1]). Let B be an interpolating Blaschke product with positive zeros. Then there exist two positive constants C_1 and C_2, and $\rho_0 \in (0, 1)$ such that

$$(2.10) \qquad C_1 n(r, B) \le M_1(r, B') \le C_2 n(r, B), \quad \rho_0 < r < 1,$$

where, for $0 < r < 1$, $n(r, B)$ denotes the number of zeros of B in the disk $\{|z| < r\}$.

We remark that Girela and Peláez [**18**] have recently improved this result showing that the conclusion holds for any infinite Blaschke product whose sequence of zeros is exponential.

Proof of Theorem 2.1. Set $\varphi(r) = \sqrt{\phi(r)}$, $0 < r < 1$. Using Theorem 1.2, we see that there exists a decreasing sequence of positive numbers $\Gamma = \{\gamma_n\}_{n=1}^\infty$ with $\gamma_1 < 1$ such that, if F_Γ is the conformal mapping from $\mathbb{D}$ onto Ω_Γ with $F_\Gamma(0) = 0$ and $F_\Gamma'(0) > 0$, then we have that $F_\Gamma(r) \to \infty$, as $r \to 1$ and

$$(2.11) \qquad M_\infty(r, F_\Gamma) = F_\Gamma(r) \le \varphi(r), \quad 0 < r < 1.$$

Using the results of Section 1.5, it follows that we also have

$$(2.12) \qquad M_1(r, F_\Gamma') = O(\varphi(r)), \quad \text{as } r \to 1.$$

Now, using Theorem 2.2, it follows easily (see [**15**, pp. 7-8] for the details) that there exists an interpolating Blaschke product B whose zeros $\{a_n\}$ are positive and such that

$$(2.13) \qquad M_1(r, B') = O(\varphi(r)), \quad \text{as } r \to 1.$$

For simplicity set $F_\Gamma = F$ and define $f(z) = B(z)F(z)$ $(z \in \mathbb{D})$. Bearing in mind that B is bounded by 1 and using (2.12), (2.13) and (2.11), we see that

$$M_1(r, f') \le M_1(r, F') + M_1(r, B')M_\infty(r, F) = O\left(\varphi^2(r)\right), \quad \text{as } r \to 1.$$

Hence, $f \in \mathcal{L}(1, \phi)$.

On the other hand, we have

$$(2.14) \qquad (1 - |a_n|^2)\frac{|f'(a_n)|}{1 + |f(a_n)|^2} = (1 - |a_n|^2)|B'(a_n)||F(a_n)|.$$

Since B is an interpolating Blaschke product, it follows that there exists $\delta > 0$ such that $(1 - |a_n|^2)|B'(a_n)| \ge \delta$, for all n. The fact that F is unbounded together with (2.11) implies that $F(a_n) \to \infty$, as $n \to \infty$. Then (2.14) implies that $(1 - |a_n|^2)\frac{|f'(a_n)|}{1 + |f(a_n)|^2} \to \infty$, as $n \to \infty$ and, hence, f is not a normal function. $\square$

Similar arguments can be used to prove the sharpness of (2.7). Indeed, we have:

THEOREM 2.3. ([**15**, Theorems 1.4 and 1.5]). Let $1 \le p < \infty$ and let $\omega :$ $[0, 1] \to [0, \infty)$ be a continuous and increasing function with $\omega(0) = 0$ and

$$\frac{\omega(\delta)}{\delta^{1/p}} \to \infty, \quad \text{as } \delta \to 0.$$

and set $\phi(r) = \omega(1 - r)/(1 - r)$ $(0 < r < 1)$, then there exists $f \in \mathcal{L}(p, \phi)$ which is not a normal function.

If we assume in addition that ω is a Dini weight and satisfies the condition b_1 then we can assert that there exists $f \in \Lambda(p, \omega)$ which is not a normal function.

The functions f constructed to prove this theorem are also of the form $f = FB$ where F is a conformal mapping from $\mathbb{D}$ onto one of the domains Ω_Γ considered in Section 1.5 and B can be taken to be the Blaschke product whose sequence of zeros is $\{1 - 2^{-n}\}_{n=1}^\infty$.

2.3. An open question. All the examples of non-normal functions f constructed in [**5, 12, 13, 15**] to prove the sharpness of (2.7) and (2.9) are of the form $f = FB$ where F is a certain analytic function and B is a Blaschke product. It is natural to ask whether or not Blaschke products are really needed. More precisely, we may raise the following questions.

QUESTION 2.4. If ϕ is a positive continuous function defined in $[0, 1)$ with $\phi(r) \to \infty$, as $r \to 1$, does there exist an **outer function** $f \in \mathcal{L}(1, \phi)$ which is not a normal function?

QUESTION 2.5. Suppose that $1 \le p < \infty$ and let $\omega : [0, 1] \to [0, \infty)$ be a continuous and increasing function with $\omega(0) = 0$ and $\omega(\delta)\delta^{-1/p} \to \infty$, as $\delta \to 0$. Set $\phi(r) = \omega(1 - r)/(1 - r)$ $(0 < r < 1)$. Does there exist an **outer function** $f \in \mathcal{L}(p, \phi)$ which is not a normal function?

We do not know the answer to these questions. We can just prove that there exist non-Bloch outer functions in the desired spaces $\mathcal{L}(p, \phi)$.

THEOREM 2.6. *(i) Let ϕ be a positive continuous function defined in $[0, 1)$ with $\phi(r) \to \infty$, as $r \to 1$. Then, there exists an outer function $f \in \mathcal{L}(1, \phi)$ which is not a Bloch function.*

(ii) Suppose that $1 \le p < \infty$ and let $\omega : [0, 1] \to [0, \infty)$ be a continuous and increasing function with $\omega(0) = 0$ and $\omega(\delta)\delta^{-1/p} \to \infty$, as $\delta \to 0$. Set $\phi(r) = \omega(1 - r)/(1 - r)$ $(0 < r < 1)$. Then there exists an outer function $f \in \mathcal{L}(p, \phi)$ which is not a Bloch function

Proof. We shall just prove (i), the same argument works for (ii). Let F and B be the functions constructed in the proof of Theorem 2.1. Set $G(z) = F(z)/z$ $(z \in \mathbb{D})$. Since F is a conformal mapping from $\mathbb{D}$ onto a certain domain Ω with $F(0) = 0$, it follows that G is an outer function (see [**9**, Theorem 3.17]). Also, it is clear that

(2.15) $M_\infty(r, G) \asymp M_\infty(r, F)$ and $M_1(r, G') \asymp M_1(r, F')$,   as $r \to 1$.

On the other hand, $\text{Re}(B(z) + 1) > 0$, for all z, which implies that $B + 1$ is also an outer function. Since the product of two outer functions is an outer function, it follows that the function $g = G(B + 1)$ is outer. Using (2.15), it follows easily that $g \in \mathcal{L}(1, \phi)$.

Recall that B is an interpolating Blaschke product with positive zeros $\{a_n\}$ and that $F \in BMOA \subset \mathcal{B}$. Then, also $G \in \mathcal{B}$ and

$$(2.16) \qquad (1 - |a_n|)|G'(a_n)(B(a_n) + 1)| = (1 - |a_n|)|G'(a_n)| = O(1).$$

Since $(1 - |a_n|)|B'(a_n)| \geq \delta$, $n \geq 1$, for a certain $\delta > 0$, and $|G(a_n)| \to \infty$, as $n \to \infty$, it follows that $(1 - |a_n|)|B'(a_n)G(a_n)| \to \infty$, as $n \to \infty$, which, together with (2.16), yields $(1 - |a_n|)|g'(a_n)| \to \infty$, as $n \to \infty$. Thus, $g \notin \mathcal{B}$. $\square$

3. Bloch functions and division by inner functions

In this section we shall consider another problem on analytic function spaces which can be solved making use of the conformal mappings introduced in Section 1.5.

We recall that a function I, analytic in $\mathbb{D}$, is said to be an inner function if $I \in H^\infty$ and I has a radial limit $I(e^{i\theta})$ of modulus one for almost every $e^{i\theta} \in \partial\mathbb{D}$.

DEFINITION 3.1. A subspace X of H^1 is said to have the f-property (or property of division by inner functions) if $h/I \in X$ whenever $h \in X$ and I is an inner function with $h/I \in H^1$.

These notion was introduced by Havin [21]. Lots of spaces including the Hardy spaces H^p ($1 \leq p \leq \infty$), the Dirichlet space $\mathcal{D}$ and several spaces of Dirichlet type, the spaces $BMOA$ and $VMOA$, the Lipschitz spaces Λ_α ($0 < \alpha \leq 1$), the disc algebra $\mathcal{A}$ and many other are known to have the f-property. We mention the recently published paper [17] as a source to find these and many other results and references about the f-property and the stronger K-property introduced by Korenblum in [25].

The first example of a space not possessing the f-property was given by Gurarii [19] who proved that the space of analytic functions in $\mathbb{D}$ with an absolutely convergent power series does not have the f-property. Later on Anderson proved in [1] that $\mathcal{B}_0 \cap H^\infty$ does not have the f-property. Consequently, the same is true for $\mathcal{B}_0 \cap H^p$ for every $p \in [1, \infty)$.

Since $H^\infty \subset \mathcal{B}$, we see that $\mathcal{B} \cap H^\infty = H^\infty$ which has the f-property. However, for $1 \leq p < \infty$ the question of whether or not the space $\mathcal{B} \cap H^p$ has the f-property remained open. In a joint work with C. González and J. A. Peláez (cf. [16]) we have solved this problem, proving the following result:

THEOREM 3.2. ([16, Theorem 1.4]). If $1 \leq p < \infty$ then the space $\mathcal{B} \cap H^p$ does not have the f-property.

Sketch of the proof. Let B be an infinite Blaschke product in $\mathcal{B}_0$, whose sequence of zeros has a subsequence accumulating at 1 (the existence of such a product was first proved by Sarason [30]). Set

$$\mu(r) = \sup_{r \leq |z| < 1} (1 - |z|^2)|B'(z)|, \quad \varphi(r) = \frac{1}{\mu(r)}, \quad 0 < r < 1.$$

Since $B \in \mathcal{B}_0$, we have that $\varphi(r) \uparrow \infty$, as $r \uparrow 1$. Let $F = F_\Gamma$ be the conformal mapping constructed in Theorem 1.2 for this φ and set $f = BF$. It is easy to see that

(3.1) $$f \in \mathcal{B} \cap H^p, \quad \text{for all } p < \infty.$$

Let now B_1 be an interpolating Blaschke product, subproduct of B, whose sequence of zeros tends to 1 and set

(3.2) $$B_2 = \frac{B}{B_1}, \quad g = \frac{f}{B_2} = B_1 F.$$

It is clear that B_2 is a Blaschke product and that $g \in H^p$ for all $p < \infty$. On the other hand, using that B_1 is an interpolating Blaschke product, we can deduce that $g \notin \mathcal{B}$ which, together with (3.1) and (3.2), implies that $\mathcal{B} \cap H^p$ $(1 \leq p < \infty)$ does not have the f-property. $\qquad\square$

Acknowledgements

The author wishes to thank the referee for his/her comments and remarks.

References

[1] J. M. Anderson, *On division by inner factors*, Comment. Math. Helv. **54**, 2 (1979), 309–317.

[2] J. M. Anderson, J. Clunie and Ch. Pommerenke, *On Bloch functions and normal functions*, J. Reine Angew. Math. **270** (1974), 12–37.

[3] A. Baernstein II, *Integral means, univalent functions and circular symmetrization*, Acta Math. **133** (1974), 139–169.

[4] A. Baernstein II, *Analytic functions of bounded mean oscillation*, in *Aspects of Contemporary Complex Analysis*, D. Brannan and J. Clunie (editors), Academic Press (1980), 3–36.

[5] O. Blasco, D. Girela and M. A. Márquez, *Mean growth of the derivative of analytic functions, bounded mean oscillation, and normal functions*, Indiana Univ. Math. J. **47** (1998), 893–912.

[6] O. Blasco and G. Soares de Souza, *Spaces of analytic functions on the disc where the growth of $M_p(F, r)$ depends on a weight*, J. Math. Anal. Appl. **147**, 2 (1990), 580–598.

[7] P. Bourdon, J. Shapiro and W. Sledd, *Fourier series, mean Lipschitz spaces and bounded mean oscillation*, in *Analysis at Urbana 1, Proceedings of the Special Year in Modern Analysis at the University of Illinois, 1986-87*, E. R. Berkson, N. T. Peck and J. Uhl (editors), London Math. Soc. Lecture Notes Series 137, Cambridge Univ. Press (1989), 81–110.

[8] J. A. Cima and K. E. Petersen, *Some analytic functions whose boundary values have bounded mean oscillation*, Math. Z. **147** (1976), 237–347.

[9] P. L. Duren, *Theory of H^p Spaces*, Academic Press, New York-London (1970). Reprint: Dover, Mineola, New York (2000).

[10] P. L. Duren, *Univalent Functions*, Springer Verlag, New York, Berlin, Heidelberg, Tokyo, (1983).

[11] J. B. Garnett, *Bounded Analytic Functions*, Academic Press, New York, etc. (1981).

[12] D. Girela, *On a theorem of Privalov and normal funcions*, Proc. Amer. Math. Soc. **125**, no. 2 (1997), 433–442.

[13] D. Girela, *Mean Lipschitz spaces and bounded mean oscillation*, Illinois J. Math. **41**, No. 2 (1997), 214–230.

[14] D. Girela, *Analytic functions of bounded mean oscillation*, in *Complex functions spaces*, R. Aulaskari (editor), Univ. Joensuu Dept. Math. Report Series No. 4 (2001), 61–171.

[15] D. Girela and C. González, *Mean growth of the derivative of infinite Blaschke products*, Complex Variables Theory Appl. **45**, No. 1 (2001), 1–10.

[16] D. Girela, C. González and J. A. Peláez, *Multiplication and division by inner functions in the space of Bloch functions*, to appear in Proc. Amer. Math. Soc.

[17] D. Girela, C. González and J. A. Peláez, *Toeplitz operators and division by inner functions*, in *Proceedings of the First Advanced Course in Operator Theory and Complex Analysis*, A. Montes-Rodríguez (editor), Secretariado de Publicaciones de la Universidad de Sevilla (2005), 85–103.

[18] D. Girela and J. A. Peláez, *On the derivative of infinite Blaschke products*, Illinois. J. Math. **48**, No. 1 (2004), 121–130.

[19] V. P. Gurarii, *The factorization of absolutely convergent Taylor series and Fourier integrals.* (in Russian). Investigations on linear operators and the theory of functions, III, Zap. Naučn. Sem. Leningrad. Otdel. Mat. Inst. Steklov. (LOMI) **30** (1972), 15–32.

[20] G. H. Hardy and J. E. Littlewood *Some properties of fractional integrals, II*, Math. Z. **34** (1932), 403–439.

[21] V. P. Havin, *On the factorization of analytic functions smooth up to the boundary* (in Russian), Zap. Nauch. Sem. LOMI. **22** (1971), 202–205.

[22] W. K. Hayman, *On functions with positive real part*, J. London Math. Soc. **36** (1961), 35–48.

[23] J. A. Jenkins *On circularly symmetric functions*, Proc. Amer. Math. Soc. **6** (1955), 620–624.

[24] F. R. Keogh, *Some theorems on conformal mapping of bounded star-shaped domains*, Proc. London Math. Soc. **3**, 9 (1959), 481–491.

[25] B. I. Korenblum, *An extremal property of outer functions* (in Russian), Mat. Zamet. **10** (1971), 53–56; translation in Math. Notes **10** (1971), 456–458 (1972).

[26] K. I. Oskolkov, *Uniform modulus of continuity of summable functions on sets of positive measure*, (in russian), Dokl. Akad. Nauk SSSR **229**, 2 (1976), 304–306; English transl. in Sov. Math. Dokl. **17**, 4 (1976/1977), 1028–1030.

[27] K. I. Oskolkov, *Approximation properties of summable functions on sets of full measure*, (in russian), Mat. Sbornik, n. Ser. **32**, 4 (1977), 563–589; English transl. in Math. USSR Sbornik **32**, 4 (1978), 489–514.

[28] K. I. Oskolkov, *On Luzin's C-property for a conjugate function*, (in russian), Trudy Mat. Inst. Stelklova, **164** (1983), 124–135; English transl. in Proc. Steklov Inst. Math. **164**, (1985), 141–153.

[29] Ch. Pommerenke, *Univalent Functions*, Vandenhoeck und Ruprecht, Göttingen, (1975).

[30] D. Sarason, *Blaschke products in $\mathcal{B}_0$*, Linear and Complex Analysis problem Book, Lecture Notes in Math. **1043** (Springer-Verlag, Berlin, Heidelberg, New York, Tokyo 1984), 337–338.

DEPARTAMENTO DE ANÁLISIS MATEMÁTICO, FACULTAD DE CIENCIAS, UNIVERSIDAD DE MÁLAGA, 29071 MÁLAGA, SPAIN

E-mail address: `girela@uma.es`

Contemporary Mathematics
Volume **393**, 2006

Composition operators between Bergman spaces of functions of several variables

Hyungwoon Koo and Wayne Smith

ABSTRACT. We study the composition operator induced by a vector-valued holomorphic function φ from the unit ball B^m in $\mathbf{C}^m$ to B^n, where n and m are positive integers. Our main result is that every such composition operator maps each Hardy or weighted Bergman space on B^n into an associated Hardy or weighted Bergman space on B^m. In certain cases we are able to show our result is sharp, in the sense that the target space is optimal.

1. Introduction and statement of result

Let $B^n = \{z = (z_1, \ldots, z_n) \in \mathbf{C}^n : \sum |z_j|^2 < 1\}$ be the open unit ball centered at origin in $\mathbf{C}^n$. We shall write $H(B^n)$ for the space of all holomorphic functions on B^n. For $0 < p < \infty$ and $\alpha > -1$, the weighted Bergman space $A_\alpha^p(B^n)$ is the space of all $f \in H(B^n)$ for which

$$||f||_{A_\alpha^p}^p = \int_{B^n} |f(z)|^p (1 - |z|^2)^\alpha \, dV(z) < \infty,$$

where dV is normalized volume measure on B^n. Also, for $0 < p < \infty$, the Hardy space $H^p(B^n)$ is the space of all $g \in H(B^n)$ for which

$$||g||_{H^p}^p = \sup_{0<r<1} \int_{\partial B^n} |g(r\zeta)|^p d\sigma(\zeta) < \infty$$

where $d\sigma$ is normalized surface measure on ∂B^n. If $g \in H^p(B^n)$, then the radial limit $g(\zeta) = \lim_{r \to 1^-} g(r\zeta)$ exists for almost all $\zeta \in \partial B^n$ and

$$||g||_{H^p}^p = \int_{\partial B^n} |g(\zeta)|^p d\sigma(\zeta).$$

We will often use the following notation to allow unified statements:

$$A_{-1}^p(B^n) = H^p(B^n).$$

Let φ be a vector-valued holomorphic function from B^m to B^n for some positive integers n and m. That is,

$$\varphi = (\varphi_1, \ldots, \varphi_n) : B^m \to B^n$$

2000 *Mathematics Subject Classification.* Primary 47B33, Secondary 30D55, 46E15.
Key words and phrases. Composition operator, ball, Bergman space.
This research was partially supported by a Korea University Grant.

where each φ_j is holomorphic on B^m. Then φ induces the composition operator C_φ, defined on $H(B^n)$ by

$$C_\varphi f = f \circ \varphi.$$

In the case that $m = n = 1$, it is a well known consequence of Littlewood's Subordination Principle that every composition operator C_φ is bounded on each of the spaces $A_\alpha^p(B^1)$, $p > 0$, $\alpha \geq -1$; see for example [**CM**]. This result does not extend to the case that $m = n > 1$, where even such a simple function as $\varphi(z_1, z_2) = (2z_1 z_2, 0)$ is known to induce an unbounded composition operator on $H^p(B^2)$; see section 3.5 in [**CM**]. A generalization has been found by enlarging the target space for C_φ:

> If φ is any vector-valued holomorphic self-map of B^n, then C_φ maps A_α^p boundedly into $A_{n+\alpha-1}^p(B^n)$.

This result was given for $\alpha = -1$ by B. MacCluer and P. Mercer in [**MM**], and subsequently extended to $\alpha > -1$ by J. Cima and P. Mercer in [**CiMe**]. When $n = 1$ the choice $\varphi(z) = z$ (which makes C_φ the identity operator) shows it is sharp in the sense that the target space can not be replaced by a smaller Bergman or Hardy space. When $n > 1$, however, it is not known if the result is sharp in this sense. The main result of this paper is an extension of this theorem that allows $m \neq n$, and which we can show is sharp in certain cases.

THEOREM 1.1. *Let n and m be positive integers, and let $\alpha \geq -1$. Let φ be a vector-valued holomorphic function from B^m to B^n. Then C_φ maps $A_\alpha^p(B^n)$ boundedly into $A_{n+\alpha-1}^p(B^m)$:*

$$C_\varphi : A_\alpha^p(B^n) \to A_{n+\alpha-1}^p(B^m).$$

Moreover, there is a constant C independent of φ such that

$$\|C_\varphi\| \leq C \left(\frac{1 + |\varphi(0)|}{1 - |\varphi(0)|} \right)^{\frac{n+\alpha+1}{p}}.$$

After background material is provided in §2, the proof of Theorem 1.1 will be given in §3. In §4 we give examples that show our result is sharp when either $(n, \alpha) = (1, -1)$ or $m = 1$.

Michael Stessin and Kehe Zhu [**SZ**] have a recent preprint with results similar to ours for $m = 1$. They determine the optimal range spaces and characterize compactness of composition operators $C_\varphi : A_\alpha^p(\Omega) \to A_\beta^p(B^1)$ where Ω is either B^n or the polydisk in $\mathbf{C}^n$.

2. Background: Carleson Type Criteria

Our work uses a characterization of measures μ on B^n which satisfy $A_\alpha^p(B^n) \subset L^q(B^n, d\mu)$. This characterization is given using what are called Carleson measure criteria. We must first introduce some notation. We write $\langle z, w \rangle = z_1 \bar{w}_1 + \cdots + z_n \bar{w}_n$ for the Hermitian inner product of $z = (z_1, ..., z_n)$ and $w = (w_1, ..., w_n)$ in $\mathbf{C}^n$. For $\zeta \in \partial B^n$ and $\delta > 0$, let $S^n(\zeta, \delta) = \{z \in B^n : |1 - \langle z, \zeta \rangle| < \delta\}$. Finally, for $\alpha > -1$ we use $dV_\alpha(z) = (1 - |z|^2)^\alpha dV(z)$ to denote weighted volume measure on B^n. Then we have (see p. 67 of [**R**])

$$(2.1) \qquad V_\alpha\left(S^n(\zeta, \delta)\right) \approx \delta^{n+1+\alpha} \quad \text{and} \quad \sigma\left(\overline{S^n(\zeta, \delta)} \cap \partial B^n\right) \approx \delta^n,$$

where α and n are fixed, and $0 < \delta < 2$. Here and below the relation $A \approx B$ means that $A \lesssim B$ and $B \lesssim A$, where $A \lesssim B$ means that $A \leq C \cdot B$ for some inconsequential constant $C > 0$.

For $\beta \geq n$, we say finite positive Borel measure μ on $\overline{B^n}$ is a β-Carleson measure if

$$||\mu||_\beta = \sup \mu(\overline{S^n(\zeta, \delta)})/\delta^\beta < \infty,$$

where the supremum is taken over all $\zeta \in \partial B^n$ and all $\delta \in (0, 2)$, and μ is a compact β-Carleson measure if

$$\lim_{\delta \to 0} \sup_{\zeta \in \partial B^n} \mu(\overline{S^n(\zeta, \delta)})/\delta^\beta = 0.$$

We note that the support of an n-Carleson measure may intersect ∂B^n, but if $\beta > n$, then it follows from (2.1) that a β-Carleson measure is supported on the open ball B^n. We warn the reader that the terminology "β-Carleson measure" is not standard in the literature. A different exponent on δ is sometimes found.

The following Carleson measure criteria for the inclusion $A^p_\alpha(B^n) \subset L^q(B^n, d\mu)$ are known.

LEMMA 2.1. *Assume that $0 < p \leq q < \infty$.*

(1) *Let μ be a finite positive Borel measure on $\overline{B^n}$. Then $H^p(B^n) \subset L^q(B^n, d\mu)$ if and only if μ is an nq/p - Carleson measure. In this case the identity operator $I : H^p(B^n) \to L^q(B^n, d\mu)$ is bounded with $||I|| \lesssim ||\mu||^{1/q}_{nq/p}$, and I is compact if and only if μ is a compact nq/p - Carleson measure.*

(2) *Let μ be a finite positive Borel measure on B^n and let $\alpha > -1$. Then $A^p_\alpha(B^n) \subset L^q(B^n, d\mu)$ if and only if μ is an $(n + 1 + \alpha)q/p$ - Carleson measure. In this case the identity operator $I : A^p_\alpha(B^n) \to L^q(B^n, d\mu)$ is bounded with $||I|| \lesssim ||\mu||^{1/q}_{(n+1+\alpha)q/p}$, and I is compact if and only if μ is a compact $(n + 1 + \alpha)q/p$ - Carleson measure.*

The work of many authors has been summarized in this lemma, going back to L. Carleson in [C] where the case $\alpha = -1$, $n = 1$, $p = q$ and I bounded was considered. A good source for the history of subsequent developments is [L]. Theorem 2.2 in [L] corresponds to $\alpha > -1$, $n = 1$ and I bounded in the lemma, and the extensions to $n > 1$ are discussed in Section 6 of that paper. Another reference for the case $p = q$ and I bounded in the lemma is [CM], Theorems 2.37 and 2.38. The method for going from I bounded to I compact is well known; see the proof Theorem 4.3 in [MS], where $n = 1$ and $p = q$. The same method applies in the general situation.

The next proposition uses the notation φ^* for the radial limit function for $\varphi : B^m \to B^n$, i.e. $\varphi^*(\zeta) = \lim_{r \to 1_-} \varphi(r\zeta)$, $\zeta \in \partial B^m$, and $\varphi^* : \partial B^m \to \overline{B^n}$. Note that $\varphi^*(\zeta)$ exists almost everywhere on ∂B^m, since each φ_j is a bounded holomorphic function and so has radial limits almost everywhere.

PROPOSITION 2.2. *Let $0 < p \leq q < \infty$, $\alpha \geq -1$, and $\beta > -1$. Suppose that $\varphi : B^m \to B^n$ is a holomorphic vector-valued function and define the Borel measures μ on B^n and μ^* on $\overline{B^n}$ by $\mu(E) = V_\beta(\varphi^{-1}(E))$ and $\mu^*(E) = \sigma(\varphi^{*-1}(E))$. Then*

(1) *$C_\varphi : A^p_\alpha(B^n) \to A^q_\beta(B^m)$ is bounded (respectively compact) if and only if μ is an (compact) $(n + 1 + \alpha)q/p$ - Carleson measure on B^n;*

(2) $C_\varphi : H^p(B^n) \to H^q(B^m)$ is bounded (respectively compact) if and only if μ^* is an (compact) nq/p - Carleson measure on $\overline{B^n}$.

PROOF. For $f \in A_\alpha^p(B^n)$,

$$\|C_\varphi(f)\|_{A_\beta^q(B^m)}^q = \int_{B^m} |f(\varphi(z))|^q dV_\beta(z) = \int_{B^n} |f(w)|^q d\mu(w),$$

and for $f \in H^p(B^n)$

$$\|C_\varphi(f)\|_{H^q(B^m)}^q = \int_{\partial B^m} |f(\varphi^*(\zeta))|^q d\sigma(\zeta) = \int_{\overline{B^n}} |f(w)|^q d\mu^*(w),$$

where the last equality in each displayed line used a change of variables formula from measure theory (see p. 163 in [H]). (In the second display the symbol f represents both an element of $H^p(B^n)$ and its extension via radial limits to a function on $\overline{B^n}$.) Thus the results follow from Lemma 2.1. □

The observation that the Carleson measure criteria in Proposition 2.2 depend only on the ratio q/p yields the following immediate corollary:

COROLLARY 2.3. Let $\alpha \geq -1$, $\beta > -1$, and $s \geq 1$. Then
  (1) $C_\varphi : A_\alpha^p(B^n) \to A_\beta^{ps}(B^m)$ is bounded (respectively compact) for some $p \in (0, \infty)$ if and only if it is bounded (compact) for all $p \in (0, \infty)$.
  (2) $C_\varphi : H^p(B^n) \to H^{ps}(B^m)$ is bounded (respectively compact) for some $p \in (0, \infty)$ if and only if it is bounded (compact) for all $p \in (0, \infty)$.

As mentioned in the introduction, the case $m = n$ in Theorem 1.1 is known. This result, with uniform bounds for the norm of C_φ and in the Carleson criteria and when $\varphi(0) = 0$, is the final piece of background needed to prove Theorem 1.1.

THEOREM 2.4. Let $0 < p < \infty$, $\alpha \geq -1$ and $\varphi : B^n \to B^n$ be a holomorphic vector-valued function with $\varphi(0) = 0$. Then $C_\varphi : A_\alpha^p(B^n) \to A_{\alpha+n-1}^p(B^n)$ is bounded, and moreover there exists a constant C independent of $\zeta \in \partial B^n$, $\delta > 0$, and φ (assuming only that $\varphi(0) = 0$) such that $\|C_\varphi\| \leq C$ and

$$V_{\alpha+n-1}\varphi^{-1}(S^n(\zeta, \delta)) \leq C\delta^{\alpha+n+1} \qquad if \quad \alpha + n > 0,$$

or

$$\sigma\varphi^{*-1}(\overline{S^n(\zeta, \delta)}) \leq C\delta \qquad if \quad \alpha + n = 0.$$

When $n = 1$, this theorem has been long known; see for example [MS] or [CM]. When $n > 1$, it is proved for B^n and also for more general domains in [MM] for $\alpha = -1$ and in [CiMe] for $\alpha > -1$, but the uniform estimates for $\|C_\varphi\|$ and the measure of $S^n(\zeta, \delta)$ when $\varphi(0) = 0$ are not stated explicitly. In [CM] it is proved for B^n, and there the proof clearly establishes the uniform estimates; see [CM], p. 227.

3. Proof of Main Theorem

We now give the proof of Theorem 1.1. Let ψ be an automorphism of B^n with $\psi(0) = \varphi(0)$. It is easy to see from Proposition 2.2 that $A_\alpha^p(B^n)$ is invariant under C_ψ with $\|C_\psi\| \leq C \left(\frac{1+|\varphi(0)|}{1-|\varphi(0)|}\right)^{\frac{n+\alpha+1}{p}}$ (see Exercise 3.5.4 on p. 172 of [CM]). Therefore, it suffices to show that there exists a constant C such that $\|C_\varphi\| \leq C$ for all φ with $\varphi(0) = 0$.

Proof of Theorem 1.1 for $m \leq n$:

The case $m = n$ is covered by Theorem 2.4, so we may assume that $m < n$. Let $\varphi(0) = 0$ and $e_1^n = (1, 0, ..., 0) \in \partial B^n$. Let R_ζ be a unitary transformation of B^n with $R_\zeta(\zeta) = e_1^n$ and let $P : B^n \to B^m$ be the projection defined by $P(z) = (z_1, ..., z_m)$. Let $\eta_\zeta = P \circ R_\zeta \circ \varphi$. Then $\eta_\zeta : B^m \to B^m$ and $\eta_\zeta(0) = 0$. Since R_ζ is a unitary transformation, $\langle R_\zeta(z), R_\zeta(w) \rangle = \langle z, w \rangle$ which leads to $S^n(\zeta, \delta) = R_\zeta^{-1} S^n(e_1^n, \delta)$. Moreover, it is readily verified that

$$S^n(\zeta, \delta) = (P \circ R_\zeta)^{-1}(S^m(e_1^m, \delta)).$$

Therefore, we have

$$\varphi^{-1}(S^n(\zeta, \delta)) = \varphi^{-1}(P \circ R_\zeta)^{-1}(S^m(e_1^m, \delta)) = \eta_\zeta^{-1}(S^m(e_1^m, \delta)).$$

Since $\eta_\zeta(0) = 0$ for each $\zeta \in \partial B^n$, from Theorem 2.4 there exists a constant C independent of ζ and δ such that

$$V_{\beta+m-1}\eta_\zeta^{-1}(S^m(e_1^m, \delta)) \leq C\delta^{\beta+m+1},$$

for all $\beta > -1$. Let $\beta = \alpha + n - m$, and note that $\beta > -1$ since $m < n$. Then

$$(3.1) \qquad V_{\alpha+n-1}\varphi^{-1}(S^n(\zeta, \delta)) = V_{\alpha+n-1}\eta_\zeta^{-1}(S^m(e_1^m, \delta)) \leq C\delta^{\alpha+n+1},$$

where again C is independent of $\zeta \in \partial B^n$ and $\delta > 0$. We also note for use below that the same constant C works for all φ satisfying $\varphi(0) = 0$. Proposition 2.2(1) now shows that $C_\varphi : A_\alpha^p(B^n) \to A_{n+\alpha-1}^p(B^m)$ is bounded, and the proof is complete.

Proof of Theorem 1.1 for $m > n$:

For $\zeta \in \partial B^m$ and $\lambda \in B^1$, define the function $\varphi_\zeta : B^1 \to B^n$ by $\varphi_\zeta(\lambda) = \varphi(\lambda\zeta)$. Then, from the just proved case $m \leq n$ of the theorem, for each fixed $\zeta \in \partial B^m$ and $\beta \geq -1$ we have

$$C_{\varphi_\zeta} : A_\beta^p(B^n) \to A_{n+\beta-1}^p(B^1)$$

is bounded. Since $\varphi(0) = 0$ we have $\varphi_\zeta(0) = 0$, and so from (3.1) we see that the Carleson criteria holds uniformly in $\zeta \in \partial B^m$. Thus there exists a constant C such that

$$(3.2) \qquad \|C_{\varphi_\zeta}\| \leq C, \quad \text{for all } \zeta \in \partial B^n.$$

We first consider the case that $n + \alpha > 0$, so $n + \alpha - 1 > -1$. Then using the method slice integration (see [**R**], p. 15) we have

$$\|C_\varphi(f)\|_{A_{n+\alpha-1}^p(B^m)}^p = \int_0^1 \int_{\partial B^m} |f(\varphi(r\zeta))|^p d\sigma(\zeta) r^{2m-1}(1-r^2)^{n+\alpha-1} dr$$

$$= \int_0^1 \left(\int_{\partial B^m} \int_0^{2\pi} |f(\varphi_\zeta(re^{i\theta}))|^p \frac{d\theta}{2\pi} d\sigma(\zeta) \right) r^{2m-1}(1-r^2)^{n+\alpha-1} dr$$

$$= \int_{\partial B^m} \int_{B^1} |C_{\varphi_\zeta}(f)(\lambda)|^p |\lambda|^{2m-2}(1-|\lambda|^2)^{n+\alpha-1} dV(\lambda) d\sigma(\zeta)$$

$$\leq \int_{\partial B^m} \|C_{\varphi_\zeta}(f)\|_{A_{n+\alpha-1}^p(B^1)}^p d\sigma(\zeta)$$

$$\leq \int_{\partial B^m} \|C_{\varphi_\zeta}\|^p \cdot \|f\|_{A_\alpha^p(B^n)}^p d\sigma(\zeta).$$

Use of (3.2) now completes the proof.

If $n + \alpha = 0$, then $n = 1$ and $\alpha = -1$. So $C_{\varphi_\zeta} : H^p(B^1) \to H^p(B^1)$, with $\|C_{\varphi_\zeta}\|^p \leq \frac{1+|\varphi_\zeta(0)|}{1-|\varphi_\zeta(0)|} = 1$; see, for example, Corollary 3.7 in [**CM**]. Therefore, the same argument shows that

$$\|C_\varphi(f)\|^p_{H^p(B^m)} = \int_{\partial B^m} |f(\varphi(\zeta))|^p d\sigma(\zeta)$$

$$= \int_{\partial B^m} \int_0^{2\pi} |f(\varphi_\zeta(e^{i\theta}))|^p \frac{d\theta}{2\pi} d\sigma(\zeta)$$

$$\leq \int_{\partial B^m} \|C_{\varphi_\zeta}(f)\|^p_{H^p(B^1)} d\sigma(\zeta)$$

$$\leq \|f\|^p_{H^p(B^1)}.$$

4. When $m = 1$ or $n = 1$

In this section we restrict ourselves to the case when either $m = 1$ or $n = 1$. First assume that $n = 1$, so $\varphi : B^m \to B^1$ and $C_\varphi : A^p_\alpha(B^1) \to A^p_\alpha(B^m)$. The following proposition gives a necessary condition for $C_\varphi : H^p(B^1) \to H^p(B^m)$ to be compact. This result is well known when $m = 1$, where the proof is the same.

PROPOSITION 4.1. *If $C_\varphi : H^p(B^1) \to H^p(B^m)$ is compact, then*

$$\sigma\left(\{\zeta \in \partial B^m : |\varphi^*(\zeta)| = 1\}\right) = 0.$$

PROOF. We may assume $p = 2$ by Corollary 2.3. The sequence $\{z^n\}$ converges to 0 weakly in $H^2(B^1)$, and so if C_φ is compact then the sequence $\{C_\varphi(z^n)\}$ converges to 0 in norm in $H^2(B^m)$. Let $A = \{\zeta \in \partial B^m : |\varphi^*(\zeta)| = 1\}$. Since

$$\|C_\varphi(z^n)\|^2_{H^2(B^m)} = \int_{\partial B^m} |\varphi^*(\zeta)|^{2n} d\sigma(\zeta) \geq \sigma(A),$$

it follows that $\sigma(A) = 0$. $\qquad\square$

The next example uses some properties of functions with Hadamard gaps, which we collect in the following lemma. First, recall that $f \in H(B^1)$ belongs to the Bloch space $\mathcal{B}$ if $\sup_{z \in B^1}(1 - |z|^2)|f'(z)| < \infty$.

LEMMA 4.2. *Let $f(z) = \sum_{k=0}^\infty b_k z^{n_k}$, where $n_{k+1}/n_k > \lambda > 1$. Then*

(1) *$f \in \mathcal{B}$ if and only if $\sup |b_k| < \infty$;*
(2) *If $\lim_{r \to 1^-} f(r\zeta)$ exists and is finite for some $\zeta \in \partial B^1$, then $\lim_{k \to \infty} b_k = 0$.*

A reference for Lemma 4.2 (1) is Proposition 8.12 of [**P**]. Lemma 4.2 (2) is known as the Tauberian theorem of Hardy and Littlewood for functions with Hadamard gaps; see [**HL**]. The next example also uses the inclusion $\mathcal{B} \subset A^q_\alpha(B^1)$, for all $\alpha > -1$ and $0 < q < \infty$; see Proposition 1.11 in [**HKZ**] or equation (4) on p. 73 of [**P**].

Parts (2) and (3) of the following example shows that Theorem 1.1 is sharp when $(\alpha, n) = (-1, 1)$.

EXAMPLE 4.3. *Let $\varphi : B^m \to B^1$ be an inner function and let $p > 0$. Then the following hold:*

(1) *$C_\varphi : H^p(B^1) \to H^p(B^m)$ is bounded but not compact.*
(2) *$C_\varphi\left(H^p(B^1)\right) \not\subset H^{p+\varepsilon}(B^m)$ for any $\varepsilon > 0$.*

(3) *If $\alpha > -1$ and $0 < q < \infty$, then $C_\varphi(A^q_\alpha(B^1)) \nsubseteq H^p(B^m)$.*

PROOF. Statement (1) is an immediate consequence of Theorem 1.1 and Proposition 4.1. For (2), suppose $C_\varphi \left(H^p(B^1)\right) \subset H^{p+\varepsilon}(B^m)$, and consequently $C_\varphi :$ $H^p(B^1) \to H^{p+\varepsilon}(B^m)$ is bounded by the Closed Graph Theorem. Then, from the Carleson measure criteria in Proposition 2.2, it follows that $C_\varphi : H^p(B^1) \to H^p(B^m)$ is compact. But this contradicts (1), and so completes the proof of (2).

For the proof of (3), let f be the gap series defined by $f(z) = \sum_{n=1}^{\infty} z^{2^n}$. Then, from Lemma 4.2, $f \in \mathcal{B} \subset A^q_\alpha(B^1)$ and f has no finite radial limits. If $C_\varphi(A^q_\alpha(B^1)) \subset H^p(B^m)$, then $f \circ \varphi \in H^p(B^m)$ and so $f \circ \varphi$ has a finite radial limit at almost every point $\zeta \in \partial B^m$. Also, φ has a radial limit of modulus one at almost every $\zeta \in \partial B^m$, since it is inner. Hence there exists $\zeta_0 \in \partial B^m$ such that $\lim_{r \to 1_-} \varphi(r\zeta_0) = \varphi^*(\zeta_0) \in \partial B^1$ and $\lim_{r \to 1_-} f \circ \varphi(r\zeta_0)$ exists and is finite. This means that f has a finite asymptotic value at $\varphi^*(\zeta_0)$, i.e. there is a curve in B^1 ending at ζ_0 along which f has a finite limit. Since $f \in \mathcal{B}$, the asymptotic value of f at $\varphi^*(\zeta_0)$ is also the radial limit of f at $\varphi^*(\zeta_0)$; see Theorem 4.3 in [**P**]. But this contradicts the fact that f has no finite radial limit. $\qquad\square$

When $n = 1$, taking φ to be an inner function worked well in the case of Hardy spaces, but the Bergman norm of an inner function is not amenable to calculation. When $n = 1$ and $\alpha > -1$ (and $m > 1$), we do not know whether our result is sharp. Next proposition shows how an example demonstrating Theorem 1.1 is sharp when $n = 1$ and α is sufficiently large can be modified to work for $n > 1$.

PROPOSITION 4.4. *Suppose α and $\beta \geq -1$, $j < n$ and m are positive integers, and $\psi : B^m \to B^{n-j}$. Define $\varphi = (\varphi_1, \ldots, \varphi_n) : B^m \to B^n$ by $\varphi_k = \psi_k$, $1 \leq k \leq n - j$, and $\varphi_k = 0$, $n - j < k \leq n$. If $C_\varphi : A^p_\alpha(B^n) \to A^p_\beta(B^m)$ is bounded, then $C_\psi : A^p_{\alpha+j}(B^{n-j}) \to A^p_\beta(B^m)$ is also bounded.*

PROOF. Let P be the projection defined by $P(z_1, \ldots, z_n) = (z_1, \ldots, z_{n-j})$, and notice that if $\zeta \in \partial B^{n-j}$ and $\delta > 0$, then $P^{-1} S^{n-j}(\zeta, \delta) = S^n(\hat\zeta, \delta)$, where $\hat\zeta$ is the unique point of ∂B^n satisfying $P\hat\zeta = \zeta$. Thus $V_\alpha P^{-1} S^{n-j}(\zeta, \delta) \approx \delta^{n+1+\alpha}$ from (2.1), and so $C_P : A^p_{\alpha+j}(B^{n-j}) \to A^p_\alpha(B^n)$ is bounded by Proposition 2.2. Also, $\psi = P \circ \varphi$, so $C_\psi = C_\varphi C_P$ is bounded if C_φ is bounded. $\qquad\square$

Note that as a corollary to Theorem 2.4 and this proposition, we get some cases of Theorem 1.1: $C_\psi : A^p_{\alpha+j}(B^{m-j}) \to A^p_{\alpha+m-1}(B^m)$ is bounded for every vector-valued holomorphic map $\psi : B^m \to B^{m-j}$, $0 \leq j \leq m - 1$.

Next we assume that $m = 1$, so $\varphi : B^1 \to B^n$ and $C_\varphi : A^p_\alpha(B^n) \to A^p_{\alpha+n-1}(B^1)$. For $\zeta \in \partial B^n$, define $\eta_\zeta : B^1 \to B^1$ by $\eta_\zeta(z) = <\varphi(z), \zeta>$.

THEOREM 4.5. *Suppose $C_\varphi : A^p_\alpha(B^n) \to A^p_{\alpha+n-1}(B^1)$ is compact. Then, for each $\zeta \in \partial B^n$, η_ζ does not have finite angular derivative at any point $\omega \in \partial B^1$ where $|\eta_\zeta^*(\omega)| = 1$.*

PROOF. Let $E(z_1, \ldots, z_n) = z_1$ and let R_ζ be a unitary transformation on B^n such that $R_\zeta(\zeta) = e^n_1$. Then $\langle z, \zeta \rangle = \langle R_\zeta(z), R_\zeta(\zeta) \rangle = E(R_\zeta(z))$, so $\eta_\zeta = E \circ R_\zeta \circ \varphi$ and the action of C_{η_ζ} on $A^p_{\alpha+n-1}(B^1)$ may be viewed as follows:

$$A^p_{\alpha+n-1}(B^1) \xrightarrow{C_E} A^p_\alpha(B^n) \xrightarrow{C_{R_\zeta}} A^p_\alpha(B^n) \xrightarrow{C_\varphi} A^p_{\alpha+n-1}(B^1).$$

Proposition 2.2 shows that C_E is bounded, since $E^{-1}S^1(w,\delta) = S^n((w,0,...,0),\delta)$ for all $w \in \partial B^1$ and $\delta > 0$, and $V_\alpha(S^n((w,0,...,0),\delta)) \approx \delta^{n+1+\alpha}$ from (2.1). And C_{R_ζ} is bounded, since R_ζ is a unitary transformation. Hence the assumption that C_φ is compact implies that C_{η_ζ} is compact on $A_{\alpha+n-1}^p(B^1)$ for each $\zeta \in \partial B^n$. The proof is complete since the compactness of C_ψ on any space $A_\beta^p(B^1)$ implies that ψ does not have finite angular derivative at any point $w \in \partial B^1$ where $|\psi^*(w)| = 1$ (see Theorem 2.1 of [**ST**] and Theorem 3.5 of [**MS**]).   □

We can now give an example showing that Theorem 1.1 is sharp when $m = 1$.

EXAMPLE 4.6. *Let* $p > 0$, $\alpha \geq -1$ *and* $\varphi : B^1 \to B^n$ *be defined by* $\varphi(z) = (z,0,...,0)$. *Then the following hold:*

(1) $C_\varphi\left(A_\alpha^p(B^n)\right) \not\subseteq A_{\alpha+n-1}^{p+\varepsilon}(B^1)$ *for any* $\varepsilon > 0$.
(2) $C_\varphi\left(A_{\alpha+\delta}^p(B^n)\right) \not\subseteq A_{\alpha+n-1}^p(B^1)$ *for any* $\delta > 0$.

PROOF. Let $\varepsilon > 0$ and assume to the contrary of (1) that $C_\varphi\left(A_\alpha^p(B^n)\right) \subseteq A_{\alpha+n-1}^{p+\varepsilon}(B^1)$. Then $C_\varphi : A_\alpha^p(B^n) \to A_{\alpha+n-1}^{p+\varepsilon}(B^1)$ is bounded by the Closed Graph Theorem, and the Carleson measure criteria in Proposition 2.2 can be used to show $C_\varphi : A_\alpha^p(B^n) \to A_{\alpha+n-1}^p(B^1)$ is compact. We now conclude from Theorem 4.5 that, for each $\zeta \in \partial B^n$, η_ζ does not have a finite angular derivative at any point where η_ζ has a radial limit of modulus one. This is a contradiction, since with $\zeta = e_1^n$ we have $\eta_{e_1^n}(z) = z$. This proves (1), and (2) is proved in the same way.   □

It is natural to consider whether Theorem 4.5 may be re-stated in terms of angular derivatives of the coordinate functions φ_j. In particular, the existence of a finite angular derivative for η_ζ at a point does follow from each coordinate function φ_j having a finite angular derivative. However, the converse fails.

EXAMPLE 4.7. *Let* $\zeta = (\zeta_1,\zeta_2) \in \partial B^2$, *where* $\zeta_1 \neq 0$ *and* $\zeta_2 \neq 0$. *There exists a holomorphic vector-valued function* $\varphi = (\varphi_1,\varphi_2) : B^1 \to B^2$ *such that (1)* $\eta_\zeta^*(1) = 1$, *(2)* η_ζ *has a finite angular derivative at* $z = 1$, *but (3) neither* φ_1 *nor* φ_2 *has a finite angular derivative at* $z = 1$.

PROOF. An example is given on p. 183 of [**R**] of a holomorphic vector-valued function $\psi = (\psi_1,\psi_2) : B^1 \to B^2$ such that $\psi^*(1) = (1,0)$ and ψ_1 has a finite angular derivative at $z = 1$, but ψ_2 does not. Let R_ζ be the unitary transformation of B^2 represented by the matrix $\begin{pmatrix} \overline{\zeta_1} & \overline{\zeta_2} \\ -\zeta_2 & \zeta_1 \end{pmatrix}$ so $R_\zeta(\zeta) = (1,0)$. Define $\varphi = R_\zeta^{-1}\psi$. Then $\varphi^*(1) = \zeta$, and so $\eta_\zeta^*(1) = < \varphi^*(1), \zeta > = 1$. Also $\psi = R_\zeta\varphi$, giving $\psi_1 = \varphi_1\overline{\zeta_1} + \varphi_2\overline{\zeta_2} = \eta_\zeta$, whence η_ζ has a finite angular derivative at $z = 1$. If φ_1 has a finite angular derivative at $z = 1$, then so does $\varphi_2 = (\psi_1 - \varphi_1\overline{\zeta_1})/\overline{\zeta_2}$. It would then follow that $\psi_2 = \varphi_1\zeta_2 - \varphi_2\zeta_1$ has a finite angular derivative at $z = 1$, which is a contradiction. Similar reasoning leads to a contradiction if φ_2 has a finite angular derivative at $z = 1$, and so the proof is complete.   □

References

[C]   L. Carleson, *Interpolation by bounded analytic functions and the corona problem*, Ann. of Math., 76 (1962), 547-559.

[CiMe]   J. Cima and A. Mercer, *Composition operators between Bergman spaces on convex domains in* $\mathbf{C}^n$, J. Operator Theory, 33(2) (1995), 363-369.

[CM]   C. Cowen and B. MaCluer, *Composition operators on spaces of analytic functions*, CRC Press, New York, 1995.

[H]   P. R. Halmos, *Measure Theory*, Springer-Verlag, New York, 1974.

[HKZ] H. Hedenmalm, B. Korenblum and K. Zhu, *Theory of Bergman spaces*, Springer, New York, 2000.

[HL]   G. H. Hardy and J. E. Littlewood, *A further note on the converse of Abel's theorem*, Proc. London Math. Soc. 25 (1926), 219-236.

[L]   D. Luecking, *Forward and Reverse Carleson inequalities for functions in Bergman spaces and their derivatives*, Amer. J. Math., 107 (1985), 85–111

[MS]   B. MacCluer and J.H. Shaprio, *Angular derivatives and compact composition operators on the Hardy and Bergman spaces*, Can. J. Math., XXXVIII(4)(1986), 878-906.

[MM]   B. MacCluer and P. Mercer, *Composition operators between Hardy and Weighted Bergman spaces on convex domains in* $\mathbf{C}^n$, Proc. Amer. Math. Soc., 123(7) (1995), 2093-2102.

[P]   Ch. Pommerenke, *Boundary behavior of conformal maps*, Springer-Verlag, Berlin Heidelberg New York, 1992.

[R]   W. Rudin, *Function theory in the unit bal of* $\mathbf{C}^n$, Springer-Verlag, New York, 1980.

[ST]   J. H. Shapiro and P.D. Taylor, *Compact, nuclear and Hilbert-Schmidt compostion operators on* H^2, Indiana Univ. Math. J., 23 (1973), 471-496.

[SZ]   M. Stessin and K. Zhu, *Composition operators on embedded disks*, preprint.

DEPARTMENT OF MATHEMATICS, KOREA UNIVERSITY, SEOUL 136–701, KOREA
E-mail address: koohw@korea.ac.kr

DEPARTMENT OF MATHEMATICS, UNIVERSITY OF HAWAII, HONOLULU, HAWAII 96822
E-mail address: wayne@math.hawaii.edu

Contemporary Mathematics
Volume **393**, 2006

Isometries of some classical function spaces among the composition operators

María J. Martín and Dragan Vukotić

Dedicated to Professor Joseph Cima on the occasion of his 70th birthday

ABSTRACT. We give a simple and unified proof of the characterizations of all possible composition operators that are isometries of either a general Hardy space or a general weighted Bergman spaces of the disk. We do the same for the isometries of analytic Besov spaces (containing the Dirichlet space) among the composition operators with univalent symbols.

Introduction

Throughout this note, $dm(\theta) = (2\pi)^{-1}d\theta$ will denote the normalized arc length measure on the unit circle $\mathbb{T}$. We assume that the reader is familiar with the definition of the standard Hardy spaces H^p of the disk (see [**D**], for example). We write dA for the normalized Lebesgue area measure on the unit disk $\mathbb{D}$: $dA(re^{i\theta}) = \pi^{-1}rdrd\theta$. The weighted Bergman space A_w^p is the space of all $L^p(\mathbb{D}, w\,dA)$ functions analytic in the disk, where w is a radial weight function: $w(z) = w(|z|)$, non-negative and integrable with respect to dA. Every H^p is a Banach space when $1 \leq p < \infty$, and so is A_w^p when the weight w is "reasonable" (whenever the point evaluations are bounded; roughly speaking, w should not be zero "too often" near the unit circle). The unweighted Bergman space A^p is obtained when $w \equiv 1$ (see [**DS**] for the theory of these spaces); the standard weighted space A_α^p corresponds to the case $w(z) = (\alpha + 1)(1 - |z|^2)^\alpha$, $-1 < \alpha < \infty$.

Given an analytic function φ in the unit disk $\mathbb{D}$ such that $\varphi(\mathbb{D}) \subset \mathbb{D}$, the *composition operator* C_φ with *symbol* φ defined by $C_\varphi f(z) = f(\varphi(z))$ is always bounded on any H^p or A_α^p space, in view of Littlewood's Subordination Theorem. The monographs [**S1**] and [**CM**] are standard sources for the theory of composition operators on such spaces.

2000 *Mathematics Subject Classification.* Primary 47B33; Secondary 30H05.

Key words and phrases. Hardy spaces, Bergman spaces, Besov spaces, Composition operators, Isometries.

Both authors are supported by MCyT grant BFM2003-07294-C02-01 and also partially by MTM2004-21420-E (MEC Program "Acciones Complementarias"), Spain.

Being a Hilbert space, the Hardy space H^2 has plenty of isometries. However, the only isometries of H^2 among the composition operators are the operators induced by inner functions that vanish at the origin. Nordgren ([N], p. 444) showed that if φ is inner and $\varphi(0) = 0$ then C_φ is an isometry of H^2 (alternatively, see p. 321 of [CM]). The converse follows, for example, from a result of Shapiro ([S2], p. 66). According to Cload [C], this characterization of isometries of H^2 among the composition operators had already been obtained in the unpublished thesis of Howard Schwarz [S]. Bayart [B] recently showed that every composition operator on H^2 which is similar to an isometry is induced by an inner function with a fixed point in the disk.

The surjective isometries of the more general H^p spaces have been described by Forelli [F] as weighted composition operators. A characterization of all isometries of H^p does not seem to be known. In this note we prove that the only isometries (surjective or not) of H^p, $1 \le p < \infty$, among the composition operators are again induced by inner functions that vanish at the origin (see Theorem 1.3 below). This fact may be known to some experts so our emphasis is on the method of proof, which also works for Bergman spaces.

Kolaski [K] (see also [DS], § 2.8) gave a characterization of all surjective isometries of a weighted Bergman space A_α^p similar to that of Forelli's. Again, no characterization of all isometries of these spaces seems to be known. The (Hilbert) Bergman space A_α^2, of course, possesses plenty of isometries. In a recent preprint Carswell and Hammond [CH] have shown, among other results, that the only composition operators that are isometries of the weighted (Hilbert) Bergman space A_α^2 are the rotations. We prove an analogous statement (Theorem 1.3) for an arbitrary space A_w^p with a radial weight, $p \ge 1$, where Hilbert space methods no longer work.

The surjective isometries of the Bloch space have been characterized in a well known work by Cima and Wogen [CW] while the surjective isometries of the general analytic Besov spaces B^p and some related Dirichlet-type spaces have been described more recently by Hornor and Jamison [HJ]. Recall that an analytic function in the disk is said to belong to the space B^p if its derivative is in the weighted Bergman space A_{p-2}^p. These spaces form an important scale of Möbius-invariant spaces that includes the Dirichlet space ($p = 2$) and the Bloch space (as a limit case as $p \to \infty$). They have been studied by many authors (see [AFP], [Z], [DGV] for some details).

The isometries (not necessarily surjective) among the composition operators acting on the Dirichlet space B^2 have been characterized in [MV]. Here we describe all isometries of Besov spaces B^p, $2 < p < \infty$, among the composition operators with univalent symbols (Theorem 1.4). The proof follows a similar pattern to that of the proofs for Hardy and Bergman spaces, with some variations typical of analytic Besov spaces.

1. Main results and their proofs

1.1. Hardy and Bergman spaces. We characterize all isometries among the composition operators on the general H^p and A^p spaces by giving an essentially unified proof. The crucial point in both statements is that the symbol φ of any isometry C_φ must fix the origin. Once this has been established, we proceed using a simple idea that is probably known to some experts, at least in the Hilbert space context ($p = 2$). In order to prove the claim about fixing the origin, we first establish an auxiliary result similar to several others that are often used in the theory of best approximation.

LEMMA 1.1. *Let μ be a positive measure on the measure space Ω, $\mathcal{M}$ a subspace of $L^p(\Omega, d\mu)$, $1 \leq p < \infty$, and let T be a linear isometry of $\mathcal{M}$ (not necessarily onto). Then*

$$\int_\Omega Tf|Tg|^{p-2}\overline{Tg}\,d\mu = \int_\Omega f|g|^{p-2}\overline{g}\,d\mu$$

for all f, g in the subspace $\mathcal{M}$.

PROOF. We apply the standard method of variation of (differentiation with respect to) the parameter. Given two arbitrary functions f, g in $\mathcal{M}$, define the function

$$N_{f,g}(t) = \int_\Omega |tf + g|^p\,d\mu, \qquad t \in \mathbb{R}.$$

Then, as described in Theorem 2.6 of [**LL**] (with f and g permuted for our convenience),

$$N'_{f,g}(0) = \frac{p}{2}\int_\Omega |g|^{p-2}(g\overline{f} + f\overline{g})\,d\mu.$$

Since T is a linear isometry of $\mathcal{M}$, we have $N_{Tf,Tg}(t) = N_{f,g}(t)$. After evaluating the derivative of each side at $t = 0$ we get

$$\int_\Omega |g|^{p-2}(g\overline{f} + f\overline{g})\,d\mu = \int_\Omega |Tg|^{p-2}(Tg\overline{Tf} + Tf\overline{Tg})\,d\mu.$$

Since this holds for arbitrary f and g we may also replace g by ig. After a cancellation, this yields

$$\int_\Omega |g|^{p-2}(g\overline{f} - f\overline{g})\,d\mu = \int_\Omega |Tg|^{p-2}(Tg\overline{Tf} - Tf\overline{Tg})\,d\mu.$$

Summing up the last two identities, we get

$$\int_\Omega |g|^{p-2}g\overline{f}\,d\mu = \int_\Omega |Tg|^{p-2}Tg\overline{Tf}\,d\mu,$$

which implies the desired formula. $\qquad\square$

From now on we assume that the weight w is not only radial but behaves "reasonably well" in the sense that A_w^p is a complete space.

PROPOSITION 1.2. *If a composition operator C_φ is an isometry (not necessarily onto) of either H^p or A_w^p, $1 \leq p < \infty$, then $\varphi(0) = 0$.*

PROOF. Consider $\mathcal{M} = H^p$, a subspace of $L^p(\mathbb{T}, dm)$, and $\mathcal{M} = A_w^p$, a subspace of $L^p(\mathbb{D}, w\,dA)$, respectively. Then set $g \equiv 1$ and $Tf = C_\varphi f$ in Lemma 1.1 and use the standard reproducing property for the origin to get

$$f(\varphi(0)) = \int_\mathbb{T} C_\varphi f\,dm = \int_\mathbb{T} f\,dm = f(0)$$

in the case of H^p. For the weighted Bergman space A_w^p, use the Mean Value Property to get

$$\int_\mathbb{D} fw\,dA = 2\int_0^1\left(\int_0^{2\pi} f(re^{i\theta})\,dm(\theta)\right)w(r)r\,dr = 2\int_0^1 f(0)w(r)r\,dr = c_w f(0)$$

(for some positive constant c_w) and, similarly,

$$\int_{\mathbb{D}} (f \circ \varphi) w dA = 2 \int_0^1 \left(\int_0^{2\pi} (f \circ \varphi)(re^{i\theta}) dm(\theta) \right) w(r) r dr = c_w f(\varphi(0)),$$

hence also $f(\varphi(0)) = f(0)$. Finally, choose the identity map: $f(z) \equiv z$ to deduce the statement in both cases.   □

Proposition 1.2 could have been established by other methods but we decided to give preference to the application of Lemma 1.1 from approximation theory.

THEOREM 1.3. *Let $1 \le p < \infty$. Then:*

(a) *A composition operator C_φ is an isometry of H^p if and only if φ is inner and $\varphi(0) = 0$.*

(b) *A composition operator C_φ is an isometry of A_w^p if and only if φ is a rotation.*

PROOF. (a) Since C_φ is an isometry, we have $\|z\|_{H^p} = \|\varphi\|_{H^p}$, hence

$$0 = \|z\|_{H^p}^p - \|\varphi\|_{H^p}^p = \int_{\mathbb{T}} (1 - |\varphi|^p) \, dm.$$

Since $|\varphi| \le 1$ almost everywhere on $\mathbb{T}$, it follows that $1 - |\varphi|^p = 0$ almost everywhere on $\mathbb{T}$ and, thus, φ is inner. We already know from the Corollary that $\varphi(0) = 0$.

(b) In view of the Corollary ($\varphi(0) = 0$) and the Schwarz Lemma, we get that $|\varphi(z)| \le |z|$ for all z in $\mathbb{D}$. Since w is assumed to be a nontrivial weight, it must be strictly positive on a set of positive measure in $\mathbb{D}$, hence the equality

$$0 = \|z\|_{A_w^p}^p - \|\varphi\|_{A_w^p}^p = \int_{\mathbb{D}} (|z|^p - |\varphi|^p) \, w(z) \, dA$$

is only possible if $|\varphi(z)| = |z|$ throughout $\mathbb{D}$, that is, when φ is a rotation.   □

1.2. Analytic Besov spaces. The *analytic Besov space B^p*, $1 < p < \infty$, is defined as the set of all analytic functions in the disk such that

$$\|f\|_{B^p}^p = |f(0)|^p + s_p(f) = |f(0)|^p + (p-1) \int_{\mathbb{D}} |f'(z)|^p (1 - |z|^2)^{p-2} dA(z) < \infty.$$

These spaces are Banach spaces with the *Möbius-invariant* seminorm s_p, in the sense that $s_p(f \circ \phi) = s_p(f)$ for every disk automorphism ϕ. It is well known that $B^p \subset B^q$ when $p < q$. For the theory of B^p and other conformally invariant spaces of analytic functions in the disk, we refer the reader to [**AFP**], [**DGV**], and [**Z**], for example.

A general composition operator on B^p is not necessarily bounded, roughly speaking because "too many points of the disk can be covered too many times" by φ (for an exact condition for the boundedness in terms of the counting function, see [**AFP**]). However, the boundedness of C_φ is guaranteed for every univalent symbol φ when $p \ge 2$. Indeed, after applying the Schwarz-Pick Lemma and the change of variable $w = \varphi(z)$, we get

$$
\begin{aligned}
s_p^p(f \circ \varphi) &= (p-1) \int_{\mathbb{D}} |f'(\varphi(z))|^p |\varphi'(z)|^p (1 - |z|^2)^{p-2} dA(z) \\
&\leq (p-1) \int_{\mathbb{D}} |f'(\varphi(z))|^p (1 - |\varphi(z)|^2)^{p-2} |\varphi'(z)|^2 dA(z) \\
&= (p-1) \int_{\varphi(\mathbb{D})} |f'(w)|^p (1 - |w|^2)^{p-2} dA(w) \\
&\leq (p-1) \int_{\mathbb{D}} |f'(w)|^p (1 - |w|^2)^{p-2} dA(w),
\end{aligned}
$$

showing that $s_p(f \circ \varphi) \leq s_p(f)$ for all f in B^p, whenever φ is univalent and $p \geq 2$.

THEOREM 1.4. *Let φ be a univalent self-map of the disk and $2 < p < \infty$. Then the induced composition operator C_φ is an isometry of B^p if and only if φ is a rotation.*

PROOF. The sufficiency of the condition is trivial.

For the necessity, suppose C_φ is an isometry of B^p, $2 < p < \infty$. We first show that again we must have $\varphi(0) = 0$. Since $s_p(f \circ \varphi) \leq s_p(f)$ in this case and

$$
\|f \circ \varphi\|_{B^p}^p = |f(\varphi(0))|^p + s_p(f \circ \varphi) = |f(0)|^p + s_p(f) = \|f\|_{B^p}^p,
$$

it follows that $|f(\varphi(0))| \geq |f(0)|$ for all f in B^p. Writing $\varphi(0) = a$ and choosing f to be the standard disk automorphism

$$
\varphi_a(z) = \frac{a - z}{1 - \bar{z}a}
$$

that interchanges the origin and the point a, we get

$$
0 = |\varphi_a(a)| = |\varphi_a(\varphi(0))| \geq |\varphi_a(0)| = |a|,
$$

hence $\varphi(0) = 0$. This proves the claim.

Thus, if C_φ is an isometry of B^p, it must satisfy $s_p(f \circ \varphi) = s_p(f)$ for all f in B^p. Choose $f(z) \equiv z$. Using the definition of s_p, the Schwarz-Pick Lemma (note that $p - 2 > 0$ by assumption), and the change of variable $\varphi(z) = w$, we get

$$
\begin{aligned}
0 &= s_p(z)^p - s_p(\varphi)^p \\
&= (p-1) \int_{\mathbb{D}} (1 - |z|^2)^{p-2} dA(z) - (p-1) \int_{\mathbb{D}} |\varphi'(z)|^p (1 - |z|^2)^{p-2} dA(z) \\
&\geq 1 - (p-1) \int_{\mathbb{D}} (1 - |\varphi(z)|^2)^{p-2} |\varphi'(z)|^2 dA(z) \\
&= 1 - (p-1) \int_{\varphi(\mathbb{D})} (1 - |w|^2)^{p-2} dA(w) \\
&\geq 1 - (p-1) \int_{\mathbb{D}} (1 - |w|^2)^{p-2} dA(w) \\
&= 0,
\end{aligned}
$$

hence equality must hold throughout. In particular, it must hold in the Schwarz-Pick Lemma (in the third line of the chain above) and so φ must be a disk automorphism. Since it also fixes the origin, it follows that φ is actually a rotation. $\square$

Note that the requirement that φ be univalent was crucial near the end of the proof. Also, the requirement that $p > 2$ was fundamental because the key inequality above becomes reversed when $p < 2$. The above proof still works when $p = 2$ (the Dirichlet space) as we no longer need the Schwarz lemma but in this case it only follows that the normalized area of $\varphi(\mathbb{D})$ equals the area of $\mathbb{D}$ itself, that is, φ must be a (univalent) map of $\mathbb{D}$ onto a subset of full measure. All such maps that fix the origin generate isometries, a fact already proved without the univalence assumption in [**MV**]. It would be interesting to know which composition operators acting on B^p are isometries when $1 < p < 2$.

References

[AFP]   J. Arazy, S. D. Fisher, and J. Peetre, *Möbius invariant function spaces*, J. Reine Angew. Math. **363** (1985), 110–145.

[B]   F. Bayart, *Similarity to an isometry of a composition operator*, Proc. Amer. Math. Soc. **131** (2003), no. 6, 1789–1791.

[CH]   B. J. Carswell and C. Hammond, *Composition operators with maximal norm on weighted Bergman spaces*, preprint, January 2005.

[CW]   J. A. Cima and W. R. Wogen, *On isometries of the Bloch space*, Illinois J. Math. **24** (1980), no. 2, 313–316.

[C]   B. A. Cload, *Composition operators: hyperinvariant subspaces, quasi-normals and isometries*, Proc. Amer. Math. Soc. **127** (1999), no. 6, 1697–1703.

[CM]   C. Cowen and B. MacCluer, *Composition Operators on Spaces of Analytic Functions*, Studies in Advanced Mathematics, CRC Press, Boca Raton, 1995.

[DGV]   J. J. Donaire, D. Girela, and D. Vukotić, *On univalent functions in some Möbius invariant spaces*, J. Reine Angew. Math. **553** (2002), 43–72.

[D]   P. L. Duren, *Theory of H^p Spaces*, Academic Press, New York, 1970. Reprinted by Dover, Mineola, NY 2000.

[DS]   P. L. Duren and A. P. Schuster, *Bergman Spaces*, Mathematical Surveys and Monographs, vol. 100, American Mathematical Society, Providence, RI, 2004.

[F]   F. Forelli, *The isometries of H^p*, Canad. J. Math. **16** (1964), 721–728.

[HJ]   W. Hornor and J. E. Jamison, *Isometries of some Banach spaces of analytic functions*, Integral Equations Operator Theory 41 (2001), no. 4, 410–425.

[K]   C. J. Kolaski, *Isometries of weighted Bergman spaces*, Canad. J. Math. **34** (1982), no. 4, 910–915.

[LL]   E. H. Lieb and M. Loss, *Analysis*, Second Edition, Graduate Studies in Mathematics **14**, American Mathematical Society, Providence, RI, 2001.

[MV]   M. J. Martín and D. Vukotić, Isometries of the Dirichlet space among the composition operators, *Proc. Amer. Math. Soc.*, to appear.

[N]   E. A. Nordgren, Composition operators, *Canad. J. Math.* **20** (1968), 442–449.

[S]   H. Schwarz, *Composition Operators on H^p*, Ph. D. Thesis, University of Toledo, Ohio, 1969.

[S1]   J. H. Shapiro, *Composition Operators and Classical Function Theory*, Springer-Verlag, New York, 1993.

[S2]   J. H. Shapiro, *What do composition operators know about inner functions?*, Monatsh. Math. **130** (2000), 57–70.

[Z]   K. Zhu, *Analytic Besov spaces*, J. Math. Anal. Appl. **157** (1991), 318–336.

DEPARTAMENTO DE MATEMÁTICAS, UNIVERSIDAD AUTÓNOMA DE MADRID, 28049 MADRID, SPAIN

E-mail address: mjose.martin@uam.es

DEPARTAMENTO DE MATEMÁTICAS, UNIVERSIDAD AUTÓNOMA DE MADRID, 28049 MADRID, SPAIN

E-mail address: dragan.vukotic@uam.es

Contemporary Mathematics
Volume **393**, 2006

New results on a classical operator

Alfonso Montes-Rodríguez and Stanislav A. Shkarin

ABSTRACT. It is remarkable how the classical Volterra integral operator, which was one of the first operators which attracted mathematicians' attention, is still worth of being studied. In this essentially survey work, by collecting some of the very recent results related to the Volterra operator, we show that there are new (and not so new) concepts that are becoming known only at the present days. Discovering whether the Volterra operator satisfies or not a given operator property leads to new methods and ideas that are useful in the setting of Concrete Operator Theory as well as the one of General Operator Theory. In particular, a wide variety of techniques like summability kernels, theory of entire functions, Gaussian cylindrical measures, approximation theory, Laguerre and Legendre polynomials are needed to analyze different properties of the Volterra operator. We also include a characterization of the commutator of the Volterra operator acting on $L^p[0,1]$, $1 \le p \le \infty$. Only the cases $p = 2$ and $p = 1$ were previously known.

1. INTRODUCTION

For $1 \le p < \infty$, we consider the space $L^p[0,1]$ of complex measurable functions f on $[0,1]$ for which the norm

$$\|f\|_p^p = \int_0^1 |f(x)|^p dx$$

is finite. For $p = \infty$, the space $L^\infty[0,1]$ is the one of measurable functions on $[0,1]$ for which the essential supremum norm denoted by $\|f\|_\infty$ is finite. The Volterra operator is built upon one of the oldest concepts in mathematics: the integral. For $1 \le p \le \infty$, the classical Volterra operator is defined as

$$(Vf)(x) = \int_0^x f(t)\, dt, \qquad \text{for each } f \in L^p[0,1].$$

Although other operators like the Fourier transform or the derivative operator were previously studied, the work on integral equations of Volterra and Fredholm

---

2000 *Mathematics Subject Classification.* Primary 47G10, 47B38; Secondary 47A10.

Key words and phrases. Volterra operator, cyclic operators, supercyclic operators, antisupercyclic operators, norms, power bounded operators, compact operators, commutator.

Partially supported by Ministerio de Ciencia y Tecnología I+D+I ref. BFM2003-00034 and Junta de Andalucía, ref. FQM-260.

around the end of the nineteenth century has greatly influenced the development of Operator Theory during the twentieth century. It is a striking fact that at the beginning of the twenty-first century, new and interesting results related to the classical Volterra operator go on appearing.

The Volterra operator acting on $L^p[0,1]$, $1 \leq p \leq \infty$, is clearly bounded and linear. Indeed, V acts compactly on $L^p[0,1]$, see Conway's book [7], for instance. In this note we discuss some recent work that has the Volterra operator as the main protagonist. Our goal is not to survey the large amount of work on Volterra and Volterra-type operators that is being carrying out in the last few years. That would be too ambitious. Thus we will mainly remain concerned about the work in which the authors have been recently involved.

Section 2 is devoted to the studying of the cyclic properties of the Volterra operator.

In section 3, we review other related problems like norms of iterates of V as well as of $I - V$ and power boundedness. A computation of the value of $\|V\|_p$ is also included. The latter only differs in minor details from the one provided by Bennewitz and Saitō, see [4]. It is curious that while the fact that $\|V\|_2 = 2/\pi$ has been known for several decades, see [15], the exact value $\|V\|_p$, $1 < p < \infty$ and $p \neq 2$, has become known only so recently.

In Section 4 we deal with the commutator of V. For $1 \leq p \leq \infty$, we denote by $[V]_p$ the commutator of V acting on $L^p[0,1]$, that is, the set of linear operators T that acts boundedly on $L^p[0,1]$ and $TV = VT$. In [34], Sarason characterized $[V]_2$. Let $\mathcal{H}^\infty(\mathbb{R})$ be the space of boundary functions on the real line of bounded analytic functions on the upper half-plane. Let J denote the canonical embedding from $L^2[0,1]$ into $L^2(\mathbb{R})$ and P the orthogonal projection from $L^2(\mathbb{R})$ onto $L^2[0,1]$, regarded as a subspace of the latter. Let M_ϕ denote the multiplication operator by ϕ and let $\mathcal{F}$ denote the Fourier transform.

THEOREM (Sarason). *A linear operator bounded on $L^2[0,1]$ commutes with V if and only if there exists $\phi \in \mathcal{H}^\infty(\mathbb{R})$ such that*

$$T = P\mathcal{F}M_\phi\mathcal{F}^{-1}J$$

Furthermore, ϕ can be chosen such that $\|\phi\|_\infty = \|T\|$.

From this characterization it follows that the commutator $[V]_2$ is the closure in the strong operator topology of the algebra generated by V. An elementary proof of the last statement was obtained by J. Erdos [10]. Mincheva [23] characterized $[V]_1$. Here, we will characterize the commutator $[V]_p$ for each p, $1 \leq p \leq \infty$, in terms of the belonging of certain convolutions to the Sobolev space $W^{1,p}[0,1]$. Although our characterization is somewhat difficult to handle, some known results can be derived and for most important cases $p = 1$, $p = 2$ and $p = \infty$, simpler characterizations are provided. The proofs are based on standard techniques of real and functional analysis.

2. CYCLIC PROPERTIES OF THE VOLTERRA OPERATOR

Cyclicity

An operator T acting on a Banach space $\mathcal{B}$ is said to be cyclic if there is a vector x in $\mathcal{B}$, called a cyclic vector for T, such that the linear span of the orbit

$\{T^n x\}_{n \geq 0}$ is dense in $\mathcal{B}$. Clearly, $\mathcal{B}$ must be separable. A typical example of a cyclic operator is the Volterra operator V acting on $L^p[0,1]$, $1 \leq p < \infty$. Indeed, since the orbit of the constant function 1 under V is $\{x^n/n!\}_{n \geq 0}$, it follows, from Weierstrass's Theorem, that 1 is a cyclic vector for V. If one applies Müntz-Satz's Theorem, see [30, p. 313], then one easily sees that 1 is also a cyclic vector for any positive power of V. Much more is known, the Volterra operator is unicellular, that is, the lattice of invariant subspaces for V is totally ordered. Indeed, Donoghue's Theorem [8] asserts that the only invariant subspaces of the Volterra operator are $L^p[\alpha, 1]$ with $0 \leq \alpha \leq 1$. The result is also true for $L^p[0,1]$, $1 \leq p < \infty$, see [21]. As a consequence of Donoghue's Theorem, the Volterra operator is one of the few operators for which the cyclic vectors are characterized (the operators for which their lattices of invariant subspaces are known are very scarse): a *function f is cyclic for V if and only if f does not vanish on any neighborhood of zero*. An excellent approach to Donoghue's Theorem, that has influenced a good deal of later work in Operator Theory, is due to Sarason, see [33] and [35]. In the latter references it is shown that under suitable similarities, the invariant subspaces of the Volterra operator are related to those of multiplication by z in the Hardy space.

A final interesting remark about the cyclicity of the Volterra operator is that the direct sum of V with itself is not cyclic. Indeed, $V \oplus \lambda V$ acting on $L^p[0,1] \oplus L^q[0,1]$, $1 \leq p, q < \infty$ is cyclic if and only if λ is in $\mathbb{C} \setminus [0, \infty)$, see [28, Cor. 6.3].

Supercyclicity

A stronger concept than cyclicity is supercyclicity, which means that the scalar multiples of the orbit of some vector under an operator are dense in the underlying space, see [26] for a survey on this subject. Salas [32] asked whether the Volterra operator is supercyclic or not. Interest in this question arises from the fact that the left inverse of the Volterra operator, that is, the derivative operator is hypercyclic in many spaces where it is boundedly defined and the well known fact that an invertible operator is hypercyclic/supercyclic if and only if the inverse is, see [16]. Recall that hypercyclic means that the orbit itself of some vector is dense. Indeed, although not boundedly defined, the derivative operator is hypercyclic on $L^2[0,1]$, see [13, Theorem 2.1]. Furthermore, the authors have shown that, extended in a natural way, the Volterra operator becomes hypercyclic, see [28, Prop. 2.1]. In [10], Gallardo and the first named author proved

THEOREM 2.1. *The Volterra operator acting on $L^p[0,1]$, $1 \leq p < \infty$, is not supercyclic.*

The seminal idea to prove the above theorem is the Angle Criterion for non-supercyclicity that appeared for the first time in [25]. As usual, if $\mathcal{B}$ is a Banach space, then $\langle f, g \rangle$ will denote both the linear functional $g \in \mathcal{B}^*$ acting on $f \in \mathcal{B}$ and, when x and y are vectors in a Hilbert space, their Hilbert inner product. The Angle Criterion reads as follows: *if T is a bounded operator on a Banach space $\mathcal{B}$ such that for each $f \in \mathcal{B}$ there is $g \in \mathcal{B}^*$ satisfying*

$$\sup_{n \geq 0} \frac{|\langle T^n f, g \rangle|}{\|T^n f\| \|g\|} < 1, \tag{2.1}$$

then T is not supercyclic. It is not difficult to see that it is enough to prove Theorem 2.1 in the friendly confines of $L^2[0,1]$. Then to apply the Angle Criterion it is needed to obtain lower estimates of the norm $\|V^n f\|_2$ for $f \in L^2[0,1]$. This is accomplished by checking how the Volterra operator acts on the Legendre polynomials $\{P_n\}$, defined by $P_n = h^{(n)}$, where $h(x) = x^n(1-x)^n$ for each non-negative integer n, see [29, Chap. 10]. It is not difficult to show that if V is supercyclic, then there must be a supercyclic f in $L^2[0,1]$ which is continuous and not vanishing at $1/2$. But in such a case, it can be proved that

$$\frac{(2n+1)!}{(n!)^2}\langle V^{\star n} P_n, f\rangle \to f(1/2), \qquad \text{as } n \to \infty. \tag{2.2}$$

Essentially this means that the sequence above is a summability kernel at $1/2$. Now, using (2.2), one easily checks that $\|V^n f\|_2 \geq \|P_n\|^{-1}|\langle V^{\star n} P_n, f\rangle|$. Since $\|P_n\|_2 = (2n+1)^{-1/2}$, it is enough to take g equal to the characteristic function of $[1/2,1]$ to apply the Angle Criterion.

Antisupercyclicity

The second named author [36] has proved that the Volterra operator fails to satisfy the Angle Criterion in a very drastic way. The operators that do not satisfy in the strongest way the Angle Criterion are those for which $T^n f/\|T^n\|$ tends weakly to zero for any f in the underlying space. If the underlying space is a Hilbert space, this means that the limit in (2.1) exists and is equal to 0 for any $f \neq 0$ and any $g \neq 0$, or what is the same, the angles that form the elements of the orbit of $f \neq 0$ under T with any g tend to $\pi/2$. In [36], these operators are called antisupercyclic. It is a "good" operator property in the sense that it is preserved under similarities and, therefore, it would be interesting to characterize the norm closure of the antisupercyclic operators. A trivial example of antisupercyclic operator is the shift acting on the space $\ell^2(\mathbb{Z})$ of bilateral sequences of square summable modulus. In [36], the antisupercyclic bilateral weighted shifts on $\ell^p(\mathbb{Z})$, $1 \leq p < \infty$ are characterized: *A weighted bilateral shift acting on $\ell^p(\mathbb{Z})$, $1 < p < \infty$, if and only if it is not supercyclic*, while there is no antisupercyclic operator on $\ell^1(\mathbb{Z})$. The characterization of supercyclic bilateral weighted shifts on $\ell^p(\mathbb{Z})$, $1 \leq p < \infty$, can be found in [32]. In [36], the antisupercyclic normal operators are also characterized. In addition, in [36, Theorem 1.1 and Proposition 1.6], it is proved that

THEOREM 2.2. *The Volterra operator acting on $L^p[0,1]$, $1 < p < \infty$, is antisupercyclic and acting on $L^1[0,1]$ is not.*

Indeed, in [36] it is proved that for any complex number z with $\Im z > 0$, the Riemann-Liouville operator V^z, which is boundedly defined as

$$(V^z f)(x) = \frac{1}{\Gamma(z)} \int_0^x (x-t)^{z-1} f(t)\,dt, \quad f \in L^p[0,1], 1 \leq p \leq \infty,$$

is antisupercyclic for $1 < p < \infty$. The proof is based on a very sharp analysis of the behavior of $\|V^z f\|_p$ for $f \in L^p[0,1]$. The proof uses deep techniques of the theory of entire functions such as Carlson's Theorem or Cartwright's Theorem. In fact, a very accurate formula for the behavior of $V^z f$ for $\Im z > 0$ and $f \in L^p[0,1]$ is proved. For each Lebesgue measurable function $f : [0,1] \to \mathbb{C}$, we set

$$\text{supp}\,(f) = \{x \in [0,1] : \mu(y \in (x-\varepsilon, x+\varepsilon) \cap [0,1] : f(y) \neq 0) > 0 \text{ for each } \varepsilon > 0\}.$$

Theorem 1.3 in [36] reads as follows.

THEOREM 2.3. *For each f in $L^p[0,1]$, $1 \le p \le \infty$ and $\theta \in (-\pi/2, \pi/2)$, we have*

$$\lim_{r \to \infty} (|\Gamma(z)| \|V^z\|)^{1/r} = (1 - \inf \operatorname{supp}(f))^{\cos(\theta)}, \qquad where \ z = re^{i\theta}.$$

In particular, the above theorem asserts that the limit does exist!

Weak supercyclicity

If norm density is replaced by density with respect to the weak topology, one obtains the concept of weak supercyclicity. Since norm closures and weak closures of convex sets coincide, weakly supercyclic operators are still cyclic. Furthermore, Hilden and Wallen [17] proved that no isometry on Hilbert space can be supercyclic. In a strong contrast with this, Bayart and Matheron [3] have recently proved that there are weakly supercyclic isometries, in fact unitary operators, in Hilbert space. On the other hand, weakly supercyclic operators share many properties with the supercyclic ones. For instance, if T is supercyclic, then so is T^n for every positive integer n, see [2]. As a consequence, if one proves that an operator is weakly supercyclic, then one is proving that all the powers are cyclic, as it happens with the powers of the Volterra operator. Therefore, there still is the possibility that the Volterra operator is weakly supercyclic. This is not the case, indeed, the following theorem has recently been shown by the authors [28, Theorem 1.1].

THEOREM 2.4. *The Volterra operator acting on $L^p[0,1]$, $1 \le p < \infty$, is not weakly supercyclic.*

The proof of Theorem 2.4 is much more subtle that the one of Theorem 2.1. To begin with, the Angle Criterion cannot be used in the setting of weakly supercyclic operators. Instead, we use the fact that if $h, g \in L^2[0,1]$ with $h \ne 0$, then

$$\{F \in L^2[0,1] \text{ such that } |\langle F, h \rangle| > |\langle F, g \rangle|\} \tag{2.3}$$

is weakly open. The advantage of a set like the one above is that, as for the angle criterion, the scalar multiples are also absorbed. In the proof, F is taken of the form $\lambda V^n f$. This is not, of course, the only thing needed to prove Theorem 2.4. Apart from the summability kernel idea, we also need an analysis of how the orbits of $V^n f$ goes to zero. This analysis is accomplished by introducing certain Gaussian cylindrical measures and, then, an existence probabilistic argument is used to show that the Volterra operator cannot compress too much the space in all directions at the same time.

The same weakly open sets as in (2.3) are used in [28] to show that certain composition operators are not weakly supercyclic. On the other hand, the methods of the proof of Theorem 2.4 lead to a complete general theorem that shows that certain operators cannot be weakly supercyclic and hopefully will be useful in future work. To state it we need the following definition.

DEFINITION 2.5. Let T be a bounded linear operator on a Banach space $\mathcal{B}$ and for each non-negative integer n let $\mathcal{B}_n$ denote the Banach $T^n(\mathcal{B})$ endowed with the norm $\|y\|_{\mathcal{B}_n} = \|y\|_{\mathcal{B}} + \inf\{\|x\|_{\mathcal{B}} : x \in \mathcal{B}, T^n x = y\}$. A set $X \subset \mathcal{B}$ is said T-*big* if there exists a non-negative integer n such that X contains a non-empty, weakly open subset of $\mathcal{B}_n$.

Observe that the topology of $\mathcal{B}_n$ is stronger than the one it inherits from $\mathcal{B}$.

REMARK 2.6. It is clear that any weakly open set is T-big. Of course, the converse is not true, as for the Volterra operator.

The following Theorem can be found in [28].

THEOREM 2.7. *Let T be a bounded linear operator on a Banach space $\mathcal{B}$. Assume that there exists a T-big set $X \subset \mathcal{B}$ such that for any f in X either zero belongs to the orbit $\{T^n f\}_{n \geq 0}$ or there exist a continuous linear operator R from $\mathcal{B}$ into a separable Hilbert space $\mathcal{H}$, a Hilbert-Schmidt operator S on $\mathcal{H}$ and a nonzero vector $x \in \mathcal{B}^\star$ such that*

$$\sum_{n=0}^{\infty} \frac{\|T^{\star n} x\|_{\mathcal{B}^\star}}{\|SRT^n f\|_{\mathcal{H}}}$$

is finite. Then T is not weakly supercyclic.

The corollary below has a simpler appearance.

COROLLARY 2.8. *Let T be a Hilbert–Schmidt operator on a separable Hilbert space $\mathcal{H}$. Assume that there exists a T-big set $X \subset \mathcal{H}$ such that for each f in X either zero belongs to the orbit $\{T^n f\}_{n \geq 0}$ or there exists a nonzero vector y in $\mathcal{H}$ such that*

$$\sum_{n=0}^{\infty} \frac{\|T^{\star n} y\|}{\|T^n f\|}$$

is finite. Then T is not weakly supercyclic.

In a work in preparation [5], the theorem above has been used by Bermudo and the authors to show that no rational function in V is weakly supercyclic.

We end this section with an open question.

Question. Does antisupercyclicity imply not weak supercyclicity?

3. THE ASYMPTOTIC BEHAVIOR OF THE NORM OF THE ITERATES

The iterates of V

As mentioned in the introduction $\|V\|_2 = 2/\pi$. Indeed, $\|V\|_2$ is exactly the square root of the reciprocal of the maximum of the eigenvalues of the operator $V^\star V$, see [15]. The question of the value of $\|V^n\|_2$ for positive integer n seems to have been posed by Halmos. The same methods as for $\|V\|_2$, although more involved, can be used to prove that $\|V^2\|_2$ equals to the square root of the inverse of the smallest positive zero of $\cos(x) \cosh(x) + 1$, a fact that seems to be due to A. Brown, see [39]. Lao and Whitley [20] computed $\|V^n\|_2$ for some values of n and conjectured that $\lim_{n \to \infty} \|n! V^n\|_2 = 1/2$. This conjecture was proved by Kershaw [19]. The proof uses the fact that the powers of the Volterra operator preserves the cone of positive functions. Under this situation the Kreĭn-Rutman Theorem guarantees the existence of a positive eigenvalue for $V^{\star n} V^n$, which is estimated. Little and Read [22] gave a simpler proof of Kershaw's result using test functions. The following theorem is due to Thorpe, see [39].

THEOREM 3.1. *Let V denote the Volterra operator. Then*

$$\|V^n\|_2 = \frac{1}{\sqrt{\lambda_0(n)}},$$

where $\lambda_o(n)$ is the first eigenvalue of the differential equation $(-1)^n f^{(2n)} = \lambda f$ subject to the restrictions $f^{(k)}(0) = 0$, $0 \leq k \leq n$, and $f^{(k)}(1) = 0$, $n \leq k \leq 2n - 1$.

Most recently, Eveson [12] has obtained sharp estimates on the $L^p[0,1]$ norm for certain Volterra kernel operators. In particular, for the Volterra operator he proves

THEOREM 3.2. *Let V denote the Volterra operator. Then*

$$\|V^n\|_p \sim \frac{C_p}{n!},$$

where $C(p) = 1$ if $p = 1$ or $p = \infty$ and $C(p) = p^{-1/p}(1 - 1/p)^{(1-1/p)}$ if $1 < p < \infty$.

The estimate in the above Theorem is achieved by comparing $\|V^n\|_p$ with an asymptotically equivalent sequence of rank one operators.

The operator $I - V$

Completely different methods apply to the operator $I - V$. It was known that $I - V$ is power bounded on $L^2[0,1]$. This follows from the fact that $I - V$ is similar to $(I + V)^{-1}$, a fact due to Pedersen [1], and $\|(I + V)^{-1}\|_2 = 1$ see [15]. In 1945, Hille [18] proved that $\|(I - V)^n\|_1 = O(n^{1/4})$. In [27, Thm. 2.2], Sánchez, Zemánek and the first named author proved

THEOREM 3.3. *Let V denote the Volterra operator, then there are positive constants C_1 and C_2 depending only on p such that*

$$C_1 n^{|1/4 - 1/(2p)|} \leq \|(I - V)^n\|_p \leq C_2 n^{|1/4 - 1/(2p)|}.$$

As a consequence of the above theorem, $I - V$ is power bounded if and only if $p = 2$ and, in this sense, the Volterra operator characterizes $L^2[0,1]$. The method of proof of Theorem 3.3 is completely different from those involving the norms of V^n. It is derived by using sharp asymptotic estimates on the Laguerre polynomials due to Szegö [38, Chapter 8] and the fact that if a real sequence $a_n \to \infty$ as $n \to \infty$, κ is real and $f \in L^1[0,1]$, then

$$\int_0^1 |f(x)||\cos(a_n x + \kappa)|^p \, dx \to \frac{1}{\pi} \int_0^\pi \sin^p(x) \, dx \int_0^1 |f(x)| \, dx \quad as \quad n \to \infty.$$

The same methods applies to show that

$$\|(I - V)^{n+1} - (I - V)^n\|_p \asymp n^{-1/2 + |1/4 - 1/(2p)|},$$

see [27, Thm. 2.5]. It is interesting to observe that the quantity $|1/4 - 1/2p|$ is exactly the Banach-Mazur distance from $L^p[0,1]$ to $L^2[0,1]$.

The fact that $I - V$ is not power bounded for $p \neq 2$ has an important application. It allows us to show that power boundedness is not equivalent to $\mathcal{C}^1$ boundedness, see [27]. An operator is $\mathcal{C}^1$ bounded if there is $C > 0$ such that

$$\left\| \sum_{k=0}^n \frac{\lambda^k T^k}{n+1} \right\| \leq C \quad \text{for each } n \geq 0 \text{ and for each } |\lambda| = 1, \tag{3.1}$$

which, in turn, is equivalent to the fact that the norm of the partial sums of the resolvent $(T - \lambda)^{-1}$ are uniformly bounded by $C(|\lambda| - 1)^{-1}$, see [27, Thm. 3.1]. On finite dimensional spaces, the latter condition is equivalent to power boundedness and this, in turn, is equivalent to the fact that the norm of the resolvent itself is bounded by $C(|\lambda| - 1)^{-1}$, see [27] and the survey by Strikwerda and Wade [37]. The fact that $I - V$ is C^1 bounded on $L^p[0, 1]$, $1 \le p \le \infty$, but power bounded only on $L^2[0, 1]$ shows an extremal behavior of $I - V$. The question of whether C^1 boundedness implies power boundedness on Hilbert spaces remains open, see [27].

The exact norm of the Volterra operator on $L^p[0,1]$

In this subsection we compute the norm of V acting on $L^p[0, 1]$, $1 < p < \infty$. The proof only differs in minor details from the one in [4]. We observe that for $p = 1$ or $p = \infty$, it is well known and easy to see that the norm is equal to 1. In contrast with the case $p = 2$, the calculation required to compute $\|V\|_p$, $1 < p < \infty$, and $p \ne 2$ is quite involved.

THEOREM 3.4. *Let V denote the Volterra operator. If $1 < p < \infty$, then*

$$\|V\|_p = \frac{p \sin(\pi/p)}{\pi(p-1)^{1/p}}.$$

Furthermore, there is a unique non-negative function f such that $\|f\|_p = 1$ and $\|Vf\|_p = \|V\|_p$ and any other function on the unit sphere in which the norm is attained is equal to αf, with $|\alpha| = 1$.

We need the following well known fact. For the sake of completeness we include the proof.

LEMMA 3.5. *Let T be a compact operator acting on a reflexive Banach space B. Then there is x in B such that $\|x\| = 1$ and $\|Tx\| = \|T\|$.*

PROOF. Consider a sequence $\{x_n\}$ in B such that $\|x_n\| = 1$ for every n and $\|Tx_n\| \to \|T\|$. Since B is reflexive, its unit ball is weakly sequentially compact and, therefore, we can choose a subsequence $\{x_{n_k}\}$ which converges weakly to $x \in B$. Since T is compact, $\|Tx_{n_k}\| \to \|Tx\|$ as $k \to \infty$ and, therefore, $\|Tx\| = \|T\|$, which proves the result.   □

In the computations below appears the Beta function which is defined for $\Re p > 0$ and $\Re q > 0$ as

$$B(p, q) = \int_0^1 t^{p-1}(1 - t)^{q-1}dt,$$

see [29, pp. 18-19]. In particular, we will use Euler's formula, see [29, p. 21],

$$B(t, 1 - t) = \Gamma(t)\Gamma(1 - t) = \frac{\pi}{\sin(\pi t)}.$$

PROOF OF THEOREM 3.4. Upon applying Lemma 3.5, there is g in $L^p[0, 1]$ such that $\|g\|_p = 1$ and $\|Vg\|_p = \|V\|_p$. Since $\|V|g\|_p \ge \|Vg\|_p$, it follows that $\|V|g\|_p = \|Vg\|_p$ if and only if g is a scalar multiple of a non-negative function. Thus, the norms acting on real and complex spaces $L^p[0, 1]$ coincide. From now on we assume that the space $L^p[0, 1]$ is over the field of real numbers. In particular, we have shown that there is a non-negative function f in $L^p[0, 1]$ such that $\|f\|_p = 1$ for which the norm is attained. Thus f is a solution of the extremal problem

$$\max \|V\varphi\|_p^p \quad \text{subject to the restriction } \|\varphi\|_p^p = 1.$$

Since $L^p[0,1]$, $1 < p < \infty$, is uniformly smooth, the functions from $L^p[0,1]$ into $\mathbb{R}$ defined by $F(\varphi) = \|V\varphi\|_p^p$ and $F_0(\varphi) = \|\varphi\|_p^p$ are Fréchet differentiable. Upon using the infinite dimension version of the Lagrange Multiplier Theorem, we find that there is $\lambda \in \mathbb{R}$ such that the Fréchet derivative of $\mathcal{L}(\varphi) = F(\varphi) - \lambda F_0(\varphi)$ at f must be zero.

The Fréchet derivative $\mathcal{L}'(\varphi)$ is an element of the dual space of $L^p[0,1]$, that is, $L^q[0,1]$, where $1/q + 1/p = 1$, and for any h in $L^p[0,1]$,

$$\langle \mathcal{L}', h \rangle = \lim_{t \to 0} \frac{\mathcal{L}(\varphi + th) - \mathcal{L}(\varphi)}{t}.$$

Hence, we must compute

$$\lim_{t \to 0} \left[\int_0^1 \frac{|(V(\varphi + th))(x)|^p - |(V\varphi)(x)|^p}{t} \, dx - \lambda \int_0^1 \frac{|\varphi(x) + th(x)|^p - |\varphi(x)|^p}{t} \, dx \right].$$

Upon applying Lebesgue's bounded convergence theorem, we obtain $\langle \mathcal{L}'(\varphi), h \rangle$ is equal to

$$p \int_0^1 \mathrm{sgn}\,((V\varphi)(x))((V\varphi)(x))^{p-1}(Vh)(x) \, dx - p\lambda \int_0^1 \mathrm{sgn}\,(\varphi(x))(\varphi(x))^{p-1} h(x) \, dx,$$

where $\mathrm{sgn}(\alpha) = 1$ if $\alpha \geq 0$ and $\mathrm{sgn}\,(\alpha) = -1$ if $\alpha < 0$. Since $\mathcal{L}'(f) = 0$ and $f \geq 0$, we obtain

$$\int_0^1 ((Vf)(x))^{p-1} \int_0^x h(y) \, dy \, dx - \lambda \int_0^1 (f(x))^{p-1} h(x) \, dx = 0 \quad \text{for each } h \in L^p[0,1].$$

Upon reversing the order of integration in the first integral and writing the two integrals in one we obtain

$$\int_0^1 h(x) \left(\int_x^1 ((Vf)(y))^{p-1} \, dy - \lambda(f(x))^{p-1} \right) dx = 0 \quad \text{for each } h \in L^p[0,1].$$

Therefore,

$$\int_x^1 ((Vf)(y))^{p-1} \, dy = \lambda(f(x))^{p-1} \quad \text{almost everywhere on } [0,1].$$

Since f is non-negative and $\|f\|_p = 1$, the function

$$h(x) = \int_x^1 (V(f(y))^{p-1} \, dy$$

is continuous and decreasing with $h(0) > 0$ and $h(1) = 0$. In particular, the last display implies that $\lambda > 0$ and f is continuous and decreasing with $f(0) > 0$ and $f(1) = 0$. Therefore $(Vf)(y) > 0$ for $0 < y \leq 1$. It also follows from the display above, that f is infinitely differentiable on $[0,1)$. Thus taking derivatives, we obtain

$$-((Vf)(x))^{p-1} = \lambda(p-1)(f(x))^{p-2} f'(x).$$

Upon setting $F(x) = (Vf)(x)$ we end up with the boundary problem

$$\begin{cases} (F(x))^{p-1} = -\lambda(p-1)(F'(x))^{p-2} F''(x), \\ F(0) = 0, \\ F'(1) = 0. \end{cases} \tag{3.2}$$

In addition, since, $F'(x) = f(x) > 0$ for $0 \le x < 1$, we find that F is strictly increasing. Therefore, the inverse function $G = F^{-1} : [0, F(1)] \to [0, 1]$ is well defined. Upon substituting $x = G(y)$ in (3.2), we obtain

$$\begin{cases} y^{p-1} = \lambda(p-1)G''(y)(G'(y))^{-p-1}, \\ G(0) = 0, \\ G'(F(1)) = +\infty. \end{cases}$$

Setting $G'(y) = H(y)$, we obtain a first order differential equation with separated variables

$$y^{p-1}dy = \lambda(p-1)H^{-p-1}dH$$

whose solution is

$$G'(y) = H(y) = \left(\frac{\lambda(p-1)}{c - y^p} \right)^{1/p},$$

where c is a real constant. Since $G(0) = 0$, it follows that

$$G(y) = \int_0^y \left(\frac{\lambda(p-1)}{c - t^p} \right)^{1/p} dt.$$

Since $G'(F(1)) = H(F(1)) = +\infty$, we have $c - (F(1))^p = 0$, or equivalently, $G(c^{1/p}) = 1$. Hence,

$$\int_0^{c^{1/p}} \left(\frac{\lambda(p-1)}{c - t^p} \right)^{1/p} dt = 1.$$

Upon performing the change of variables $t = (cu)^{1/p}$ and using Euler's formula, we obtain

$$1 = \frac{\lambda^{1/p}(p-1)^{1/p}}{p} \int_0^1 u^{1/p-1}(1-u)^{-1/p}du$$

$$= \frac{\lambda^{1/p}(p-1)^{1/p}}{p} B(1/p, 1 - 1/p)$$

$$= \frac{\lambda^{1/p}(p-1)^{1/p}\pi}{p \sin(\pi/p)}.$$

Therefore,

$$\lambda = \frac{1}{p-1} \left(\frac{\sin(\pi/p)}{\pi/p} \right)^p.$$

To find the value of c, we use that $\|f\|_p = 1$, or what is the same,

$$\int_0^1 (F'(x))^p \, dx = 1.$$

Performing the change of variables $x = G(y)$ in the last integral, we obtain

$$\int_0^{c^{1/p}} (G'(y))^{1-p}dy = \int_0^{c^{1/p}} (H(y))^{1-p}dy = 1.$$

Thus using the formulas for H and λ, we have

$$\left(\frac{\sin(\pi/p)}{\pi/p} \right)^{1-p} \int_0^{c^{1/p}} (c - y^p)^{(p-1)/p}dy = 1.$$

Upon performing again the change of variables $y = (cu)^{1/p}$ and using Euler's formula, we obtain

$$
\begin{aligned}
1 &= \frac{c}{p} \left(\frac{\sin(\pi/p)}{\pi/p} \right)^{1-p} \int_0^1 (1-u)^{1-1/p} u^{1/p-1} du \\
&= \frac{c}{p} \left(\frac{\sin(\pi/p)}{\pi/p} \right)^{1-p} B(2-1/p, 1/p) \\
&= \frac{c}{p} \left(\frac{\sin(\pi/p)}{\pi/p} \right)^{1-p} \Gamma(2-1/p)\Gamma(1/p) \\
&= \frac{c}{p} \left(1 - \frac{1}{p} \right) \left(\frac{\sin(\pi/p)}{\pi/p} \right)^{1-p} \Gamma(1-1/p)\Gamma(1/p) \\
&= c \frac{p-1}{p} \left(\frac{\sin(\pi/p)}{\pi/p} \right)^{-p}.
\end{aligned}
$$

Thus,

$$
c = \frac{p}{p-1} \left(\frac{\sin(\pi/p)}{\pi/p} \right)^p.
$$

Now, we can compute $\|V\|_p$. We have

$$
\|V\|_p^p = \|Vf\|_p^p = \int_0^1 (F(x))^p \, dx.
$$

Again we perform the change of variables $x = G(y)$ in the last integral. We obtain

$$
\|V\|_p^p = \int_0^{c^{1/p}} y^p G'(y) \, dy = \int_0^{c^{1/p}} y^p \left(\frac{\lambda(p-1)}{c-y^p} \right)^{1/p} dy.
$$

Upon performing the usual change of variables $y = (cu)^{1/p}$ once more, we obtain

$$
\begin{aligned}
\|V\|_p^p &= c \frac{(\lambda(p-1))^{1/p}}{p} \int_0^1 u^{1/p}(1-u)^{-1/p} du \\
&= c\lambda^{1/p} \frac{(p-1)^{1/p}}{p} B(1+1/p, 1-1/p) \\
&= \frac{c\lambda^{1/p}(p-1)^{1/p}}{p^2} B(1/p, 1-1/p).
\end{aligned}
$$

Using Euler's formula, we obtain

$$
\|V\|_p^p = \frac{c\lambda^{1/p}(p-1)^{1/p}}{p} \frac{\pi/p}{\sin(\pi/p)}.
$$

Upon replacing the values of λ and c, the required formula follows.

Finally, observe that λ and c are uniquely determined and, therefore, so are G and F. Since $F = Vf$ and V is one-to-one, we find that the non-negative function f is unique. The proof is complete. $\qquad\square$

REMARK 3.6. A different procedure to solve the differential equation for G is to perform the substitution $K(y) = G((cy)^{1/p})/y^{1/p}$. In such a case, an easy calculation shows that K satisfies the hypergeometric diferential equation, see [29, Chap. 4]. Then one can obtain that

$$G(y) = (p-1)^{1/p} y F_{21}(1/p, 1/p; 1 + 1/p; y^4/c),$$

where $F_{21}(a, b; c, z)$ stands for the confluent hypergeometric function with parameters a, b and c.

4. THE COMMUTATOR OF THE VOLTERRA OPERATOR

The general case

Before stating the main result of this section, we need to introduce and discuss some additional notation. Let $\mathcal{M}_0$ be the space of finite σ-additive complex-valued Borel measures on $\mathbb{R}$, whose supports are in the interval $[0, 1)$, endowed with the variation norm $\|\mu\| = |\mu|([0, 1))$. Here the positive measure $|\mu|$ is the variation of μ. It is well-known that $\mathcal{M}_0$ is a Banach space. For $f \in L^1[0, 1]$ we can consider the extension Jf defined on $\mathbb{R}$ that coincides with f on $[0, 1]$ and vanishes outside of $[0, 1]$. Let μ_f be the absolutely continuous measure on $\mathbb{R}$ with density Jf. Let $\chi = \chi_{[0,1]}$ denote the characteristic function of $[0, 1]$. For $\mu \in \mathcal{M}_0$ and $f, g \in L^1[0, 1]$ we define the following operations

$$\mu \circ f = \left. \frac{d(\mu * \mu_f)}{d\lambda} \right|_{[0,1]} \quad \text{and} \quad f \star g = (Jf * Jg)\chi,$$

where $*$ stands for convolution and $d\nu/d\lambda$ is the density of the absolutely continuous measure ν, that is, the Radon-Nykodim derivative with respect to the Lebesgue measure. The next proposition summarizes some elementary properties of the operations $\circ$ and $\star$ that follow immediately from the well-known properties of the convolution, see [14], for instance.

PROPOSITION 4.1. *Assume that $1 \le p \le \infty$. If μ is in $\mathcal{M}_0$, f is in $L^1[0, 1]$ and g is in $L^p[0, 1]$, then $\mu \circ g$ as well as $f \star g$ are in $L^p[0, 1]$ with $\|\mu \star g\|_p \le \|\mu\| \|g\|_p$ and $\|f \star g\|_p \le \|f\|_1 \|g\|_p$. Furthermore, $\star$ is commutative and associative and $\mu \circ (f \star g) = (\mu \circ f) \star g$.*

It should be observed that $\star$ can also be represented as

$$(f \star g)(x) = \int_0^x f(t)g(x-t)\, dt \tag{4.1}$$

for almost every $0 \le x \le 1$. Clearly, $Vf = f \star 1$ for each $f \in L^1[0, 1]$. According to Proposition 4.1 for each $\mu \in \mathcal{M}_0$ the operator

$$S_\mu f = \mu \circ f$$

acts boundedly on $L^p[0, 1]$ for each $1 \le p \le \infty$ and $VS_\mu = S_\mu V$. As it was pointed out by Sarason [34], Sarason's Theorem implies that $[V]_2$ contains operators, which are not of the form S_μ for any $\mu \in \mathcal{M}_0$. It was shown by Mincheva, [23] that $[V]_1 = \{S_\mu : \mu \in \mathcal{M}_0\}$. For a sort of generalization of this result for operators acting on $L^1([0,1] \times [0,1])$ and commuting with partial integrations see [24]. We shall show that both observations follow from our characterization.

For $1 \leq p \leq \infty$, we denote by $W^{1,p}[0,1]$ the space of absolutely continuous complex valued functions defined on $[0,1]$ such that $f' \in L^p[0,1]$ for which the norm

$$\|f\|_{1,p} = |f(0)| + \|f'\|_p$$

is finite. Clearly, $W^{1,p}[0,1]$ is a Banach space isomorphic to $L^p[0,1]$. The commutator of V can be characterized in terms of the following set

$$X_p = \{a \in L^1[0,1] : a \star f \in W^{1,p}[0,1] \text{ for each } f \in L^p[0,1]\}. \tag{4.2}$$

THEOREM 4.2. *Let V denote the Volterra operator acting on $L^p[0,1]$, $1 \leq p \leq \infty$. Then $[V]_p$ is the set of the operators T for which there is $a \in X_p$ such that $Tf = (a \star f)'$ for each $f \in L^p[0,1]$.*

REMARK 4.3. Although the characterization above is not a very manageable one, Theorems 4.4 and 4.7 below will provide a simpler equivalent formulation of the condition $a \in X_p$ in the three most important cases: $p = 1$, $p = 2$ and $p = \infty$.

PROOF OF THEOREM 4.2. First we prove that if $a \in X_p$, then $T_a f = (a \star f)'$ is bounded on $L^p[0,1]$. To this end, we take $g \in L^p[0,1]$ and a sequence $\{f_n\}$ in $L^p[0,1]$ such that that $\|f_n\|_p \to 0$ and $\|(a \star f_n)' - g\|_p \to 0$. Then

$$(a \star f_n)(x) = \int_0^x (a \star f_n)'(t)\, dt \to \int_0^x g(t)\, dt, \text{ for each } 0 \leq x \leq 1.$$

On the other hand, by Proposition 4.1 we see that $\|a \star f_n\|_p \leq \|a\|_1 \|f_n\|_p \to 0$, which along with the last display implies that $\int_0^x g(t)\, dt = 0$ almost everywhere on $[0,1]$. Therefore, g is the zero function and by the Closed Graph Theorem, see [31], it follows that T_a is bounded on $L^p[0,1]$.

Next we show that T_a commutes with V. To this end, we first take $f \in L^\infty[0,1]$. From Proposition 4.1, it follows that

$$\|f \star a\|_{L^\infty[0,x]} \leq \|a\|_{L^1[0,x]} \|f\|_{L^\infty[0,x]} \leq \|f\|_\infty \|a\|_{L^1[0,x]} \to 0 \quad \text{as } x \to 0.$$

Since $a \in X_p$, we find that $a \star f$ is continuous. Hence according to the last display $(a \star f)(0) = 0$. It follows that $VT_a f = V((a \star f)') = a \star f - (a \star f)(0) = a \star f$. On the other hand, by Proposition 4.1, we have $T_a V f = T_a(1 \star f) = (a \star 1 \star f)' = (1 \star a \star f)' = (V(a \star f))' = a \star f$. Hence $T_a V f = V T_a f$ for each $f \in L^\infty[0,1]$. Since $L^\infty[0,1]$ is dense in $L^p[0,1]$ for each $1 \leq p \leq \infty$, we find that T_a commutes with V.

It remains to prove that if T is bounded on $L^p[0,1]$ and commutes with V, then $T = T_a$ for some $a \in X_p$. Let u_0 be the function identically 1 and for each positive integer n we set $u_n = u_0 \star u_{n-1}$. We take $a = Tu_0$ which is clearly in $L^p[0,1]$. Then for each positive integer n, we have $TVV^{n-1}u_0 = V^n Tu_0 = V^n a$. The fact that $Vf = u_0 \star f$ along with Proposition 4.1 implies that

$$TVu_n = a \star u_n \qquad \text{for each } n = 1, 2, \ldots.$$

Since the operator TV is bounded from $L^1[0,1]$ into $L^p[0,1]$ and the linear span of $\{u_n = x^n/n!\}$ is dense in $L^1[0,1]$, we see that $TVf = VTf = a \star f$ for each $f \in L^p[0,1]$. Since $Tf \in L^p[0,1]$ and $VTf = a \star f$, it follows that $a \star f$ is absolutely continuous and $(a \star f)' \in L^p[0,1]$ for each $f \in L^p[0,1]$, that is, $a \in X_p$. Since $VTf = a \star f$, taking derivatives we find that $Tf = (a \star f)'$ for each $f \in L^p[0,1]$, that is, $T = T_a$. The proof is complete. $\qquad \square$

The commutator of the Volterra operator on $L^1[0,1]$ and $L^\infty[0,1]$

Let $\mathbb{V}[0,1]$ denote the space of functions $f : [0,1] \to \mathbb{C}$ of bounded variation. The following theorem contains a characterization of the commutator of $[V]_1$ as well as $[V]_\infty$. We have

THEOREM 4.4. *For each* $1 \le p \le \infty$, *we have* $\mathbb{V}[0,1] \subseteq X_p$. *Furthermore,* $X_1 = X_\infty = \mathbb{V}[0,1]$.

As usual, the equality in Theorem 4.4 means that each function in $X_1 = X_\infty$ coincides almost everywhere with a function of bounded variation.

PROOF OF THEOREM 4.4. First, we note that a complex valued function a defined on $[0,1]$ has bounded variation if and only if there is a (unique) measure $\mu = \mu_a \in \mathcal{M}_0$ such that a coincides, except possibly for a countable set, with the restriction to $[0,1]$ of the distribution function of μ. In addition, it is straightforward to check that if $f \in L^1[0,1]$, then $a \star f$ is absolutely continuous and $(a \star f)' = \mu \circ f$. As a consequence, by Proposition 4.1, we see that $a \star f \in W^{1,p}[0,1]$ for each $f \in L^p[0,1]$, for each $1 \le p \le \infty$. Thus $\mathbb{V}[0,1]$ is contained in X_p for each $1 \le p \le \infty$.

Next we prove that X_1 is contained in $\mathbb{V}[0,1]$. Thus suppose now that $a \in X_1$. According to Theorem 4.2, we have that $T_a f = (a \star f)'$ acts boundedly on $L^1[0,1]$. Therefore, there exists a constant $c > 0$ such that $\|(a \star f)'\|_1 \le c\|f\|_1$ for each $f \in L^1[0,1]$. Now, for each $0 < \varepsilon < 1$, let $f_\varepsilon = \varepsilon^{-1}\chi_{[0,\varepsilon]}$. Clearly, $\|f_\varepsilon \star a - a\|_1$ tends to 0 as ε tends to 0. On the other hand, $\|f_\varepsilon\|_1 = 1$ and, therefore, $\|(f_\varepsilon \star a)'\|_1 \le c$ for each $0 < \varepsilon < 1$. Let μ_ε be the absolutely continuous measure on $[0,1]$ whose density is $(f_\varepsilon \star a)' \in L^1[0,1]$. Of course, $\mu_\varepsilon \in \mathcal{M}_0$ and $\|\mu_\varepsilon\| = \|(f_\varepsilon \star a)'\|_1 \le c$. According to the Radon Theorem, we may regard $\mathcal{M}_0$ as the dual $\mathcal{C}^*$ of the space $\mathcal{C}$ consisting of continuous functions on $[0,1]$ that vanish at 1 endowed with the supremum norm. Indeed, the corresponding dual pairing is

$$\langle h, \nu \rangle = \int_{[0,1)} h \, d\nu.$$

By Alaoglu's theorem [31] the closed ball of radius c of $\mathcal{M}_0$ is a metrizable compact set with respect to the weak star topology $\sigma(\mathcal{M}_0, \mathcal{C})$. Hence there exist $\mu \in \mathcal{M}_0$ and a sequence of $\{\varepsilon_n\}$, $0 < \varepsilon_n < 1$, tending to 0 such that $\|\mu\| \le c$ and $\mu_{\varepsilon_n} \to \mu$ in the weak star topology. According to the well known property of weak star convergence of measures, the distribution functions $(f_{\varepsilon_n} \star a)(x)$ of measures μ_{ε_n} converge to the distribution function $F_\mu(x)$ at each point x where F_μ is continuous. Since F_μ has bounded variation, it is continuous everywhere except possibly for a countable set. Hence $f_{\varepsilon_n} \star a \to F_\mu$ almost everywhere. On the other hand, $f_{\varepsilon_n} \star a \to a$ in $L^1[0,1]$. Therefore, it follows that $a = F_\mu$ almost everywhere and, consequently, a has bounded variation. Therefore $X_1 = \mathbb{V}[0,1]$.

It remains to prove that X_∞ is contained in $\mathbb{V}[0,1]$. Thus suppose that $a \in X_\infty$. According to Theorem 4.2 we have that $T_a f = (a \star f)'$ acts boundedly on $L^\infty[0,1]$. Therefore, there exists a constant $c > 0$ such that

$$\|(a \star f)'\|_\infty \le c\|f\|_\infty \quad \text{for each} \quad f \in L^\infty[0,1]. \tag{4.3}$$

Since $(a \star 1)' = (Va)' = a$, we immediately obtain that $a \in L^\infty[0,1]$ and $\|a\|_\infty \le c$. We need the following technical lemma whose proof is delayed.

LEMMA 4.5. *Assume that for $f \in L^{\infty}[0,1]$ there is a positive constant c such that for each finite sequence $0 \le x_0 < \cdots < x_n < 1$, we have*

$$\sum_{j=1}^{n} |f(x_{j+1} + \lambda) - f(x_j + \lambda)| \le c \quad \text{for almost all } \lambda \in [0, 1 - x_n]. \quad (4.4)$$

Then there exists $g \in \mathbb{V}[0,1]$ such that $f(x) = g(x)$ for almost all $x \in [0,1]$.

Let $\mathbb{T}$ denote the unit circle in the complex plane. For each finite sequence $0 \le x_0 < \cdots < x_n < 1$ and $\gamma = (\gamma_1, \ldots, \gamma_n) \in \mathbb{T}^n$, we consider

$$g_{x,\gamma}(t) = \begin{cases} 0, & \text{if } t \in [0, x_0) \cup [x_n, 1]; \\ \gamma_k, & \text{if } t \in [x_{k-1}, x_k), \ 1 \le k \le n. \end{cases}$$

A direct computation shows that

$$(a \star g_{x,\gamma})'(u) = \sum_{j=1}^{n} \gamma_j(f(u - x_j) - f(u - x_{j-1})) \quad \text{for almost all } u \in [x_n, 1].$$

By (4.3) there is a null Lebesgue set $A(\gamma)$ contained in $[x_n, 1]$ such that

$$\left| \sum_{j=1}^{n} \gamma_j(f(u - x_j) - f(u - x_{j-1})) \right| \le c \quad \text{for each } u \in [x_n, 1] \setminus A(\gamma).$$

Let $\{\gamma^m\}$ be a dense sequence in $\mathbb{T}^n$ and let $A = \bigcup_{m=1}^{\infty} A(\gamma^m)$. Then $\mu(A) = 0$ and

$$\sup_{m \in \mathbb{N}} \left| \sum_{j=1}^{n} \gamma_j^m(f(u - x_j) - f(u - x_{j-1})) \right| = \sum_{j=1}^{n} |f(u - x_j) - f(u - x_{j-1})| \le c$$

for each $u \in [x_n, 1] \setminus A$. Thus, f satisfies (4.4) and, by Lemma 4.5, it follows that f has bounded variation and, therefore, $X_{\infty} = \mathbb{V}[0,1]$. The proof will be complete once we have proved Lemma 4.5. □

PROOF OF LEMMA 4.5. First, we shall prove that for each $0 < x < 1$, there is a set A_x with $\mu(A_x) = 1$, here μ is the Lebesgue measure, such that the following limits exist

$$f_-(x) = \lim_{\substack{y \to x \\ y \in (0,x) \cap A_x}} f(y) \quad \text{and} \quad f_+(x) = \lim_{\substack{y \to x \\ y \in (x,1) \cap A_x}} f(y).$$

Indeed, if this is not the case, there exist $x \in (0,1)$, $\varepsilon > 0$ and Lebesgue measurable sets A and B contained in $[0,1]$ such that $|f(y) - f(u)| \ge \varepsilon$ for each $y \in A$ and $u \in B$ and either $\mu(A \cap (x - \delta, x)) > 0$ and $\mu(B \cap (x - \delta, x)) > 0$ for each $\delta > 0$ or $\mu(A \cap (x, x + \delta)) > 0$ and $\mu(B \cap (x, x + \delta)) > 0$ for each $\delta > 0$. We choose a positive integer n such that $n\varepsilon > c$. Since almost every point of a set of positive Lebesgue measure is a density point, in any case we can choose $0 < x_0 < \cdots < x_n < 1$ such that x_j is a density point of A for even j and x_j is a density point of B for odd j. From the definition of a density point, it immediately follows that

$$C = \{\lambda \in (0, 1 - x_n) : x_j + \lambda \in A \text{ for even } j \text{ and } x_j + \lambda \in B \text{ for odd } j\}$$

has positive Lebesgue measure.

Since $|f(y) - f(u)| \geq \varepsilon$ for $y \in A$ and $u \in B$, we have that

$$\sum_{j=1}^{n} |f(x_{j+1} + \lambda) - f(x_j + \lambda)| \geq n\varepsilon > c$$

for any $\lambda \in C$, which contradicts (4.4) and the existence of the limits defining the functions f_- and f_+ is proved.

Now, the derivative of $F = Vf$ is $F'(x) = f(x)$ almost everywhere on $[0,1]$ and one easily checks that if F is differentiable at x, then $f_-(x) = f_+(x) = F'(x)$. Hence f_+ and f_- both coincide with f almost everywhere. Thus, it is enough to prove that f_- has bounded variation, or what is the same,

$$\sum_{j=1}^{n} |f_-(x_{j+1}) - f_-(x_j)| \leq c \quad \text{for } 0 < x_0 < \cdots < x_n < 1.$$

To this end, we consider $\Omega = \{x \in \mathbb{R}^{n+1}, \ 0 < x_0 < \cdots < x_n < 1\}$ and

$$\Phi(x) = \sum_{j=1}^{n} |f_-(x_{j+1}) - f_-(x_j)| \qquad \text{for } x \in \Omega.$$

Since $f_- = f$ almost everywhere, (4.4) implies that Φ is bounded by c almost everywhere with respect to the one-dimensional Lebesgue measure on the intersection of the convex set Ω with each straight line whose direction is $(1, \ldots, 1)$. By Fubini's Theorem Φ is bounded by c almost everywhere on Ω with respect to $(n+1)$-dimensional Lebesgue measure. Using that f_- is left continuous, we find that almost everywhere boundedness of Φ by c implies boundedness of Φ by c. Thus f_- is of bounded variation. The proof of Lemma 4.5 and that of Theorem 4.4 is now complete.   □

As a corollary of Theorem 4.4, we have

COROLLARY 4.6. $[V]_1 = \{S_\mu : \mu \in \mathcal{M}_0\} = [V]_\infty$.

PROOF. As already shown, the operators S_μ act boundedly on $L^p[0,1]$ and commute with V for each $1 \leq p \leq \infty$. Suppose now that $p = 1$ or $p = \infty$ and T is in $[V]_p$. By Theorem 4.4 there exists a function $a : [0,1] \to \mathbb{C}$ of bounded variation such that $Tf = (a \star f)'$ for each $f \in L^p[0,1]$. Since a has bounded variation, there exists $\mu \in \mathcal{M}_0[0,1]$ such that a coincides, except for a countable set, with the restriction to $[0,1]$ of the distribution function of μ. Now, it is straightforward to check that if $f \in L^p[0,1]$, then $(a \star f)' = \mu \circ f$ and the result follows.   □

Observe that for the first equality in the above corollary the operators S_μ act on $L^1[0,1]$, while for the second they act in $L^\infty[0,1]$. The first equality in Corollary 4.6 is exactly the mentioned result of Mincheva, while the second equality shows that her description of the commutator of the Volterra operator acting on $L^1[0,1]$ remains valid when V acts on $L^\infty[0,1]$.

The commutator of the Volterra on $L^2[0,1]$.

We end with a simpler characterization for $[V]_2$. If f is in $L^2(\mathbb{R})$, its Fourier transform will be denoted $\widehat{f}$.

THEOREM 4.7. *A function $a \in L^1[0,1]$ is in X_2 if and only if there exists b in $L^2(\mathbb{R})$ that coincides with a almost everywhere on $[0,1]$, vanishes on $(-\infty,0)$ and satisfies $\widehat{b}(x) = O(x^{-1})$ as $x \to \pm\infty$.*

PROOF. Suppose first that there is b in $L^2(\mathbb{R})$ satisfying the conditions in the statement. We take any $f \in L^2[0,1]$ and set $h = b * Jf$. Since b coincides with a on $[0,1]$ and vanishes on $(-\infty,0]$, we see that $a \star f$ coincides with h on $[0,1]$. On the other hand, $\widehat{h}(x) = \widehat{f}(x)\widehat{a}(x) = O(x^{-1}\widehat{f}(x))$ as $x \to \pm\infty$. Thus $G(x) = -ix\widehat{h}(x)$ is in $L^2(\mathbb{R})$. It follows that G is the Fourier transform of some function $g \in L^2(\mathbb{R})$. Using the well known property of the Fourier transform, we find that h is absolutely continuous and $h' = g \in L^2(\mathbb{R})$. Hence $a \star f = h$ coincides with h in $[0,1]$ and belongs to $W^{1,2}[0,1]$, that is, $a \in X_2$.

Conversely, if $a \in X_2$, then, by Theorem 4.2, $T_a f = (a \star f)'$ acts boundedly on $L^2[0,1]$ and commutes with V. Sarason's Theorem implies that there is $\phi \in H^\infty(\mathbb{R})$ such that $T_a = PFM_\phi\mathcal{F}^{-1}J$. Using the latter formula for T_a along with the properties of the Fourier transform, one easily checks that if f is in $\mathcal{D}[0,1] = \{f \in C^\infty(\mathbb{R}) : \mathrm{supp}(f) \subseteq [0,1]\}$, then $(T_a f)(x) = (2\pi)^{-1}(f \star \widehat{\phi})$ for each $0 \le x \le 1$ and where $\widehat{\phi}$ is regarded as a Schwarz distribution. Set $R_a f = a \star f$, observe that $R_a = VT_a$ and $(Vf)(x) = f * \chi_{[0,1]}(x)$, for $0 \le x \le 1$. Thus we obtain that

$$(R_a f)(x) = (2\pi)^{-1}(f * \psi)(x) \text{ for each } 0 \le x \le 1, \text{ where } \psi = \widehat{\phi} \star \chi. \text{ Hence } \widehat{\psi} = \widehat{\widehat{\phi}}\widehat{\chi}.$$

Since $\widehat{\widehat{\phi}}(x) = 2\pi\phi(-x)$ and $\widehat{\chi}(x) = (1-e^{-ix})/(ix)$, we conclude that $\widehat{\psi} \in L^2(\mathbb{R})$ and $\widehat{\psi}(x) = O(x^{-1})$ as $x \to \pm\infty$. On the other hand, since $\phi \in H^\infty(\mathbb{R})$, we immediately see that $\widehat{\psi}$ belongs to the Hardy space of the lower half-plane. Hence ψ vanishes on $(-\infty,0)$. If we set $b = (2\pi)^{-1}\psi$, we see that $(R_a f)(x) = (b * f)(x)$ for each $0 \le x \le 1$ and for each $f \in \mathcal{D}[0,1]$, which implies that b coincides with a on $[0,1]$. Thus b satisfies all the requirements of the statement. The proof of Theorem 4.7 is complete. $\qquad\square$

Further properties and remarks

PROPOSITION 4.8. *If $1 \le p, q \le \infty$ with $1/p + 1/q = 1$. Then $X_p = X_q$.*

PROOF. The case $p = 1$ or $p = \infty$ is part of the statement of Theorem 4.4. Thus suppose that $1 < p < \infty$ and $a \in L^1[0,1]$. Clearly, it is enough to prove that if $a \in X_p$, then $a \in X_q$. Thus suppose that $a \in X_p$. Recall that the dual space of $L^p[0,1]$ is isometrically isomorphic to $L^q[0,1]$ under the following dual pairing

$$\langle f, g \rangle = \int_0^1 f(t)g(t)\,dt, \qquad f \in L^p[0,1] \text{ and } g \in L^q[0,1].$$

By Theorem 4.2 the operator $T_a f = (a \star f)'$ acts boundedly on $L^p[0,1]$. Hence, the adjoint T_a^* acts boundedly on $L^q[0,1] = (L^p[0,1])^*$. Since $R_a = T_a V$, where $R_a f = a \star f$, we have $R_a^* = V^* T_a^*$. Now, if $f \in L^p[0,1]$ and $g \in L^q[0,1]$, then

$$\langle f, R_a^* g \rangle = \langle R_a f, g \rangle = \langle a \star f, g \rangle = \int_0^1 g(x) \int_0^x f(t)a(x-t)\,dt\,dx.$$

Now, Fubini's Theorem along with a linear change of variables shows that

$$\langle f, R_a^* g \rangle = \int_0^1 f(t) \int_0^{1-t} g(x)a(1-t-x)\,dx\,dt = \langle f, UR_a Ug \rangle,$$

where $(Ug)(x) = g(1 - x)$. Hence $R_a^* = U R_a U$ (observe that R_a in the right hand side acts on $L^q[0, 1]$). Since $V = R_1$, we find that $V^* = UVU$. Taking into account that $R_a^* = V^* T_a^*$ and $U^2 = I$, we obtain that the operator R_a acting on $L^q[0, 1]$ satisfies the equality $R_a = VUT_a^*U$. Since UT_a^*U acts boundedly on $L^q[0, 1]$ and V is bounded from $L^q[0, 1]$ into $W^{1,q}[0, 1]$, we obtain that $a \star f = R_a f \in W_q^1[0, 1]$ for any $f \in L^q[0, 1]$, that is, $a \in X_q$. The proof is complete.   □

Erdos [11] proved that, for each $1 \le p < \infty$, the commutant $[V]_p$ is commutative, that is, $TS = ST$ for each $T, S \in [V]_p$. This fact can easily be derived from Theorem 4.2. In passing, we show that the same statement remains true for $p = \infty$.

PROPOSITION 4.9. *Assume that $1 \le p \le \infty$ and two bounded operators on $L^p[0, 1]$ commute with V, then they also commute with each other.*

PROOF. Suppose that T and S act boundedly on $L^p[0, 1]$ and commute with V. From Theorem 4.2, it follows that there exist $a, b \in L^1[0, 1]$ such that $TVf = a \star f$ and $SVf = b \star f$ for any $f \in L^p[0, 1]$. By Proposition 4.1, we find that that $V^2 STf = V^2 TSf = a \star b \star f$ for any $f \in L^p[0, 1]$. Since $\ker V^2 = \{0\}$, we obtain that $TSf = STf$ for each $f \in L^p[0, 1]$, that is, S and T commute.   □

REMARK 4.10. As mentioned in Section 2 the invariant subspaces for V acting in $L^p[0, 1]$, $1 \le p < \infty$ are $L^p[\alpha, 1]$, $0 \le \alpha \le 1$. From Theorem 4.2, it immediately follows that the latter subspaces are also invariant for each operator in the commutant $[V]_p$.

REMARK 4.11. Since there exist essentially unbounded (and therefore having unbounded variation) functions $a \in L^2[0, 1]$ such that $\widehat{Ja}(x) = O(1/x)$ as $x \to \pm\infty$, Theorems 4.2 and 4.7 imply that $X_2 \neq \mathbb{V}[0, 1]$, that is, there exists $T \in [V]_2$, which is not of the form S_μ.

REMARK 4.12. Biswas, Lambert, and Petrovic [6] study a related problem called the generalized eigenvalue problem. They search for complex numbers λ for which there exists a nonzero operator T that acts boundedly on $L^2[0, 1]$ such that $TV = \lambda VT$ for certain subclasses of operators T.

References

[1]   G. R. Allan, *Power-bounded elements and radical Banach algebras*, Linear Operators, Banach Center Publ., vol. 38, Inst. Math., Polish Acad. Sci., Warsaw, 1997, pp. 9–16.

[1]   S. I. Ansari, *Hypercyclic and cyclic vectors*, J. Funct. Anal. **128** (1995), 374–383.

[3]   F. Bayart and E. Matheron, *Hyponormal operators, weighted shifts and weak forms of supercyclicity*, Proceeding of the Edimbourgh Mathematical Society (to appear).

[4]   C. Bennewitz and Y. Saitō, *Approximation numbers of Sobolev embedding operators on an interval*, J. London Math Soc. (2) **70** (2004), 244–260.

[5]   S. Bermudo, A. Montes-Rodríguez and S. A. Shkarin, *Analytic functions of the Volterra operator, in preparation.*

[6]   A. Biswas, A. Lambert and S. Petrovic, *Extended eigenvalues and the Volterra operator*, Glasg. Math. J., **44** (2002), 521–534.

[7]   J. B. Conway, *A course in Functional Analysis*, Springer-Verlag, New York, 1985.

[8]   W. F. Donoghue, Jr., *The lattice of invariant subspaces of a completely continuous quasi-nilpotent transformation*, Pacific J. Math. **9** (1957), 1031–136.

[10] J. Erdos, *The commutant of the Volterra operator*, Integral Equations and Operator Theory **5** (1982), 127–130.

[11] J. Erdos, *Operators with abelian commutant*, J. Lond. Math. Soc. **9** (1974/75), 637–640.

[12] S. P. Eveson, *Asymptotic behavivour of iterates of Volterra operators on $L^p[0,1]$*. Preprint.

[13] E. Gallardo-Gutiérrez and A. Montes-Rodríguez, *The Volterra operator is not supercyclic*, Integral Equations and Operator Theory **50** (2004), 211–216.

[14] C. Graham and O. McGehee, *Essays in commutative harmonic analysis*, Springer-Verlag, Berlin, 1979.

[15] P. R. Halmos, *A Hilbert Problem Book*, Van Nostrand Inc., New York, 1967.

[16] D. A. Herrero and C. Kitai, *On invertible hypercyclic operators*, Proc. Amer. Math. Soc. **116** (1992), 873–875.

[17] H. M. Hilden and L. J. Wallen, *Some cyclic and non-cyclic vectors of certain operators*, Indiana Univ. Math. J. **23** (1973), 557–565.

[18] E. Hille, *Remarks on ergodic theorems*, Trans. Amer. Math. Soc. **57** (1945), 246–269.

[19] D. Kershaw, *Operator norms of powers of the Volterra operator*, J. Integral Equations Appl. **11** (1989), 351–362.

[20] N. Lao and R. Whitley, *Norms of powers of the Volterra operator*, Integral Equations Operator Theory **27** (1997), 419-425.

[21] B. Ya. Levin, *Distribution of Zeros of Entire Functions*, American Mathematical Society, Providence, R.I., 1980.

[22] G. Little and J. B. Reade, *Estimates for the norm of the n-th indefinite integral*, Bull. London Math. Soc. **30** (1998), 539-542.

[23] S. Mincheva, *Automorphisms in the commutant of the integration operator in the space of Lebegue integrable functions*, Transform Methods and Special Functions, Varna, Bulgarian Acad. Sci., Sofia 1998, 1996, pp. 335–344.

[24] S. Mincheva, *Automorphisms of C commuting with partial integration operators in a rectangle*, Algebraic analysis and related topics, Warsaw, vol. 53, Polish Acad. Sci. Banach Center Publ. Warsaw 2000, 1999, pp. 167–176.

[25] A. Montes-Rodríguez and H. N. Salas, *Supercyclic subspaces: spectral theory and weighted shifts*, Adv. Math. **163** (2001), 74–134.

[26] A. Montes-Rodríguez and H. N. Salas, *Supercyclic subspaces*, Bull. London Math. Soc. **35** (2003), 721–737.

[27] A. Montes-Rodríguez, J. Sánchez-Álvarez and J. Zemánek, *Uniform Abel-Kreiss bounded and the extremal behaviour of the Volterra operator*, Proc. London Math. Soc. (to appear).

[28] A. Montes-Rodríguez and S. A. Shkarin, *The Volterra operator is not weakly supercyclic*. Preprint.

[29] E. D. Rainville, *Special functions*, Chelsea Publishing Company, New York, 1971.

[30] W. Rudin, *Real and Complex Analysis*, McGraw-Hill, New York, 1987.

[31] W. Rudin, *Functional Analysis*, McGraw-Hill, New York, 1991.

[32] H. N. Salas, *Supercyclicity and weighted shifts*, Studia Math. **135** (1999), 55–74.

[33] D. Sarason, *A remark on the Volterra operator*, J. Math. Anal. Appl. **12** (1965), 244–246.

[34] D. Sarason, *Generalized interpolation in H^∞*, Trans. Amer. Math. Soc., **127** (1967), 179–203.

[35] D. Sarason, *Invariant subspaces*, Topics in operator theory, vol. 13, Amer. Math. Soc., Providence, R.I., 1974, pp. 1–47. Math. Surveys.

[36] S. A. Shkarin, *Antisupercyclic operators and orbits of the Volterra operator.*, London Math. Soc. (to appear).

[37] J. C. Strikwerda and B. A. Wade, *A survey of the Kreiss matrix theorem for power bounded families of matrices and its extensions*, Linear Operators, Banach Center Publ., vol. 38, Inst. Math., Polish Acad. Sci., Warsaw, 1997, pp. 339–360.

[38] G. Szegö, *Orthogonal Polynomials*, American Mathematical Society, Providence, 1959.

[39] B. Thorpe, *The norms of powers of the indefinite integral operator on $(0,1)$*, Bull. London Math. Soc. **30** (1998), 543-548.

DEPARTAMENTO DE ANÁLISIS MATEMÁTICO, FACULTAD DE MATEMÁTICAS, UNIVERSIDAD DE SEVILLA, APTDO. 1160, SEVILLA 41080, SPAIN
E-mail address: `amontes@us.es`

FACULTY OF MATHEMATICS AND MECHANICS, MOSCOW STATE UNIVERSITY, VOROBJOVY GORY, 119899 MOSCOW, RUSSIA
E-mail address: `shkarin@math.uni-wuppertal.de`

Contemporary Mathematics
Volume **393**, 2006

Size conditions to be in a finitely generated ideal of H^∞

Jordi Pau

Introduction

Let H^∞ be the Banach algebra of all bounded analytic functions on the unit disc $\mathbb{D}$. Our starting point is Carleson's famous Corona theorem that says that the unit disc is dense in the space M_{H^∞} of maximal ideals of H^∞ with the weak-star topology. This can be reformulated by saying that, given functions $f_1, \dots, f_N \in H^\infty$, the condition

$$\sum_{j=1}^{N} |f_j(z)| \geq \delta > 0 \quad (z \in \mathbb{D})$$

implies that the ideal $I = I(f_1, \dots, f_N)$ generated by the functions $f_j \in H^\infty$ equals the whole algebra, that is, there exist functions $g_1, \dots, g_N \in H^\infty$ such that $\sum_{j=1}^{N} g_j f_j = 1$.

In order to generalize this result it is natural to ask if given $g \in H^\infty$, the condition

(1) $$|g(z)| \leq C \sum_{j=1}^{n} |f_j(z)|, \quad \forall z \in \mathbb{D},$$

for some constant C implies that the function g is in the ideal I generated by the functions f_j. Condition (1) is clearly a necessary condition, but a well known example due to Rao (see below or [**Ra**]) shows that the answer is, in general, negative. On the other hand, Wolff proved that this condition implies that the function g^3 belongs to the ideal I and also that the condition

$$|g(z)| \leq C \left(\sum_{j=1}^{N} |f_j(z)|^2 \right)^{3/2} \quad z \in \mathbb{D}$$

for some constant C implies that the function g is in the ideal I.

From now, given functions $f_1, \dots, f_N \in H^\infty$ we denote $|f| = (\sum_{j=1}^{\infty} |f_j|^2)^{1/2}$.

1991 *Mathematics Subject Classification.* Primary 30D55; Secondary 46J15.
Key words and phrases. Corona problems, Carleson measure, interpolating Blaschke product.
The author is partially supported by SGR grant 2001SGR00431 and DGICYT grant BFM2002-00571.

In view of the previous results, the following problem arises naturally: for which positive continuous functions h on $[0, \infty)$ (increasing in a neighbourhood of zero) the condition $|g| \leq h(|f|)$ on $\mathbb{D}$, $g \in H^\infty$ implies that the function g is in the ideal generated by the functions f_j?

Therefore we are looking for sufficient size conditions for a function $g \in H^\infty$ to be in a finitely generated ideal of H^∞.

For functions of the form $h(s) = s^\alpha$, with $\alpha \geq 1$, the problem is completely solved. For $1 \leq \alpha < 2$, a variation of Rao's example (see below) shows that the answer is negative, and for $\alpha > 2$, work of Wolff, Cegrell and others gives an affirmative answer (see [Ce1], [Ga]). The problem for $\alpha = 2$ was an old question of Wolff, which remained open for twenty years. However, Treil (see [Tr]) has recently shown (using a connection with the best estimates of the solutions of the corona theorem) that the answer is, in general, negative.

Example 1: Let B_1 and B_2 be Blaschke products without common zeros satisfying

$$\inf_{z \in \mathbb{D}} (|B_1(z)| + |B_2(z)|) = 0.$$

Let m be a positive integer greater than 1 and consider the functions $g = (B_1 B_2)^{m-1}$, $f_1 = B_1^m$ and $f_2 = B_2^m$. Using the elementary inequality $2ab \leq a^2 + b^2$ we obtain

$$|g| = (|B_1|^{m/2} |B_2|^{m/2})^{2(m-1)/m} \leq \{(|B_1|^m + |B_2|^m)/2\}^{2(m-1)/m}.$$

Hence, given $\varepsilon > 0$, we can choose $m = m(\varepsilon)$ with $2m/(m+1) > 2 - \varepsilon$. Since $(f_1 + f_2)/2 \leq 1$ we have

$$|g(z)| \leq (|f_1(z)| + |f_2(z)|)^{2-\varepsilon}.$$

Suppose now that there exist functions $g_1, g_2 \in H^\infty$ with

$$B_1^{m-1} B_2^{m-1} = g_1 B_1^m + g_2 B_2^m.$$

Let $\{z_n\}$ be the zeros of the Blaschke product B_1, then $g_2(z_k) B_2^m(z_k) = 0$. Since B_1 and B_2 have no common zeros there exists a function $k_2 \in H^\infty$ such that $g_2 = k_2 B_1^{m-1}$. Hence $B_2^{m-1} = g_1 B_1 + k_2 B_2^m$. In an analogous way there is a function $k_1 \in H^\infty$ such that $g_1 = k_1 B_2^{m-1}$. Hence $1 = k_1 B_1 + k_2 B_2$ that contradicts the corona theorem since $\inf_{z \in \mathbb{D}}(|B_1(z)| + |B_2(z)|) = 0$. Hence g does not belong to the ideal generated by f_1 and f_2.

In [Li], Lin shows that $|g| \leq h(|f|)$ implies that g is in the ideal I for the function

$$h(s) = s^2(-\log s)^{-(3/2+\varepsilon)}$$

with $\varepsilon > 0$, and in [Ce2] Cegrell gives an affirmative answer for this problem for the function

$$h(s) = \frac{s^2}{(-\log s)^{3/2}(\log(-\log s))^{3/2} \log\log(-\log s)}.$$

In [Pa1], the author established the following strongest known positive case for this problem.

THEOREM A. *Let $k : (0,1) \to [0,\infty)$ be a nondecreasing bounded continuous function such that $k(x)/x$ is nonincreasing and*

$$\int_0^1 \frac{k(x)}{x} |\log x| \, dx < +\infty,$$

and let $H(x) = \sqrt{k(x) \int_0^x \frac{k(s)}{s} \, ds}$. Further, let $g, f_1, \ldots, f_n \in H^\infty$, where $0 < |f|^2 := \sum_{j=1}^n |f_j|^2 < 1$. Then the condition

$$|g| \leq |f|^2 H(|f|^2)$$

implies the existence of solutions $g_1, \ldots, g_n \in H^\infty$ of the equation

$$g = f_1 g_1 + \cdots + f_n g_n.$$

For example, if we take $k(x) = |\log x|^{-2} (\log |\log x|)^{-3/2}$, we see that the problem stated before has an affirmative answer for the function

$$h(s) = s^2 (-\log s)^{-3/2} (\log(-\log s))^{-1},$$

and this clearly improves the previous results.

Treil's example shows that, in general, for $g \in H^\infty$ the condition $|g| \leq |f|^2$ does not imply that the function g belongs to the ideal I generated by the functions $f_j \in H^\infty$, $1 \leq j \leq N$, but under some extra assumptions on the structure of the ideal, things can go much better. To explain this we make first some definitions.

A sequence $\{z_n\}$ in $\mathbb{D}$ is said to be an *interpolating sequence* if for every bounded sequence (w_n) of complex numbers there exists $f \in H^\infty$ with $f(z_n) = w_n$ for every $n \in \mathbb{N}$. A Blaschke product

$$B(z) = \prod_{j=1}^\infty \frac{\bar{z}_j}{|z_j|} \cdot \frac{z_j - z}{1 - \bar{z}_j z}$$

whose zero sequence is an interpolating sequence is called an *interpolating Blaschke product*.

By Carleson's interpolating theorem $\{z_n\}$ is interpolating if and only if

$$\inf_{n \in \mathbb{N}} \prod_{j \neq n} \rho(z_j, z_n) \geq \delta > 0,$$

where $\rho(z, w) = \left| \dfrac{z - w}{1 - \bar{z}w} \right|$ denotes the pseudohyperbolic distance in $\mathbb{D}$. The extension of ρ to the maximal ideal space of H^∞ is defined by

$$\rho(x, y) = \sup\{|x(f)| : f \in H^\infty, \|f\|_\infty \leq 1, y(f) = 0\}.$$

If $f \in H^\infty$, then $Z(f) = \{m \in M_{H^\infty} : m(f) = 0\}$ denotes its zero set; $Z_{\mathbb{D}}(f) = Z(f) \cap \mathbb{D}$. If I is an ideal, then

$$Z(I) = \bigcap_{f \in I} Z(f)$$

is the hull or zero set of the ideal I.

For every $m \in M_{H^\infty}$ and $f \in H^\infty$ with $m(f) = 0$ the order of the zero m of f is defined by

$$\mathrm{ord}(f, m) = \sup\{n \in \mathbb{N} : f = f_1 \ldots f_n,\; f_j \in H^\infty,\; m(f_j) = 0\,(1 \le j \le n)\}.$$

If $m(f) \ne 0$, then $\mathrm{ord}(f, m) = 0$.

Using the fact that for every $m \in M_{H^\infty}$ there exists an analytic map L_m of $\mathbb{D}$ onto the Gleason part $P(m) = \{x \in M_{H^\infty} : \rho(m, x) < 1\}$ of m with $L_m(0) = m$, we see that $\mathrm{ord}(f, m)$ is the usual multiplicity of the zero of the analytic function $f \circ L_m$ at the origin.

If I is an ideal of H^∞ and $m \in Z(I)$, then we can define the order of the ideal I at the zero m as

$$\mathrm{ord}(I, m) = \inf_{f \in I} \mathrm{ord}(f, m).$$

A well known theorem due to Tolokonnikov (see [**To**]) says that the condition $\mathrm{ord}(I, m) \le k$ for any $m \in Z(I)$ is equivalent to the fact that the ideal I contains a product of k interpolating Blaschke products. Hence the condition $\mathrm{ord}(I, m) < \infty$ for each $m \in Z(I)$ is equivalent to saying that the ideal I contains a finite product of interpolating Blaschke products.

The following result was proved in [**GMN**].

THEOREM B. *Let $f_1, \ldots, f_N \in H^\infty$ and let $I = I(f_1, \ldots, f_N)$ the ideal generated by the functions f_j. If $\mathrm{ord}(I, m) < \infty$ for each $m \in Z(I)$ then for $g \in H^\infty$ the condition*

$$|g(z)| \le C|f(z)|^2 \quad z \in \mathbb{D}$$

implies that $g \in I$.

This result can be improved in the following way.

THEOREM 1. *Let $g, f_1, \ldots, f_N \in H^\infty$, and let $I = I(f_1, \ldots, f_N)$ be the ideal generated by $f_1, \ldots, f_N$. If $\mathrm{ord}(I, m) \le k$, for all $m \in Z(I)$, then for $0 \le \alpha < 1/k$, the condition*

$$|g(z)| \le C|f(z)|^{2-\alpha} \quad \forall z \in D$$

implies $g \in I$.

One is going to ask if the same result remains true for $\alpha = 1/k$ since it is well known that for $k = 1$ it is true (if the ideal I contains an interpolating Blaschke product then $|g| \le C|f|$ implies that $g \in I$, see for example [**GMN**]). The proof of Theorem 1 given here does not allow the case $\alpha = 1/k$, so it is an interesting question to ask if for $k > 1$, the condition $|g| \le |f|^{2-1/k}$ implies that the function g belongs to the ideal I. It seems that new ideas must be developed.

Finally we want to say that something similar happens when one looks for conditions to be in the closure of the ideal I generated by the functions f_j. If $\mathrm{ord}(I, m) < \infty$ for all $m \in Z(I)$, then the condition $g \le C|f|$ implies that g is in the closure of the ideal I (see [**NP**] or [**GMI**]), while it is known that, in general, this result is not true (see [**Bg**]).

1. Preliminary results

The methods of the proof of Theorem 1 are the standard ones developed by Wolff to prove the corona theorem. First let us recall that a positive Borel measure μ on the unit disc $\mathbb{D}$ is called a *Carleson measure* if there exists a constant C so that

$$\int_{\mathbb{D}} |f|^2 \, d\mu \leq C \|f\|_2^2$$

for every function f in the Hardy space H^2 of analytic functions g in $\mathbb{D}$ with $\|g\|_2^2 = \sup_{0<r<1} \int_0^{2\pi} |g(re^{i\theta})|^2 \frac{d\theta}{2\pi} < \infty$.

It is well known that Carleson measures are those positive Borel measures μ for which there exists a constant K such that

$$\mu(Q) \leq K \, l(Q)$$

for every Carleson cube Q defined by

$$Q = \{z \in \mathbb{D} : 1 - |z| < l(Q), \, |\arg z - \theta_0| < l(Q)\}.$$

Let $N(\mu) = \sup\{\frac{\mu(Q)}{l(Q)} : Q$ Carleson cube$\}$ denote the *Carleson norm* of the measure μ.

We will use two results, due to Carleson and Wolff, on the existence of bounded solutions of the $\bar{\partial}$ equation (see, for example, [**Ga**] p. 321-322).

PROPOSITION 2. *(a) Let G be a bounded continuous function on $\mathbb{D}$ and assume that $d\mu = |G| \, dx \, dy$ is a Carleson measure. Then there exists $u \in C(\overline{\mathbb{D}}) \cap C^1(\mathbb{D})$ such that $\bar{\partial}u = G$ with $\|u\|_{L^\infty(\partial \mathbb{D})} \leq C$, where C depends only on the Carleson norm of μ.*

(b) Let $G(z)$ be bounded and C^1 on the disc $\mathbb{D}$ and assume that the measures $d\sigma_1(z) = |G(z)|^2 (1 - |z|^2) \, dx \, dy$ and $d\sigma_2(z) = |\partial G(z)| (1 - |z|^2) \, dx \, dy$ are Carleson measures. Then there exists $u \in C(\overline{\mathbb{D}}) \cap C^1(\mathbb{D})$ such that $\bar{\partial}u = G$ and

$$\|u\|_{L^\infty(\partial \mathbb{D})} \leq C_1 N(\sigma_1)^{1/2} + C_2 N(\sigma_2).$$

The following result can be found for example in [**ABN**] or [**Pa1**].

LEMMA 3. *Let $g \in H^\infty$. Then for each $\varepsilon > 0$ the measure*

$$\frac{|g'(z)|^2}{|g(z)|^{2-\varepsilon}} (1 - |z|^2) \, dx \, dy$$

is a Carleson measure.

Let us also recall that a Blaschke product B is said to be of *Carleson-Newman type* (a *CN-Blaschke product*, for short) if B is a finite product of interpolating Blaschke products.

A well known theorem (for example, see [**McK**]) tells us that a Blaschke product with zero sequence (z_n) is a CN-Blaschke product if and only if the measure $\mu = \sum_{n=1}^\infty (1 - |z_n|^2)$ is a Carleson measure. Any such sequence will be called a *CN-sequence*.

By Carleson's interpolation theorem, a Blaschke product B is an interpolating Blaschke product if and only if $\inf_{z \in Z_{\mathbb{D}}(B)}(1 - |z|^2)|B'(z)| > 0$. Hence, given a CN-Blaschke product $B = \prod_{j=1}^{k} B_j$ where B_j are interpolating Blaschke products, we can define

$$\delta_B = \min_{1 \le j \le k} \inf_{z \in Z_{\mathbb{D}}(B_j)} (1 - |z|^2)|B_j'(z)| > 0.$$

Recall that the pseudohyperbolic distance is defined by $\rho(z, w) = \left|\frac{z-w}{1-\bar{z}w}\right|$. Let $D(z, r) = \{w \in \mathbb{D} : \rho(z, w) < r\}$ be the pseudohyperbolic disc of center z and radius $r > 0$.

We shall also need the following results.

LEMMA 4. *(see* [**Ga**], *p. 310) Let* $\{z_n\}$ *be an interpolating sequence in* $\mathbb{D}$ *with* $\inf_n \prod_{j \ne n} \rho(z_j, z_n) \ge \delta > 0$. *Let* $0 < \eta < (1 - \sqrt{1 - \delta^2})/\delta$, *and let* $w_n \in \mathbb{D}$ *such that* $\rho(z_n, w_n) < \eta$ *for every* $n \in \mathbb{N}$. *Then* $\{w_n\}$ *is an interpolating sequence with*

$$\inf_n \prod_{j \ne n} \rho(w_j, w_n) \ge \frac{\delta - \frac{2\eta}{1+\eta^2}}{1 - \delta\frac{2\eta}{1+\eta^2}}.$$

LEMMA 5. *Let* B *be an interpolating Blaschke product with zero sequence* $\{z_n\}$. *Let* $0 < \eta < (1 - \sqrt{1 - \delta_B^2})/\delta_B$. *Then there is a constant* $c = c(\eta, \delta_B)$ *such that*

$$(1 - |z|^2)|B'(z)| \ge c > 0$$

for each z *in the pseudohyperbolic disc* $D(z_n, \eta)$.

PROOF. Just apply Lemma 4 to the sequence $\{w_j\}$ defined by

$$w_j = \begin{cases} z_j & if \quad j \ne n \\ z & if \quad j = n \end{cases},$$

and note that $(1 - |z_j|^2)|B'(z_j)| = \prod_{k \ne j} \rho(z_k, z_j)$. $\square$

LEMMA 6. *Let* B *be an interpolating Blaschke product with zero sequence* $\{z_n\}$. *Let* $0 < \eta < (1 - \sqrt{1 - \delta_B^2})/\delta_B$ *and* $0 \le \alpha < 1$. *Then*

$$\int_{D(z_n,\eta)} \frac{|b_k'(z)|}{|b_k(z)|^{1+\alpha}} \, dx \, dy \le C \, (1 - |z_n|^2).$$

PROOF. Note that $D(z_n, \eta) \subset cQ(z_n)$ for some constant c, where

$$Q(z_n) = \{w \in \mathbb{D} : 1 - |w| \le 1 - |z_n|, |\arg w - \arg z_n| < 1 - |z_n|\}.$$

By Lemma 5 we have

$$I = \int_{D(z_n,\eta)} \frac{|b_k'(z)|}{|b_k(z)|^{1+\alpha}} \, dx \, dy \le C \int_{D(z_n,\eta)} \frac{|b_k'(z)|^2}{|b_k(z)|^{1+\alpha}} \, (1 - |z|^2) \, dx \, dy$$

$$\le C \int_{cQ(z_n)} \frac{|b_k'(z)|^2}{|b_k(z)|^{1+\alpha}} \, (1 - |z|^2) \, dx \, dy$$

$$\le C \, (1 - |z_n|^2)$$

by Lemma 3. $\square$

LEMMA 7. *Let (X,μ) be a measure space. For $j = 1,\ldots,n$, consider functions $f_j \in L^1(X,\mu)$. Then*

$$\Big(\int_X \prod_{j=1}^n |f_j|^{1/n}\, d\mu\Big)^n \le \prod_{j=1}^n \int_X |f_j|\, d\mu.$$

PROOF. For $n = 1$, this is an equality. For $n > 1$, Hölder's inequality gives

$$\int_X \prod_{j=1}^n |f_j|^{1/n}\, d\mu \le \Big(\int_X |f_n|\, d\mu\Big)^{1/n}\Big(\int_X \prod_{j=1}^{n-1} |f_j|^{1/(n-1)}\, d\mu\Big)^{(n-1)/n},$$

that by induction is bounded by

$$\Big(\int_X |f_n|\, d\mu\Big)^{1/n}\Big(\prod_{j=1}^{n-1} \int_X |f_j|\, d\mu\Big)^{1/n}.$$

□

PROPOSITION 8. *Let b a CN Blaschke product of order less than or equal to k, that is, b is a product of k interpolating Blaschke products. Let $\{z_n\}$ be a CN sequence of distinct points such that $Z_{\mathbb{D}}(b) \subset \{z_n\}$, and let $U = \cup_n D(z_n,\eta)$, $0 < \eta < (1 - \sqrt{1-\delta_b^2})/\delta_b$. If Ψ_U denotes the characterisic function of U, then*

$$\frac{|b'(z)|}{|b(z)|^{1+\alpha}}\Psi_U(z)\, dx\, dy,$$

is a Carleson measure for all $\alpha < 1/k$.

PROOF. Let $b_1,\ldots,b_k$ be interpolating Blaschke products such that $b = \prod_{l=1}^k b_l$.

The additive property of the logarithmic derivative yields

$$b'(z)/b(z) = \sum_{j=1}^k b_j'(z)/b_j(z)$$

So, we only need to check that for fixed j, the measure

$$\frac{|b_j'(z)|}{|b_j(z)||b(z)|^\alpha}\Psi_U(z)\, dx\, dy$$

is a Carleson measure.

From now on, $c_i = c_i(\eta,\delta_b,\alpha)$ will denote different constants depending only on η,δ_b and α. Let

$$Q = \{z \in \mathbb{D} : 1 - |z| < l(Q),\ |\arg z - \theta_0| < l(Q)\}.$$

We may assume without loss of generality that $b_j = b_k$. Then $|b| = |b_k|\prod_{l=1}^{k-1}|b_l|$.

Let $A = \{z_n : D(z_n,\eta) \cap Q \neq \emptyset\}$ and let $A_j = A \cap Z_{\mathbb{D}}(b_j)$, $1 \le j \le k$. From the definition of U we have

$$\int_Q \Big|\frac{b_k'(z)}{b_k(z)}\Big|\frac{1}{|b(z)|^\alpha}\Psi_U(z)\, dx\, dy \le \sum_{z_n \in A}\int_{D(z_n,\eta)}\Big|\frac{b_k'(z)}{b_k(z)}\Big|\prod_{l=1}^k\frac{1}{|b_l(z)|^\alpha}\, dx\, dy.$$

If $z_n \in A \setminus Z_{\mathbb{D}}(b)$, then there is a constant $c_1 = c_1(\eta)$ such that $|b(z)| \geq c_1$ for $z \in D(z_n, \eta)$. Hence

$$\int_{D(z_n,\eta)} \frac{|b'_k(z)|}{|b_k(z)||b(z)|^\alpha} \, dx \, dy \leq c_1^{-\alpha} \int_{D(z_n,\eta)} \frac{|b'_k(z)|}{|b_k(z)|} \, dx \, dy$$

$$\leq c_2 \, (1 - |z_n|^2),$$

where the last inequality follows from Lemma 6.

For $z_n \in A_k$, let M be the number of indices l such that $Z_{\mathbb{D}}(b_l) \cap D(z_n, \eta) \neq \emptyset$. Clearly $M \leq k$. If $M = 1$, then $|b(z)| \geq c_3 |b_k(z)|$ for each z in $D(z_n, \eta)$. Then using Lemma 6 once again we have

$$\int_{D(z_n,\eta)} \frac{|b'_k(z)|}{|b_k(z)|} \frac{1}{|b(z)|^\alpha} \, dx \, dy \leq c_3^{-\alpha} \int_{D(z_n,\eta)} \frac{|b'_k(z)|}{|b_k(z)|^{1+\alpha}} \, dx \, dy$$

$$\leq c_4 (1 - |z_n|^2).$$

So we also may assume that $M > 1$. Reordering if necessary, we may assume that $Z_{\mathbb{D}}(b_l) \cap D(z_n, \eta) \neq \emptyset$ for $1 \leq l \leq M - 1$. Since $\eta < (1 - \sqrt{1 - \delta_b^2})/\delta_b$, by Lemma 5 there is a constant $c_5 = c_5(\eta, \delta_b) > 0$ such that for $z \in D(z_n, \eta)$ we have

$$(1 - |z|^2)|b'_l(z)| \geq c_5 \quad 1 \leq l \leq M - 1$$

and

$$(1 - |z|^2)|b'_k(z)| \geq c_5$$

Let $D_n = D(z_n, \eta)$. Hence for $z_n \in A_k$ we have

$$I = \int_{D_n} \frac{|b'_k|}{|b_k|^{1+\alpha}} \prod_{l=1}^{k-1} \frac{1}{|b_l|^\alpha} \, dx \, dy$$

$$\leq c_6 \int_{D_n} \frac{|b'_k|^{1+1/M}}{|b_k|^{1+\alpha}} \prod_{l=1}^{M-1} \frac{|b'_l|^{1/M}}{|b_l|^\alpha} (1 - |z|^2) \, dx \, dy$$

$$\leq c_6 \left(\int_{D_n} \frac{|b'_k|^2}{|b_k|^\beta} (1 - |z|^2) \, dx \, dy \right)^{\frac{M+1}{2M}} \left(\int_{D_n} \prod_{l=1}^{M-1} h_l^{\frac{1}{M-1}} (1 - |z|^2) \, dx \, dy \right)^{\frac{M-1}{2M}}.$$

by Hölder's inequality, where $\beta = 2M(1 + \alpha)/(M + 1)$ and $h_l = |b'_l|^2/|b_l|^{2M\alpha}$. We note that $\beta < 2$ and $2M\alpha < 2$ since $\alpha < 1/k$. Hence applying Lemma 7 to the functions h_l, using that $D_n \subset cQ(z_n)$ for some constant c and finally using Lemma 3 we obtain

$$I \leq c_6 \left(\int_{cQ(z_n)} \frac{|b'_k|^2}{|b_k|^\beta} (1 - |z|^2) \, dx \, dy \right)^{\frac{M+1}{2M}} \left(\prod_{l=1}^{M-1} \int_{cQ(z_n)} \frac{|b'_l|^2}{|b_l|^{2M\alpha}} (1 - |z|^2) \, dx \, dy \right)^{\frac{1}{2M}}$$

$$\leq c_7 (1 - |z_n|^2)^{(M+1)/2M} \cdot (1 - |z|^2)^{(M-1)/2M}$$

$$= c_7 (1 - |z_n|^2).$$

In a similar way it can be proved that if $z_n \in \cup_{l=1}^{k-1} A_k$ then

$$\int_{D(z_n,\eta)} \frac{|b_k'(z)|}{|b_k(z)||b(z)|^\alpha} \, dx \, dy \le c_8(1 - |z_n|^2).$$

Since (z_n) is a CN-sequence we have

$$\sum_{z_n \in A} (1 - |z_n|^2) \le c_9 \, l(Q).$$

and this finishes the proof. $\qquad\qquad\qquad\qquad\qquad\qquad\qquad\qquad\qquad\qquad\qquad$ □

2. Proof of Theorem 1

Our assumption that $\mathrm{ord}(I, m) \le k$ for all $m \in Z(I)$ is equivalent to the fact that the ideal I contains a Carleson-Newman Blaschke product $C = \prod_{j=1}^k c_j$, where c_j are interpolating Blaschke products. Then, using the same argument given in [**GMN**] we can assume that the generators f_j are Carleson-Newman Blaschke products

$$f_j = \prod_{i=1}^k b_{ij}, \quad (j = 1, \ldots, N),$$

where b_{ij} are interpolating Blaschke products. As in [**Ga**], Sec. 8, we may assume without loss of generality that $\bigcap_{j=1}^N Z_{\mathbb{D}}(f_j) = \emptyset$. As usual, using a normal families argument we can assume that $f_1, \ldots, f_N$ are analytic in a neighbourhood of the closed unit disc. Let $\varphi_j = \bar{f}_j |f|^{-2}$ and $G_{ij} = \varphi_i \bar{\partial}\varphi_j$, $(i, j = 1, \ldots, N)$. Assume that we can solve the $\bar{\partial}$-equations

$$\bar{\partial} h_{ij} = gG_{ij} \quad (i, j = 1, \ldots, N),$$

with $\|h_{ij}\|_{L^\infty(\partial\mathbb{D})} \le M$ for some finite positive constant M. Then the functions

$$g_j = g\varphi_j + \sum_{i=1}^N (h_{ji} - h_{ij}) f_i,$$

are bounded, $\sum_{i=1}^N f_j g_j = g$ and $\bar{\partial} g_j = 0$, and solves the problem.

Thus we have only to show that these $\bar{\partial}$-equations admit bounded solutions. Let $\{z_n\} = \bigcup_{j=1}^N Z_{\mathbb{D}}(f_j)$. As $\{z_n\}$ is a finite union of CN-sequences, $\{z_n\}$ is also a CN-sequence. Let $0 < \varepsilon < (1 - \sqrt{1 - \delta^2})/3\delta$ with $\delta = \min_{1 \le j \le N} \delta_{f_j} > 0$. Take a function $a \in C^\infty(\mathbb{D})$, $0 \le a \le 1$ such that $a \equiv 1$ on $\{z : \rho(z, \{z_n\}) < \varepsilon\}$, $a \equiv 0$ on $\{z : \rho(z, \{z_n\}) \ge 2\varepsilon\}$, and with $(1 - |z|)|\nabla a(z)|$ bounded.

Let $U = \bigcup_{n=1}^\infty D(z_n, 2\varepsilon)$, where $D(z_n, \varepsilon)$ denotes the pseudohyperbolic disc with center z and radius ε. Fix i and j and put $G_{ij} = G_{ij}a + G_{ij}(1 - a)$.

To solve the equation $\bar{\partial} h = gG_{ij}$, it is sufficient to show that the equations $\bar{\partial} h = gG_{ij}a$ and $\bar{\partial} h = gG_{ij}(1 - a)$ admit bounded solutions. In order to solve this, we will make use of Proposition 2 a) in the first case, and Proposition 2 b) in the second case.

To solve $\bar{\partial} h = g G_{ij} a$, we need to show that $|g G_{ij} a| \, dx \, dy$ is a Carleson measure. An elementary computation gives

$$(2) \qquad\qquad G_{ij} = \frac{\bar{f}_i}{|f|^6} \sum_{k \neq j} f_k \, (\overline{f_k f_j' - f_j f_k'}),$$

and

$$(3) \qquad\qquad \sum_{j,k=1}^{n} |f_k' f_j - f_k f_j'|^2 = |f|^2 |f'|^2 - |\partial(|f|^2)|^2,$$

where $|f'|^2 = \sum_{l=1}^{N} |f_l'|^2$. By (2) and (3), it follows that

$$(4) \qquad\qquad |G_{ij}| \leq 2 \frac{(|f|^2 |f'|^2 - |\partial(|f|^2)|^2)^{1/2}}{|f|^4} \leq 2 \frac{|f'|}{|f|^3}.$$

Using (4) and our condition on the size of $|g|$ we see that

$$|g G_{ij} a|^2 \leq c_0 a^2 \frac{\sum_{l=1}^{N} |f_l'|^2}{\left(\sum_{l=1}^{N} |f_l|^2 \right)^{1+\alpha}}.$$

Hence

$$|g G_{ij} a| \leq c_1 a \sum_{l=1}^{N} \frac{|f_l'|}{|f_l|^{1+\alpha}} \leq c_1 \sum_{l=1}^{N} \frac{|f_l'|}{|f_l|^{1+\alpha}} \Psi_U,$$

since $0 \leq a \leq 1$ and the function a vanishes outside U. Here Ψ_U denotes the characteristic function of U. By Proposition 8 we have that $|g G_{ij} a| \, dx \, dy$ is a Carleson measure, because $\alpha < 1/k$.

To solve the equation $\bar{\partial} h = g G_{ij} (1 - a)$, we note that, since $\{z_n\}$ are the zeros of the functions f_j, we have $\prod_{j=1}^{N} |f_j(z)| \geq c(\varepsilon)$ if $\rho(z, \{z_n\}) \geq \varepsilon$, and then

$$|g(z) G_{ij}(z) (1 - a(z))|^2 \, (1 - |z|) \, dx \, dy \leq \frac{c_0}{c(\varepsilon)^{2+\alpha}} \sum_{l=1}^{N} |f_l'(z)|^2 \, (1 - |z|) \, dx \, dy,$$

that, by Lemma 3 is a Carleson measure. It remains to show that

$$|\partial(g G_{ij} (1 - a))(z)| \, (1 - |z|) \, dx \, dy$$

is a Carleson measure. We note that

$$\partial(g G_{ij} (1 - a)) = g' G_{ij} (1 - a) + g(1 - a) \partial G_{ij} + g G_{ij} \partial(1 - a).$$

Since $\prod_{j=1}^{N} |f_j(z)| \geq c(\varepsilon)$ if $\rho(z, \{z_n\}) \geq \varepsilon$, we have

$$|g' G_{ij} (1 - a)| \leq |g'|^2 + |G_{ij} (1 - a)|^2 \leq |g'|^2 + \frac{c_0}{c(\varepsilon)^6} \sum_{l=1}^{N} |f_l'|^2.$$

Hence, by Lemma 3, the measure $|(g' G_{ij} (1 - a))(z)| (1 - |z|) \, dx \, dy$ is a Carleson measure. Now, an elementary computation shows that

$$|\partial G_{ij}| \leq 2 |G_{ij}| \frac{|\partial(|f|^2)|}{|f|^2} + \frac{|f|^2 |f'|^2 - |\partial(|f|^2)|^2}{|f|^6}.$$

Then, using (4), the fact that $|\partial(|f|^2)|^2 \leq |f|^2|f'|^2$ and our condition on the size of $|g|$ we have

$$|(g(1-a)\partial G_{ij}(z)|\,(1-|z|)\,dx\,dy \leq \frac{c_2}{c(\varepsilon)^{4+\alpha}} \sum_{l=1}^{N} |f_l'(z)|^2\,(1-|z|)\,dx\,dy,$$

that, by Lemma 3 is a Carleson measure.

It remains to show that $|(\partial(1-a)gG_{ij})(z)|\,(1-|z|)\,dx\,dy$ is a Carleson measure. To do this, first note that if z is in the support of $\partial(1-a)$, then $\rho(z,\{z_n\}) \geq \varepsilon$ and z is in the closure of U. Hence $z \in V = \bigcup_{n=1}^{\infty} D(z_n, 3\varepsilon)$. Then, using the fact that $(1-|z|)|\nabla a(z)|$ is bounded, we have

$$|(\partial(1-a)gG_{ij})(z)|\,(1-|z|)\,dx\,dy \leq c_3 \sum_{l=1}^{N} \frac{|f_l'(z)|}{|f_l(z)|^{1+\alpha}} \Psi_V(z),$$

that, by Proposition 8 is a Carleson measure. This completes the proof.

References

[ABN] E. Amar, J. Bruna and A. Nicolau, *On H^p-solutions of the Bezout equation*, Pacific J. Math. **171**: 2 (1995), 297-307.

[Bg] J. Bourgain, *On finitely generated closed ideals in $H^\infty(\mathbb{D})$*, Ann. Inst. Fourier (Grenoble) **35** (1985), 163-174.

[Ce1] U. Cegrell, *A generalization of the Corona Theorem in the unit disc*, Math. Z. **203** (1990), 255-261.

[Ce2] U. Cegrell, *Generalisations of the corona Theorem in the unit disc*, Proc. Roy. Irish Acad. **94** (1994), 25-30.

[Ga] J.B. Garnett, *Bounded Analytic Functions*, Academic Press, New York, 1981.

[GMI] P. Gorkin, K. Izuchi and R. Mortini, *Higher order hulls in H^∞ II*, J. Funct. Anal. **1777** (2000), 107-129.

[GMN] P. Gorkin, R. Mortini and A. Nicolau, *The generalized corona theorem*, Math. Ann. **301** (1995), 135-154.

[Li] K.C. Lin, *On the constants in the corona theorem and ideals of H^∞*, Houston J. Math. **19** (1993), 97-106.

[McK] P.J. McKenna, *Discrete Carleson measures and some interpolating problems*, Mich. Math. J. **24** (1977), 311-319.

[NP] A. Nicolau and J. Pau, *Closures of finitely generated ideals on Hardy spaces*, Ark. Mat. **39** (2001), 137-149.

[Pa1] J. Pau, *On a generalized corona problem on the unit disc*, Proc. Amer. Math. Soc. **133** (2005), 167-174.

[Ra] K.V. Rao, *On a generalized corona problem*, J. Anal. Math. **18** (1967), 277-278.

[To] V. Tolokonnikov, *Blaschke products satisfying the Carleson-Newman condition and ideals of the algebra H^∞*, Sov. J. Math. **42** (1988), 1603-1610.

[Tr] S. Treil, *Estimates in the Corona theorem and ideals of H^∞: a problem of T. Wolff*, J. Analyse Math. **87** (2002), 481-495.

DEPARTAMENT DE MATEMÀTICA APLICADA I ANALISI, UNIVERSITAT DE BARCELONA, GRAN VIA DE LES CORTS CATALANES, 585, 08007 BARCELONA, SPAIN

E-mail address: jordi.pau@ub.edu

Contemporary Mathematics
Volume **393**, 2006

The classical Dirichlet space

William T. Ross

Dedicated to J. A. Cima on the occasion of his 70th birthday and to over 45 years of dedicated service to mathematics.

1. Introduction

In this survey paper, we will present a selection of results concerning the class of analytic functions f on the open unit disk $\mathbb{D} := \{z \in \mathbb{C} : |z| < 1\}$ which have finite Dirichlet integral

$$D(f) := \frac{1}{\pi} \int_{\mathbb{D}} |f'|^2 \, dx \, dy.$$

In particular, we will cover the basic structure of these functions - their boundary values and their zeros - along with two important operators that act on this space of functions - the forward and backward shifts. This survey is by no means complete. For example, we will not cover the Toeplitz or Hankel operators on these functions, nor will we cover the important topic of interpolation. These topics are surveyed in a nice paper of Wu [**64**]. In order to make this survey more manageable, we will also restrict ourselves to this space of functions with finite Dirichlet integral and will not try to cover the many related Dirichlet-type spaces. We refer the reader to the papers [**11, 41, 47, 54**] for more on this.

2. Basic definitions

An analytic function f on the open unit disk $\mathbb{D}$ belongs to the classical *Dirichlet space* $\mathcal{D}$ if it has finite *Dirichlet integral*

$$D(f) := \frac{1}{\pi} \int_{\mathbb{D}} |f'|^2 \, dA,$$

where dA is two dimensional Lebesgue area measure. Thinking of f as a mapping from $\mathbb{D}$ to some region $f(\mathbb{D})$, one computes the Jacobian determinant J_f to be $|f'|^2$ and so the two dimensional area of $f(\mathbb{D})$, counting multiplicities, is

$$\int_{f(\mathbb{D})} 1 \, dA = \int_{\mathbb{D}} |J_f| \, dA = \int_{\mathbb{D}} |f'|^2 \, dA.$$

2000 *Mathematics Subject Classification.* Primary 31C25; Secondary 30C15.

Thus a function has finite Dirichlet integral exactly when its image has finite area (counting multiplicities).

Another interesting geometric observation to make is that for each $\zeta \in \mathbb{T}$, the quantity

(2.1)
$$L(f, \zeta) := \int_0^1 |f'(r\zeta)|\, dr$$

is the length of the curve $\{f(r\zeta) : 0 < r < 1\}$, which is the image of the ray $\{r\zeta : 0 < r < 1\}$ under the mapping f. If $D(f) < \infty$, we can apply the Cauchy-Schwartz inequality to show that

$$\int_0^{2\pi} L(f, e^{i\theta})\, d\theta \leqslant cD(f) < \infty$$

and so $L(f, e^{i\theta}) < \infty$ for almost every θ. A theorem of Beurling (see Theorem 5.5 below) says that $L(f, e^{i\theta}) < \infty$ for quasi-every θ in the sense of logarithmic capacity.

Writing

$$f(z) = \sum_{n=0}^{\infty} a_n z^n$$

as a power series and setting $z = re^{i\theta}$, one can integrate in polar coordinates to see that

$$D(f) = \sum_{n=0}^{\infty} n|a_n|^2.$$

Closely related to the Dirichlet space is the classical *Hardy space* H^2 of analytic functions f on $\mathbb{D}$ for which

$$\|f\|_{H^2}^2 := \sup_{0 < r < 1} \int_{\mathbb{T}} |f(r\zeta)|^2\, dm(\zeta) < \infty.$$

Here dm denotes Lebesgue measure on the unit circle $\mathbb{T} = \partial\mathbb{D}$ normalized so that $m(\mathbb{T}) = 1$. Observe that

$$\int_{\mathbb{T}} |f(r\zeta)|^2\, dm(\zeta) = \sum_{n=0}^{\infty} r^{2n}|a_n|^2$$

and so

$$\|f\|_{H^2}^2 = \sum_{n=0}^{\infty} |a_n|^2.$$

Thus $\mathcal{D} \subset H^2$ and so, via Fatou's theorem on radial limits [22], functions in $\mathcal{D}$ have radial boundary values almost everywhere on $\mathbb{T}$, that is to say,

(2.2)
$$f(\zeta) := \lim_{r \to 1^-} f(r\zeta)$$

exists and is finite for almost every $\zeta \in \mathbb{T}$. It will turn out (see Theorem 5.10 and Theorem 5.12) that functions in $\mathcal{D}$ have much stronger regularity near $\mathbb{T}$ than H^2 functions.

The quantity $D(f)$ is not a norm, since $D(c) = 0$ for any constant function c. However, one can endow $\mathcal{D}$ with the norm $\|f\|$, where

(2.3)
$$\|f\|^2 := \|f\|_{H^2}^2 + D(f) = \sum_{n=0}^{\infty} (1+n)|a_n|^2.$$

An easy estimate yields

$$|f(0)|^2 + D(f) \leqslant \|f\|^2 \leqslant 2\left(|f(0)|^2 + D(f)\right)$$

and several authors use the quantity

$$\sqrt{|f(0)|^2 + D(f)}$$

to define an equivalent norm on $\mathcal{D}$ which is sometimes more convenient to use. With the norm $\|\cdot\|$ in eq.(2.3) above, one defines an inner product

$$\langle f, g \rangle := \int_{\mathbb{T}} f\bar{g}\, dm + \frac{1}{\pi} \int_{\mathbb{D}} f'\overline{g'}\, dA.$$

Simple computations show that the quantities

$$\frac{1}{\pi} \int_{\mathbb{D}} (zf)'\overline{(zg)'}\, dA$$

and

$$\sum_{n=0}^{\infty} (n+1) a_n \overline{b_n},$$

where the a_n are the Taylor coefficients of f and the b_n are those for g, are both equal to $\langle f, g \rangle$.

Another power series computation shows that if we define kernels $k_z(w)$ by

$$k_z(w) := \frac{1}{\bar{w}z} \log \frac{1}{1 - \bar{w}z}, \quad z, w \in \mathbb{D},$$

then

$$f(z) = \langle f, k_z \rangle.$$

Moreover, by the Cauchy-Schwartz inequality,

$$|f(z)| \leqslant \|f\|\|k_z\| = \|f\|\langle k_z, k_z \rangle^{1/2} = \|f\| k_z(z)^{1/2}$$

and so f satisfies the pointwise estimate

$$|f(z)| \leqslant c\|f\| \left(\log \frac{1}{1 - |z|^2}\right)^{1/2}.$$

In particular, this pointwise estimate shows that if $f_n \to f$ in norm of $\mathcal{D}$, then $f_n \to f$ uniformly on compact subsets of $\mathbb{D}$. Hence $\mathcal{D}$ is a reproducing kernel Hilbert space of analytic functions on $\mathbb{D}$. Finally, it is clear from the definition of the norm that if

$$f(z) = \sum_{n=0}^{\infty} a_n z^n,$$

belongs to $\mathcal{D}$, then

$$\left\|\sum_{n=0}^{N} a_n z^n - f\right\|^2 = \sum_{n=N+1}^{\infty} (n+1)|a_n|^2$$

which goes to zero as $N \to \infty$. Thus the polynomials form a dense subset of $\mathcal{D}$.

3. The Douglas and Carleson formulas

In his investigations of the Plateau problem, Jesse Douglas [20] proved the following formula for the Dirichlet integral

$$D(f) = \int_{\mathbb{T}} \int_{\mathbb{T}} \left| \frac{f(\zeta) - f(\xi)}{\zeta - \xi} \right|^2 dm(\zeta) \, dm(\xi),$$

where we understand that the function $\zeta \mapsto f(\zeta)$ on $\mathbb{T}$ is the almost everywhere defined boundary function via radial boundary values eq.(2.2). The inner integral

$$D_\zeta(f) := \int_{\mathbb{T}} \left| \frac{f(\zeta) - f(\xi)}{\zeta - \xi} \right|^2 dm(\xi)$$

is called the *local Dirichlet integral* and one can use it to make some interesting observations [47]. First notice how

$$\int_{\mathbb{T}} D_\zeta(f) \, dm(\zeta) = D(f)$$

and so for a Dirichlet function, the local Dirichlet integral is finite almost everywhere.

PROPOSITION 3.1. *Suppose $\zeta \in \mathbb{T}$ and $f \in H^2$ with $D_\zeta(f) < \infty$. Then the oricyclic limit of f exists at ζ. Thus any $f \in \mathcal{D}$ has an oricyclic limit almost everywhere.*

Here the oricyclic approach regions with contact point ζ are

$$A_{2,c}(\zeta) := \{z \in \mathbb{D} : |\zeta - z| \leqslant c(1 - |z|)^{1/2}\}, \quad c > 0.$$

For example, $A_{2,2}(\zeta)$ contains the disk with center $\zeta/2$ and radius $1/2$. We say that f has *oricyclic limit* L at ζ if

$$\lim_{z \to \zeta, z \in A_{2,c}(\zeta)} f(z) = L$$

for every $c > 0$. Compare this to the non-tangential approach regions with contact point ζ

$$A_{1,c}(\zeta) := \{z \in \mathbb{D} : |\zeta - z| \leqslant c(1 - |z|)\}, \quad c > 0,$$

which are triangle shaped regions with vertex at ζ. We will discuss much stronger results in Section 5. We also mention that whenever $D_\zeta(f) < \infty$, the Fourier series of f converges to $f(\zeta)$.

A function $f \in H^2$ has the standard Nevanlinna factorization [22]

(3.2) $f = b s_\mu F,$

where

$$b(z) = z^m \prod_{n=1}^{\infty} \frac{\overline{a_n}}{|a_n|} \frac{a_n - z}{1 - \overline{a_n} z}$$

is the Blaschke factor with zeros at $z = 0$ and $(a_n)_{n \geqslant 1} \subset \mathbb{D} \setminus \{0\}$ (repeated according to multiplicity),

$$s_\mu(z) = \exp\left\{ -\int_{\mathbb{T}} \frac{\zeta + z}{\zeta - z} d\mu(\zeta) \right\},$$

is the singular inner factor with positive singular measure μ on $\mathbb{T}$ (i.e., $\mu \perp m$), and

$$F(z) = \exp\left\{ \int_{\mathbb{T}} \frac{\zeta + z}{\zeta - z} u(\zeta) \, dm(\zeta) \right\}, \quad u(\zeta) = \log|f(\zeta)|,$$

is the outer factor. The two inner factors b and s_μ are bounded analytic functions on $\mathbb{D}$ with unimodular boundary values almost everywhere. The outer factor F belongs to H^2.

If $f \in \mathcal{D}$, then $f \in H^2$ and as such has a factorization $f = bs_\mu F$ as in eq.(3.2). Here, the inner factors b and s_μ do not belong to $\mathcal{D}$ unless b is a finite Blaschke product and $s_\mu \equiv 1$ (i.e., $\mu \equiv 0$). However, the outer factor does belong to $\mathcal{D}$ (see Corollary 3.5 below). This following theorem of Carleson [14], which has proven to be a quite a workhorse in the subject, computes $D(f)$ in terms of the Nevanlinna factorization.

THEOREM 3.3 (Carleson). *For $f = bs_\mu F \in \mathcal{D}$ as in eq.(3.2),*

$$
\pi D(f) = \int_{\mathbb{T}} \left(m + \sum_{n=1}^{\infty} P_{a_n}(\zeta) \right) |f(\zeta)|^2 \, dm(\zeta) + \int_{\mathbb{T}} \int_{\mathbb{T}} \frac{2}{|\zeta - \xi|^2} \, d\mu(\zeta) |f(\xi)|^2 \, dm(\xi)
$$
$$
+ \int_{\mathbb{T}} \int_{\mathbb{T}} \frac{(e^{2u(\zeta)} - e^{2u(\xi)})(u(\zeta) - u(\xi))}{|\zeta - \xi|^2} \, dm(\zeta) \, dm(\xi).
$$

In the above formula,

$$
P_a(\zeta) = \frac{1 - |a|^2}{|\zeta - a|^2}, \quad a \in \mathbb{D}, \zeta \in \mathbb{T},
$$

is the usual Poisson kernel. We also agree to the understanding that if any of the factors in eq.(3.2) are missing, then those corresponding components of the Carleson formula are zero.

Carleson's formula has some nice consequences. For example, two applications of Fubini's theorem with Carleson's formula yields:

COROLLARY 3.4. *An inner function bs_μ belongs to $\mathcal{D}$ if and only if $\mu = 0$ and b is a finite Blaschke product.*

One can also use standard facts about inner functions and the Carleson formula to prove the division property for $\mathcal{D}$ (often called the F-property for $\mathcal{D}$).

COROLLARY 3.5. *Suppose $f \in \mathcal{D}$ and ϑ is an inner function that divides the inner factor of f, equivalently $f/\vartheta \in H^2$. Then $f/\vartheta \in \mathcal{D}$. Consequently, if $f \in \mathcal{D}$, its outer factor also belongs to $\mathcal{D}$.*

For a general Banach space of analytic functions $X \subset H^2$ we say that X has the F-property if whenever ϑ is inner and $f/\vartheta \in H^2$, then $f/\vartheta \in X$. When $X = H^2$, this is automatic by the Nevanlinna factorization in eq.(3.2). For other spaces of analytic functions, this becomes more involved and sometimes is false [56].

Another interesting corollary can be obtained by re-arranging the terms in Carleson's formula.

COROLLARY 3.6. *For $f \in \mathcal{D}$,*

$$
D(f) = \{D(bF) - D(F)\} + \{D(s_\mu F) - D(F)\} + D(F),
$$

where each of the individual summands is non-negative.

In [**47**], one finds an analogous decomposition of the local Dirichlet integral, namely,

$$D_\zeta(f) = \left(m + \sum_{n=1}^{\infty} P_{a_n}(\zeta)\right)|f(\zeta)|^2 + \int_{\mathbb{T}} \frac{2}{|\zeta - \xi|^2} d\mu(\xi)|f(\zeta)|^2$$

$$+ \int_{\mathbb{T}} \frac{e^{2u(\xi)} - e^{2u(\zeta)} - 2e^{2u(\zeta)}(u(\xi) - u(\zeta))}{|\zeta - \xi|^2} dm(\xi).$$

Using the local Dirichlet integral and an argument using cut-off functions, as in [**47**], one can show the following.

THEOREM 3.7. *If $f \in \mathcal{D}$, then $f = g_1/g_2$, where $g_1, g_2 \in H^\infty \cap \mathcal{D}$.*

Here H^∞ denotes the bounded analytic functions on $\mathbb{D}$. Notice how this theorem mimics the well-know fact that any H^2 function can be written as the quotient of two bounded analytic functions [**22**]. Theorem 3.7 was first proved in [**46**] in the more general setting of the Dirichlet space of a general domain. The authors in [**47**] present a new proof of this and obtain some bounds on the local Dirichlet integrals of the functions that comprise the quotient. In fact, they show that g_1 and g_2 can be chosen so that both g_1 and $1/g_2$ belong to $\mathcal{D}$. To push the analogy with H^2 further, notice how H^∞ are the multipliers of H^2 (see definitions of multipliers below) and so every H^2 function can be written an the quotient of two multipliers of H^2. Can every function in $\mathcal{D}$ be written as the quotient of two multipliers of $\mathcal{D}$?

4. Potentials

One can also realize the Dirichlet space as a space of potentials. Let

$$k(\zeta) = |1 - \zeta|^{-1/2}, \quad \zeta \in \mathbb{T},$$

and notice that the Fourier coefficients

$$\widehat{k}(n) := \int_{\mathbb{T}} k(\zeta)\overline{\zeta}^n dm(\zeta), \quad n \in \mathbb{Z},$$

satisfy the estimates

(4.1) $\delta(1 + n)^{-1/2} \leqslant |\widehat{k}(n)| \leqslant \delta^{-1}(1 + n)^{-1/2},$

for some $\delta > 0$. With the integral convolution

$$(g_1 * g_2)(\zeta) := \int_{\mathbb{T}} g_1(\overline{\xi}\zeta)g_2(\xi) \, dm(\xi),$$

one can form the space of potentials

$$L^2_{1/2} := \left\{k * g : g \in L^2\right\},$$

where L^2 is the standard Lebesgue space of measurable functions g on $\mathbb{T}$ with norm

$$\|g\|_2 := \sqrt{\int_{\mathbb{T}} |g|^2 \, dm}.$$

One places a norm on $L^2_{1/2}$ by

$$\|k * g\|_{1/2} := \|g\|_{L^2}.$$

These potentials are closely related to the standard Bessel potentials [**57**].

If $P(k * g)$ denotes the Poisson integral of $k * g$, that is,

$$P(k * g)(z) = \int_{\mathbb{T}} (k * g)(\xi) P_z(\xi)\, dm(\xi),$$

one can compute in polar coordinates to show that for $\zeta \in \mathbb{T}$ and $0 < r < 1$,

$$P(k * g)(r\zeta) = \sum_{n=-\infty}^{\infty} r^{|n|} \zeta^n \widehat{k * g}(n) = \sum_{n=-\infty}^{\infty} r^{|n|} \zeta^n \widehat{k}(n)\widehat{g}(n).$$

If one extends the definition of the Dirichlet integral to harmonic functions u by

$$D(u) := \frac{1}{\pi} \int_{\mathbb{D}} \left(|u_x|^2 + |u_y|^2 \right) dA,$$

we can define the *harmonic Dirichlet space* $\mathcal{D}_h$ to be the space of harmonic functions u on $\mathbb{D}$ with finite Dirichlet integral $D(u)$. With $u = P(k * g)$, one can compute $D(u)$ in polar coordinates to get

$$D(u) = \sum_{n=-\infty}^{\infty} |n| |\widehat{k * g}(n)|^2$$

and so by eq.(4.1), $D(u) < \infty$.

PROPOSITION 4.2. *For u harmonic on $\mathbb{D}$, the following are equivalent.*
(1) $D(u) < \infty$;
(2) $u = P(k * g)$ for some $g \in L^2$;
(3) *The boundary function*

$$u(\zeta) = \lim_{r \to 1^-} (r\zeta)$$

exists almost everywhere and satisfies

$$\int_{\mathbb{T}} \int_{\mathbb{T}} \left| \frac{u(\zeta) - u(\xi)}{\zeta - \xi} \right|^2 dm(\zeta)dm(\xi) < \infty.$$

Furthermore,

$$D(u) = \int_{\mathbb{T}} \int_{\mathbb{T}} \left| \frac{u(\zeta) - u(\xi)}{\zeta - \xi} \right|^2 dm(\zeta)dm(\xi).$$

Harmonic functions with finite Dirichlet integral must have almost everywhere defined boundary values that belong to L^2 and so we can place a norm on the harmonic Dirichlet space $\mathcal{D}_h$ by

$$\|u\| = \sqrt{\|u\|_{L^2}^2 + D(u)}.$$

From our estimates above, notice that

$$\|P(k * g)\| \asymp \|g\|_{L^2}.$$

Also observe how this norm is the same as the analytic Dirichlet space norm from eq.(2.3) when u is an analytic function.

Thinking of the Dirichlet space (harmonic or analytic) as a space of potentials allows us to discuss the fine (capacity) properties of the boundary function. We will get to this in a moment. Before doing so, let us mention the following integral representation of Dirichlet functions [38].

PROPOSITION 4.3. *An analytic function f belongs to $\mathcal{D}$ if and only if*

$$f(z) = \int_{\mathbb{T}} \frac{g(\zeta)}{(1 - \bar{\zeta}z)^{1/2}} \, dm(\zeta)$$

for some $g \in L^2$.

Potentials allow us to define a capacity. The *capacity* of a set $E \subset \mathbb{T}$ is defined to be

$$\gamma(E) := \inf\{\|g\|_{L^2}^2 : g \in L^2, g \geqslant 0, k * g \geqslant 1 \text{ on } E\}.$$

This is often called the *Bessel capacity* and is a monotone, sub-additive set function on $\mathbb{T}$. Moreover, one can show that for an arc $I \subset \mathbb{T}$ (sufficiently small),

$$\gamma(I) \sim \left(\log \frac{1}{m(I)}\right)^{-1}.$$

The capacity γ is generally larger than Lebesgue measure in that there are sets E which have zero measure but positive capacity. We say a property holds *quasi-everywhere* if it holds except possibly on a set of capacity zero. Note that if a property holds quasi-everywhere, it holds almost everywhere. We could go on at length about the Bessel capacity but we will not need it in this presentation. We refer the reader to [**3**] for a thorough treatment of all this.

We would like to mention an alternative definition and older definition of capacity which, fortunately, is equivalent to the square root of the above (Bessel) capacity. It is called the logarithmic capacity. Let M_+ denote the positive finite Borel measures on $\mathbb{T}$. For $\mu \in M_+$, define the logarithmic potential on $\mathbb{C}$ by

$$u_\mu(z) = \int_{\mathbb{T}} \log \frac{e}{|1 - \bar{\zeta}z|} \, d\mu(\zeta).$$

A computation with power series shows that

$$u_\mu(z) = \mu(\mathbb{T}) + \frac{1}{2}\sum_{n=1}^{\infty} \frac{\hat{\mu}(n)}{n} z^n + \frac{1}{2}\sum_{n=1}^{\infty} \frac{\hat{\mu}(-n)}{n}\bar{z}^n, \quad z \in \mathbb{D},$$

where

$$\hat{\mu}(n) := \int \bar{\zeta}^n d\mu(\zeta), \quad n \in \mathbb{Z},$$

are the Fourier coefficients of μ. Notice that u_μ is a non-negative harmonic function on $\mathbb{D}$. Define the *energy* of μ to be

(4.4) $$E(\mu) := \int_{\mathbb{T}} u_\mu(\zeta) d\mu(\zeta).$$

Remembering that μ is a positive measure and so

$$\hat{\mu}(-n) = \overline{\hat{\mu}(n)},$$

we can compute $E(\mu)$ to be

$$E(\mu) = \mu(\mathbb{T})^2 + \sum_{n=1}^{\infty} \frac{|\hat{\mu}(n)|^2}{n}.$$

We set

$$\mathcal{E} := \{\mu \in M_+ : E(\mu) < \infty\}$$

to be the measures of finite energy.

If $\widetilde{u_\mu}$ is the conjugate function for u_μ, normalized so that $\widetilde{u_\mu}(0) = 0$, then

(4.5)
$$f_\mu(z) := u_\mu(z) + i\widetilde{u_\mu}(z) = \mu(\mathbb{T}) + \sum_{n=1}^{\infty} \frac{\widehat{\mu}(n)}{n} z^n.$$

We also have the nice identity,

$$|f_\mu(0)|^2 + D(f_\mu) = \mu(\mathbb{T})^2 + \sum_{n=1}^{\infty} n \left| \frac{\widehat{\mu}(n)}{n} \right|^2 = E(\mu).$$

Hence $f_\mu \in \mathcal{D}$ exactly when $\mu \in \mathcal{E}$. It is also interesting to note that

$$f_\mu'(z) = \int_{\mathbb{T}} \frac{1}{\zeta - z} d\mu(\zeta),$$

the Cauchy transform of μ.

For a compact set $F \subset \mathbb{T}$, the *logarithmic capacity* of F is defined to be

$$c(F) := \sup\{\mu(F) : \mu \in \mathcal{E}, \operatorname{supp}(\mu) \subset F, u_\mu | F \leqslant 1\}.$$

Extend this definition to any set $E \subset \mathbb{T}$ by

$$c(E) := \sup\{c(F) : F \subset E, F \text{ compact}\}.$$

One can show that

$$c(E)^2 \asymp \gamma(E)$$

and so the notion of quasi-everywhere (i.e., 'except for a set of capacity zero') is the same for these capacities. Depending on the setting, various authors use different definitions of logarithmic capacity for their particular application. Fortunately, these are essentially the same.

5. Boundary values

From the observation $\mathcal{D} \subset H^2$, we know that the boundary function

$$f(\zeta) := \lim_{r \to 1^-} f(r\zeta)$$

for $f \in \mathcal{D}$ exists almost everywhere. Can we say more? The answer is a resounding yes.

If $u \in \mathcal{D}_h$, the harmonic Dirichlet space, then $u = P(k * g)$ for some $g \in L^2$ and moreover, the radial boundary function for u is $k * g$ almost everywhere. This next results begins to unpack the finer relationship between the boundary values of u and the potential $k * g$.

THEOREM 5.1. *Let* $u = P(k*g)$ *for some* $g \in L^2$. *For fixed* $\theta \in [0, 2\pi)$, *consider the following four limits:*

(1)
$$\lim_{r \to 1^-} u(re^{i\theta}),$$

   the radial limit of u;

(2)
$$\lim_{N \to \infty} \sum_{n=-N}^{N} \widehat{k * g}(n) e^{in\theta},$$

   the limit of the partial sums of the Fourier series for $k * g$;

(3)
$$\lim_{h \to 0^+} \frac{1}{2h} \int_{\theta-h}^{\theta+h} (k * g)(e^{it})\, dt;$$

(4)
$$P.V.(k * g)(e^{i\theta}).$$

If one of them exists and is finite, they all do and they are equal.

REMARK 5.2. The equivalence of (1) and (2) is Abel's theorem and an old result of Landau [**31**, p. 65 - 66]. The equivalence of (2) and (3) was pointed out by Beurling in [**6**] and uses some old results dating back to Fatou and Fejér. The equivalence of (1) and (4) can be found in [**44**]. Unfortunately, there are cases where $(k * g)(re^{i\theta})$ as a finite limit as $r \to 1$ but $(k * |g|)(e^{i\theta}) = \infty$ [**44**].

Combine the previous result with this next result of Beurling to complete the picture.

THEOREM 5.3 (Beurling [**6**]). *If $u \in \mathcal{D}_h$, then there is a set of $W \subset \mathbb{T}$ of capacity zero such that*
$$u(\zeta) := \lim_{r \to 1^-} u(r\zeta)$$
exists and is finite for every $\zeta \in \mathbb{T} \setminus W$. Thus, the four limits in Theorem 5.1 exist and are equal for these ζ.

This theorem is sharp.

THEOREM 5.4 (Carleson [**15**]). *Given any closed set $F \subset \mathbb{T}$ of capacity zero, there is an $f \in \mathcal{D}$ such that*
$$\lim_{r \to 1^-} f(r\zeta)$$
does not exist for all $\zeta \in F$.

Recall from eq.(2.1) the quantity $L(f, \zeta)$ which is the length of the arc $\{f(r\zeta) : 0 < r < 1\}$. We noted earlier that for $f \in \mathcal{D}$, that $L(f, \zeta) < \infty$ for almost every $\zeta \in \mathbb{T}$.

THEOREM 5.5 (Beurling [**6**]). *For $f \in \mathcal{D}$, $L(f, \zeta) < \infty$ for quasi-every $\zeta \in \mathbb{T}$.*

There are similar results for other classes of functions [**60**]. We point out the special nature of this result since there are examples of $f \in H^2$ for which $L(f, \zeta) = \infty$ for *every* $\zeta \in \mathbb{T}$ [**23**]. We also point out that the exceptional set in Theorem 5.5 can not be made any smaller. Indeed if $f \in \mathcal{D}$ and $L(f, \zeta) < \infty$, one can use the identity
$$f(s\zeta) - f(0) = \int_0^s \zeta f'(r\zeta)\, dr,$$
to show that $f(r\zeta)$ has a finite limit as $r \to 1$. However by Theorem 5.4, given any closet set F of capacity zero, there is an $f \in \mathcal{D}$ such that f does not have a radial limit for every $\zeta \in F$. This function must then satisfy $L(f, \zeta) = \infty$ for all $\zeta \in F$.

Before moving on to talk about other types of limits (non-tangential, oricyclic, etc.) we want to mention a nice relationship between $\mathcal{E}$, the measures of finite energy from eq.(4.4), and the inner product on $\mathcal{D}$. We already know that for f_μ defined in eq.(4.5),
$$f_\mu \in \mathcal{D} \Leftrightarrow \mu \in \mathcal{E}.$$

If we use the alternate inner product

$$(f, g) := f(0)\overline{g(0)} + \sum_{n=1}^{\infty} n a_n \overline{b_n},$$

where the a_n's are the Taylor coefficients of f and the b_n's are those for g, one can show that if p is an analytic polynomial and $\mu \in \mathcal{E}$ then

$$(p, f_\mu) = \int p(\zeta)\, d\mu(\zeta).$$

Since

$$(f_\mu, f_\mu) = |f_\mu(0)|^2 + D(f_\mu) = \mu(\mathbb{T})^2 + E(\mu),$$

we see that the linear functional

$$p \mapsto \int p(\zeta) d\mu(\zeta)$$

extends to be continuous on $\mathcal{D}$. Furthermore, we also have the following maximal-type theorem [15].

THEOREM 5.6. *For $f \in \mathcal{D}$ and $\zeta \in \mathbb{T}$, let*

$$(Mf)(\zeta) := \sup\{|f(r\zeta)| : 0 < r < 1\}$$

be the radial maximal function. For $\mu \in \mathcal{E}$,

$$\left(\int (Mf)(\zeta)\, d\mu(\zeta)\right)^2 \leqslant c(f_\mu, f_\mu) E(\mu) < \infty.$$

As a corollary to this maximal theorem, we can now prove this useful identity: For $g \in \mathcal{D}$ and $\mu \in \mathcal{E}$,

$$(5.7) \qquad\qquad (g, f_\mu) = \int g(\zeta)\, d\mu(\zeta).$$

We now move on to other types of limits considered in [29, 38, 61]. Consider the approach regions

$$A_{\gamma,c}(\zeta) := \{z \in \mathbb{D} : |\zeta - z| < c(1 - |z|)^{1/\gamma}\}, \quad \zeta \in \mathbb{T}, c, \gamma > 0.$$

These are the γ order contact regions. With a little geometry, one can see that when $\gamma = 1$, these regions are triangle shaped with vertex at ζ and are called the non-tangential approach regions. When $\gamma = 2$, these regions become (essentially) circles tangent to $\mathbb{T}$ at ζ and are called oricyclic approach regions. Observe how that when $\gamma > 1$, these domains are tangent to the circle and the degree of tangency increases as γ increases. We say, for an analytic function f on $\mathbb{D}$, that f has an A_γ-limit L at ζ if $f(z) \to L$ as $z \to \zeta$ within $A_{\gamma,c}(\zeta)$ for every $c > 0$.

THEOREM 5.8 (Beurling). *Every $f \in \mathcal{D}$ has a finite A_1-limit at quasi-every point of $\mathbb{T}$.*

From Proposition 3.1, every $f \in \mathcal{D}$ has oricyclic A_2-limits at almost every point of $\mathbb{T}$. This next result [29] improves this to higher order contact.

THEOREM 5.9 (Kinney). *If $\gamma > 0$ and $f \in \mathcal{D}$, then f has finite A_γ-limits for almost every point of $\mathbb{T}$.*

Twomey [61] improves 'almost everywhere' in the previous theorem.

THEOREM 5.10 (Twomey). *If $\gamma > 0$ and $f \in \mathcal{D}$, then f has finite A_γ-limits for quasi-every point of $\mathbb{T}$.*

The state of the art here involves the exponential contact regions

$$(5.11) \quad E_{\gamma,c}(\zeta) := \left\{ z \in \mathbb{D} : |\zeta - z| < c \left(\log \frac{1}{1 - |z|} \right)^{-1/\gamma} \right\}, \quad \zeta \in \mathbb{T}, c > 0, \gamma > 0.$$

THEOREM 5.12 (Nagel, Rudin, J. Shapiro [**38**]). *Every $f \in \mathcal{D}$ has an E_1-limit for almost every point of $\mathbb{T}$.*

Twomey improves this to the following.

THEOREM 5.13 (Twomey). *If $f \in \mathcal{D}$ and $0 < \gamma < 1$, there is a set W_γ of γ-dimensional Hausdorff content zero such that f has a finite E_γ-limit on $\mathbb{T} \setminus W_\gamma$.*

See the references in [**61**] for more on this. Twomey also points out that these results are in a sense sharp (see Theorems 6 and 7 in [**61**]). We end this section with the following theorem that points out the special nature of the existence of limits of Dirichlet function in tangential contact regions.

THEOREM 5.14. *Let C be any curve in $\mathbb{D}$ that approaches the point 1 tangent to the unit circle. Then there is a bounded analytic function f on $\mathbb{D}$ whose limit along the curve ζC does not exist for any $\zeta \in \mathbb{T}$.*

Littlewood [**32**] proved the 'almost everywhere' version of this theorem while Lohwater and Piranian [**33**] proved that f could be a Blaschke product. Aikawa [**4**] proved that f could be a bounded outer function.

6. Zeros

In this section, we address the following question: Given a sequence $Z = (z_n)_{n \geqslant 1} \subset \mathbb{D}$, what are necessary and sufficient conditions for Z to be the zeros Z_f of a function $f \in \mathcal{D} \setminus \{0\}$? To place our discussion in broader context, we review some well-known theorems about the zeros of functions from other classes.

Recall our earlier notation that H^2 denotes the classical Hardy space and H^∞ denotes the bounded analytic functions on $\mathbb{D}$. A well-known theorem of Blaschke says that if $Z = (z_n)_{n \geqslant 1}$ is a sequence of points in $\mathbb{D}$ that satisfies the *Blaschke condition*

$$\sum_{n=1}^{\infty} (1 - |z_n|) < \infty,$$

then the Blaschke product

$$b(z) = \prod_{n=1}^{\infty} \frac{\overline{z_n}}{|z_n|} \frac{z_n - z}{1 - \overline{z_n} z}$$

converges uniformly on compact subsets of $\mathbb{D}$ and forms an H^∞ function whose zeros are precisely Z. Conversely, an argument using Jensen's inequality shows that the zeros Z_f of an $f \in H^\infty \setminus \{0\}$ must satisfy the Blaschke condition. From here one can use the fact that every H^2 function is the quotient of two H^∞ functions to show that the Blaschke condition is both necessary and sufficient for a sequence $Z \subset \mathbb{D}$ to be the zeros of a function from $H^2 \setminus \{0\}$.

To look at the zeros functions which are 'smoother' up to the boundary than bounded analytic functions, we recall an old theorem of Riesz [**22**, p. 17].

THEOREM 6.1 (Riesz). *If $f \in H^\infty \setminus \{0\}$, then*

$$\int_{\mathbb{T}} \log |f| dm > -\infty.$$

For the space of analytic functions A on $\mathbb{D}$ which extend to be continuous on $\mathbb{D}^-$, called the *disk algebra*, the zeros (in $\mathbb{D}$) of an $f \in A \setminus \{0\}$ must certainly satisfy the Blaschke condition. However, if these zeros accumulate on a set $E \subset \mathbb{T}$ of positive measure then, by the continuity of f on $\mathbb{D}^-$, $f|E = 0$ making the integral

$$\int_{\mathbb{T}} \log |f| dm$$

divergent. Thus the zeros Z_f in $\mathbb{D}$ of an $f \in A \setminus \{0\}$ must satisfy the Blaschke condition as well as satisfy $m(Z_f^- \cap \mathbb{T}) = 0$. A theorem of Fatou [27, p. 80] says that if $K \subset \mathbb{T}$ is compact and of measure zero, then there is a $g \in A \setminus \{0\}$ for which $g|K = 0$. Hence, if $(z_n)_{n \geqslant 1} \subset \mathbb{D}$ satisfies

$$\sum_{n=1}^{\infty} (1 - |z_n|) < \infty \quad \text{and} \quad m\left((z_n)_{n \geqslant 1}^- \cap \mathbb{T}\right) = 0,$$

then $f := gb$, where b is the Blaschke product with zeros $(z_n)_{n \geqslant 1}$ and g is the function from Fatou's result, belongs to $A \setminus \{0\}$ with the desired zeros.

When one moves to spaces of even smoother functions, say the Lipschitz classes Λ_α, $0 < \alpha < 1$, of analytic f on $\mathbb{D}$ for which

$$\sup \left\{ \frac{|f(\zeta) - f(\xi)|}{|\zeta - \xi|^\alpha} : \zeta, \xi \in \mathbb{T}, \zeta \neq \xi \right\} < \infty,$$

the situation is even more delicate. We first notice that $f \in \Lambda_\alpha$ if and only if

$$\sup \left\{ \frac{|f(z) - f(w)|}{|z - w|^\alpha} : z, w \in \mathbb{D}^-, z \neq w \right\} < \infty.$$

Thus

$$|f(z) - f(\zeta)| \leqslant C_f |z - \zeta|^\alpha, \quad z \in \mathbb{D}, \zeta \in \mathbb{T}.$$

Taking logarithms of both sides of the above equation and replacing z by one of the zeros of f in $\mathbb{D}$ we see that

$$\log |f(\zeta)| \leqslant C_f + \alpha \log \operatorname{dist}(\zeta, Z_f).$$

By integrating both sides and using the Riesz theorem (Theorem 6.1) we see that the zeros $Z_f \subset \mathbb{D}$ of an $f \in \Lambda_\alpha \setminus \{0\}$ must satisfy

$$\int_{\mathbb{T}} \log \operatorname{dist}(\zeta, Z_f) dm(\zeta) > -\infty$$

as well as the Blaschke condition. A deep theorem of Taylor and Williams [58] (discovered independently by others [17, 40]) says the conditions

(6.2) $$\sum_{z \in Z} (1 - |z|) < \infty \quad \text{and} \quad \int_{\mathbb{T}} \log \operatorname{dist}(\zeta, Z) dm(\zeta) > -\infty$$

on a sequence $Z \subset \mathbb{D}$ are both necessary and sufficient to be the zeros of an $f \in \Lambda_\alpha \setminus \{0\}$. In fact, if the two conditions in eq.(6.2) are satisfied, there is a function f such that $f^{(n)} \in A$ for all $n \in \mathbb{N}_0$ and $Z_f = Z$.

The complete classification of the zero sets for Dirichlet functions is still unresolved. Lokki [34] mistakenly claimed that the Blaschke condition was both necessary and sufficient to be a Dirichlet zero set. But in fact, there are radii $(r_n)_{n \geqslant 1} \subset [0,1)$ which satisfy the Blaschke condition

$$\sum_{n=1}^{\infty} (1 - r_n) < \infty$$

but for which there are angles θ_n such that $z_n = r_n e^{i\theta_n}$ are not the zeros of any $f \in \mathcal{D} \setminus \{0\}$ [16]. There are also $f \in \mathcal{D} \setminus \{0\}$ whose zeros $Z_f \subset \mathbb{D}$ satisfy

$$\text{dist}(\zeta, Z_f) = 0 \text{ for all } \zeta \in \mathbb{T}.$$

Thus the two conditions in eq.(6.2) are sufficient but not necessary. There is, however, this observation of Carleson [12]: If $(r_n)_{n \geqslant 1} \subset [0,1)$ satisfy the Blaschke condition, then the Blaschke product

$$b(z) = \prod_{n=1}^{\infty} \frac{r_n - z}{1 - r_n z}$$

satisfies

$$|b'(z)| \leqslant \sum_{n=1}^{\infty} \frac{1 - r_n}{|1 - r_n z|^2} \leqslant \frac{C}{|1 - z|^2}.$$

Hence the function

$$f(z) = (1 - z)^2 b(z)$$

belongs to $\mathcal{D} \setminus \{0\}$ and has the r_n's as its zeros. A recent result of Bogdan [8] extends this observation.

THEOREM 6.3 (Bogdan). *A necessary and sufficient condition on a set $W \subset \mathbb{D}$ to have the property that every Blaschke sequence $(z_n)_{n \geqslant 1} \subset W$ is a Dirichlet zero sequence is*

(6.4) $$\int_{\mathbb{T}} \log \text{dist}(\zeta, W) dm(\zeta) > -\infty.$$

Furthermore, if eq.(6.4) holds, then there is an outer function $F \in \mathcal{D}$ such that $bF \in \mathcal{D}$ for every Blaschke product whose zeros lie in W.

For example, some simple estimates show that any Blaschke sequence lying in a single non-tangential or even a finite order contact region $A_{\gamma,c}(\zeta)$ is a Dirichlet zero set. The same is true for certain (but not all) exponential contact regions $E_{\gamma,c}(\zeta)$.

If one is hoping for a necessary and sufficient condition on the radii r_n for a sequence $z_n = r_n e^{i\theta_n}$ to be a Dirichlet zero sequence (as the Blaschke condition is for bounded analytic functions), there does not seem to be such a result. Carleson [12] proved that the radii r_n for a Dirichlet zero set must satisfy

(6.5) $$\sum_{n=1}^{\infty} \left\{ \log \frac{1}{1 - r_n} \right\}^{-1-\varepsilon} < \infty \text{ for every } \varepsilon > 0.$$

Moreover if

(6.6) $$\sum_{n=1}^{\infty} \left\{ \log \frac{1}{1 - r_n} \right\}^{-1+\varepsilon} < \infty \text{ for some } \varepsilon > 0$$

then $z_n = r_n e^{i\theta_n}$ is a Dirichlet zero set for every choice of angles θ_n. The conditions in eq.(6.5) and eq.(6.6) do not characterize the Dirichlet zero sets in that for any continuous function h on $\mathbb{R}_+$ with $h(0) = 0$ and $h(x) > 0$ for $x > 0$, and radii $(r_n)_{n\geqslant 1}$ satisfying

$$\sum_{n=1}^{\infty} \left\{ \log \frac{1}{1 - r_n} \right\}^{-1} h(1 - r_n) < \infty,$$

there is a sequence of angles θ_n so that $z_n = r_n e^{i\theta_n}$ is not a Dirichlet zero set [53]. An extension of eq.(6.5) by H. S. Shapiro and A. Shields [53] says that if

(6.7)
$$\sum_{n=1}^{\infty} \left\{ \log \frac{1}{1 - r_n} \right\}^{-1} < \infty,$$

then $z_n = r_n e^{i\theta_n}$ is a Dirichlet zero set for any choice of angles θ_n.

Unfortunately, even this sharpened condition in eq.(6.7) is not necessary. Using the exponential contact result in Theorem 5.12 one can, as was observed in [38], produce a counterexample: Beginning at $\zeta = 1$, lay down arcs $I_n \subset \mathbb{T}$ of length

$$\left\{ \log \frac{1}{1 - r_n} \right\}^{-1}$$

end-to-end (repeatedly traversing the unit circle). If we assume that

(6.8)
$$\sum_{n=1}^{\infty} \left\{ \log \frac{1}{1 - r_n} \right\}^{-1} = \infty,$$

we observe that

$$\sum_{n=1}^{\infty} m(I_n) = \infty,$$

and so each $\zeta \in \mathbb{T}$ will be contained in infinitely many of the arcs $(I_n)_{n\geqslant 1}$. Let $e^{i\theta_n}$ be the center of the arc I_n and note that simple geometry shows that for every $e^{i\theta}$, the exponential contact region $E_{1,1}(e^{i\theta})$ (see eq.(5.11)) contains infinitely many of the points $r_n e^{i\theta_n}$. Thus if $f \in \mathcal{D}$ and $f(r_n e^{i\theta_n}) = 0$ for all n, Theorem 5.12 says that the boundary function for f will vanish almost everywhere on $\mathbb{T}$, forcing f to be identically zero (see Theorem 6.1). This argument actually shows that the sequence $(r_n e^{i\theta_n})_{n\geqslant 1}$ just created can not be the zeros of any $u \in \mathcal{D}_h \setminus \{0\}$, the harmonic Dirichlet space. A relatively recent result [45] shows that one does not even have to make the zeros z_n accumulate at every point of $\mathbb{T}$, as in the above counterexample. Assuming eq.(6.8), one can arrange the angles θ_n so that the zeros $z_n = r_n e^{i\theta_n} \to 1$. In a much earlier result, Caughren [16] proved that one can have a Blaschke sequence that converges to a single point on the boundary that is not a Dirichlet zero sequence. There is also the following probabilistic version of all this [8].

THEOREM 6.9 (Bogdan). *Let* $(\theta_n)_{n\geqslant 1}$ *be a sequence of independent random variables uniformly distributed on* $(-\pi, \pi]$ *and* $(r_n)_{n\geqslant 1} \subset [0, 1)$. *If*

$$\sum_{n=1}^{\infty} \left\{ \log \frac{1}{1 - r_n} \right\}^{-1} = \infty,$$

then almost surely the sequence $(r_n e^{i\theta})_{n\geqslant 1}$ *is not a Dirichlet zero sequence.*

In another line of thought, we can also talk about boundary zero sets. If $f \in A \setminus \{0\}$, then, by Riesz's theorem (Theorem 6.1), the boundary zeros

$$E_f := \{\zeta \in \mathbb{T} : f(\zeta) = 0\}$$

must be a set of measure zero. Conversely, by Fatou's theorem mentioned earlier [27, p. 80], if E is a closed subset of $\mathbb{T}$ with $m(E) = 0$, then there is an $f \in A \setminus \{0\}$ such that $E_f = E$. For the Lipschitz classes Λ_α a necessary and sufficient condition for a closed subset of $\mathbb{T}$ to satisfy $E = E_f$ for some $f \in \Lambda_\alpha \setminus \{0\}$ is the following:

$$m(E) = 0 \quad \text{and} \quad \sum_{n=1}^{\infty} m(I_n) \log \frac{1}{m(I_n)} < \infty,$$

where I_n are the complimentary arcs of E.

For $f \in \mathcal{D}$, we know from Beurling's theorem (Theorem 5.3) that the radial limit

$$f(\zeta) = \lim_{r \to 1^-} f(r\zeta)$$

exists and is finite quasi-everywhere. For a set $E \subset \mathbb{T}$, one can define the space

$$\mathcal{D}_E := \{f \in \mathcal{D} : f|E = 0 \text{ quasi-everywhere}\}.$$

The standard capacity estimate

$$\gamma(\{\zeta \in \mathbb{T} : |f(\zeta)| > \lambda\}) \leqslant \frac{c}{\lambda^2} \|f\|^2$$

shows that $\mathcal{D}_E$ is a closed subspace of $\mathcal{D}$ [11]. Using the identity in eq.(5.7), one can show that

$$\mathcal{D}_E^{\perp} = \bigvee\{f_\mu : \mu \in \mathcal{E}, \operatorname{supp}(\mu) \subset E, u_\mu|E \leqslant 1\}.$$

The question now is: For what sets $E \subset \mathbb{T}$, is $\mathcal{D}_E = (0)$? Such sets are called *sets of uniqueness*. Certainly if $m(E) > 0$, Riesz's theorem (Theorem 6.1) says that $\mathcal{D}_E = (0)$. Carleson [13] showed that if $\gamma(E) = 0$, then $\mathcal{D}_E \neq (0)$. In fact, if E is also closed, then $\mathcal{D}_E$ contains outer functions that also belong to the disk algebra [10]. The problem is very delicate since there are sets with $\gamma(E) > 0$ and $m(E) = 0$ but $\mathcal{D}_E = (0)$. There is a complete characterization of the sets of uniqueness due to Malliavin [35] but the necessary and sufficient condition involves the 'modified logarithmic capacity' and is quite difficult to apply to particular situations. Other partial results can be found in [26]. The reference [28] contains a survey of the sets of uniqueness for several other classes of analytic functions.

7. Forward shift invariant subspaces

In this section we wish to study the forward shift operator $S : \mathcal{D} \to \mathcal{D}$

$$(Sf)(z) := zf(z).$$

In particular, we focus our attention on the invariant subspaces of S, that is, those closed linear manifolds $\mathcal{M} \subset \mathcal{D}$ such that $S\mathcal{M} \subset \mathcal{M}$. We denote the collection of these invariant subspaces by $\operatorname{Lat}(S, \mathcal{D})$. Though $\operatorname{Lat}(S, \mathcal{D})$ is not completely understood, there has been quite a lot of work on this subject. In order to place these results in some context, we mention a few classical theorems. The first, and probably one that always needs to be mentioned when talking about the shift on spaces of analytic functions, is Beurling's theorem [7, 22].

THEOREM 7.1 (Beurling).   (1) *A subspace* $\mathcal{M} \neq \{0\}$ *belongs to* $\mathrm{Lat}(S, H^2)$
if and only if $\mathcal{M} = \vartheta H^2$, *where* ϑ *is an inner function.*
(2) *If* $f \in H^2$, *then*

$$[f]_S := \bigvee\{S^n f : n \in \mathbb{N}_0\} = \vartheta_f H^2,$$

where ϑ_f *is the inner factor of* f. *Thus* f *is* S-*cyclic, that is* $[f]_S = H^2$,
if and only if f *is an outer function.*

Let us take a moment to review the description of $\mathrm{Lat}(S, X)$ for some other
well-known Banach spaces of analytic functions X. When X is the disk algebra
A, discussed earlier, one can show, by approximating every $f \in A$ with its Cesàro
polynomials, that every $\mathcal{I} \in \mathrm{Lat}(S, A)$ is a closed ideal of A and by a result of
Rudin [52] (see also [27]) takes the form $\mathcal{I} = \mathcal{I}(E, \vartheta)$, where $E \subset \mathbb{T}$ is closed with
$m(E) = 0$ and ϑ is inner such that

$$\underline{Z}(\vartheta) := \left\{ z \in \mathbb{D}^- : \varliminf_{\lambda \to z} |\vartheta(\lambda)| = 0 \right\}$$

satisfies $\underline{Z}(\vartheta) \cap \mathbb{T} \subset E$, and

$$\mathcal{I}(E, \vartheta) := \{ f \in A : f/\vartheta \in A, f|E = 0 \}.$$

Moreover, every $\mathcal{I}(E, \vartheta)$ is a non-zero closed ideal of A. From here Korenblum [30]
developed techniques, used by many others (for example [36, 56]), to discuss the
closed ideals of several other spaces of analytic functions that are smooth up to the
boundary. The results are similar to Rudin's result except that the types of closed
sets E and the inner functions ϑ have further restrictions on them. Furthermore,
in some cases, the boundary zeros of the derivatives come into play.

The Dirichlet space $\mathcal{D}$ is not an algebra of analytic functions and the functions
in the Dirichlet space need not have continuous boundary values (see Theorem 5.4).
Knowing that Dirichlet functions have radial boundary values quasi-everywhere
(Theorem 5.3), one conjectures that every $\mathcal{M} \in \mathrm{Lat}(S, \mathcal{D})$ should take the form
$\mathcal{M}(E, \vartheta)$, the space of $f \in \mathcal{D}$ such that

$$f/\vartheta \in \mathcal{D} \quad \text{and} \quad \lim_{r \to 1^-} f(r\zeta) = 0 \text{ for quasi-every } \zeta \in E.$$

Moreover, it should be the case that $f \in \mathcal{D}$ is S-cyclic, that is $[f]_S = \mathcal{D}$, if and
only if f is an outer function and the set of boundary zeros has capacity zero. We
will say more about this conjecture in a moment. Though this type of result is
unknown, there are many other things one can say.

Before discussing these results, let us first say a few words about the operator
S on $\mathcal{D}$. On the Hardy space, S is an isometry. On the Dirichlet space, S is a
two-isometry

$$\|S^2 f\|^2 - 2\|Sf\|^2 + \|f\|^2 = 0, \ \forall f \in \mathcal{D} \quad \text{and} \quad \bigcap_{n=0}^{\infty} S^n \mathcal{D} = \{0\}.$$

Moreover, Richter [43] showed that every analytic, cyclic, two-isometry on a Hilbert
space is unitarily equivalent to S on some local Dirichlet type space $\mathcal{D}(\mu)$.

By Beurling's theorem, the shift S on H^2 is *cellular indecomposable* in that if
$\mathcal{M}, \mathcal{N} \in \mathrm{Lat}(S, H^2) \setminus \{0\}$, then $\mathcal{M} \cap \mathcal{N} \neq \{0\}$. Furthermore, $\mathcal{M} \cap H^\infty \neq \{0\}$, and

the subspace

$$\mathcal{M} \ominus S\mathcal{M} := \mathcal{M} \cap (S\mathcal{M})^{\perp}$$

is one-dimensional. Richter and Shields [46] generalize this to the Dirichlet space.

THEOREM 7.2 (Richter-Shields). *Let* $\mathcal{M}, \mathcal{N} \in Lat(S, \mathcal{D}) \setminus \{0\}$. *Then*

(1) $\mathcal{M} \cap H^{\infty} \neq \{0\}$.
(2) $\mathcal{M} \cap \mathcal{N} \neq \{0\}$.
(3) $\mathcal{M} \ominus S\mathcal{M}$ *is one-dimensional.*

One way to prove Beurling's theorem is to first show that whenever $\mathcal{M} \in Lat(S, H^2) \setminus \{0\}$, then $\mathcal{M} \ominus S\mathcal{M}$ is one dimensional and $[\mathcal{M} \ominus S\mathcal{M}]_S = \mathcal{M}$. Here for a set Y, $[Y]_S$ is the smallest S-invariant subspace containing Y, or equivalently

$$[Y]_S := \bigvee \{S^n g : n \in \mathbb{N}_0, g \in Y\}.$$

As it turns out, the same technique works for the Dirichlet space [42].

THEOREM 7.3 (Richter). *If* $\mathcal{M} \in Lat(S, \mathcal{D})$, *then*

$$[\mathcal{M} \ominus S\mathcal{M}]_S = \mathcal{M}.$$

In H^2 setting, a function $\phi \in \mathcal{M} \ominus S\mathcal{M}$ is a solution (assuming that $\mathcal{M}$ contains a function that does not vanish at the origin) to the extremal problem

$$\inf \left\{ \frac{\|g\|_{H^2}}{|g(0)|} : g \in \mathcal{M} \right\}$$

and solutions to this extremal problem are constant multiplies of inner functions. Inner functions ϕ have the property that $|\phi(\zeta)| = 1$ for almost every $\zeta \in \mathbb{T}$. Furthermore, the formula

$$\phi(z) = z^m \prod_{n=1}^{\infty} \frac{\overline{z_n}}{|z_n|} \frac{z_n - z}{1 - \overline{z_n} z} \exp \left\{ - \int_{\mathbb{T}} \frac{\zeta + z}{\zeta - z} d\mu(\zeta) \right\},$$

defining an inner function on $\mathbb{D}$ is valid as a meromorphic function $\widetilde{\phi}$ on $\mathbb{D}_e := \widehat{\mathbb{C}} \setminus \mathbb{D}^-$. In fact, it is not difficult to see that

$$\widetilde{\phi}(z) = \frac{1}{\phi(1/\overline{z})}, \quad z \in \mathbb{D}_e \setminus W,$$

where $W = \{1/\overline{z} : \phi(z) = 0\}$. Also observe how ϕ and $\widetilde{\phi}$ are pseudocontinuations of each other in that

$$\lim_{r \to 1^-} \phi(r\zeta) = \lim_{r \to 1^-} \widetilde{\phi}(\zeta/r)$$

for almost every $\zeta \in \mathbb{T}$. We will say more about pseudocontinuations in the next section. By a Morera-type theorem, one can also show that if I is an arc in $\mathbb{T} \setminus \underline{Z}(\phi)$, then ϕ and $\widetilde{\phi}$ are analytic continuations of each other across I. Of course, for example by taking ϕ to be a Blaschke product whose zeros accumulate on all of $\mathbb{T}$, one can have $\underline{Z}(\phi) \supset \mathbb{T}$ and so ϕ need not have an analytic continuation across any arc of $\mathbb{T}$.

For the Dirichlet space, every $\phi \in \mathcal{M} \ominus S\mathcal{M}$ is a solution to the extremal problem

$$\inf \left\{ \frac{\|g\|_{\mathcal{D}}}{|g(0)|} : g \in \mathcal{M} \right\}$$

and has some extra regularity properties near the boundary. Here however, it is the derivative of ϕ that has the pseudo and analytic continuation properties. We start with the following.

THEOREM 7.4 (Richter-Sundberg [**48**]). *If* $\phi \in \mathcal{M} \ominus S\mathcal{M}$, *then* ϕ *is a multiplier of* $\mathcal{D}$.

Here ψ is a *multiplier* of $\mathcal{D}$ if $\psi\mathcal{D} \subset \mathcal{D}$. An easy application of the closed graph theorem shows that a multiplier ψ defines a bounded linear operator on $\mathcal{D}$ by $f \mapsto \psi f$. Moreover, multipliers are bounded analytic functions on $\mathbb{D}$. In fact

(7.5) $$\sup\{|\psi(z)| : z \in \mathbb{D}\} \leqslant \sup\{\|\psi g\| : \|g\| \leqslant 1\}.$$

This last inequality says that the H^∞ norm of a multiplier is bounded by the norm of the multiplication operator $f \mapsto \psi f$ on $\mathcal{D}$. A description of the multipliers of $\mathcal{D}$ can be found in a paper of Stegenga [**57**]. Another result of Richter and Sundberg [**49**] refines the standard estimate for multipliers in eq.(7.5) (standard in the sense that this estimate holds for most Banach spaces of analytic functions) and says that if $\phi \in \mathcal{M} \ominus S\mathcal{M}$, then

$$\sup\{|\phi(z)| : z \in \mathbb{D}\} \leqslant \|\phi\|.$$

Thus the sup norm of an extremal function is bounded by the actual norm of ϕ (which is smaller than the multiplier norm).

Extremal functions $\phi \in \mathcal{M} \ominus S\mathcal{M}$ in the Dirichlet space have even further regularity properties. Carleson [**12**] proved that if E is a 'thin set' in $\mathbb{T}$ and $\phi \in \mathcal{D}_E \ominus S\mathcal{D}_E$, then $(z\phi)'$ has an analytic continuation to $\mathbb{C} \setminus E$. Actually, Carleson proved a slightly different result since he was using another inner product on $\mathcal{D}$. Richter and Sundberg [**12**] extended this in the following way.

THEOREM 7.6 (Richter-Sundberg). *Let* $\mathcal{M} \in Lat(S, \mathcal{D})$ *and* $\phi \in \mathcal{M} \ominus S\mathcal{M}$. *Then*

(1) $(z\phi)'$ *can be written as the quotient of two bounded analytic functions on* $\mathbb{D}$, *that is to say,* $(z\phi)'$ *is a function of 'bounded type';*

(2) $(z\phi)'$ *has a meromorphic pseudocontinuation* G *of bounded type in* $\mathbb{D}_e := \widehat{\mathbb{C}} \setminus \mathbb{D}^-$. *By this we mean that*

$$\lim_{r \to 1^-} (z\phi)'(r\zeta) = \lim_{r \to 1^-} G(\zeta/r)$$

for almost every $\zeta \in \mathbb{T}$;

(3) $\psi(z) := |\phi(z)|$ *satisfies*

(a) *for every* $\zeta \in \mathbb{T}$,

$$\psi(\zeta) := \lim_{r \to 1^-} \psi(r\zeta)$$

exists.

(b) ψ *is upper semicontinuous on* $\mathbb{T}$ *and for all* $\zeta \in \mathbb{T}$,

$$\overline{\lim_{z \to \zeta}} |\phi(z)| = \psi(\zeta).$$

(4) $(z\phi)'$ *has an analytic continuation across* $\mathbb{T} \setminus \underline{Z}(\phi)$.

We will be saying more about pseudocontinuations in the next section.

REMARK 7.7.   (1) Statement (1) of the above theorem is significant in a more subtle way. The derivatives of Dirichlet functions belong to the Bergman space and it is well-known that Bergman functions need not have finite radial limits almost everywhere [**22**, p. 86]. The fact that $(z\phi)'$

belongs to the Bergman space yet has radial limits makes this extremal function distinctive.

(2) In statement (4), there are extremal functions ϕ such that $\underline{Z}(\phi) \supset \mathbb{T}$ [**49**, Thm. 4.3].

Before moving on to discuss the cyclic vectors for S, we make one final remark about multipliers. If $\mathcal{M} = \vartheta H^2$, where ϑ is inner, then $P_{\mathcal{M}}$, the projection of H^2 onto $\mathcal{M}$, is given by the formula

$$P_{\mathcal{M}} = M_\vartheta M_\vartheta^*,$$

where $M_\vartheta f = \vartheta f$. For the Dirichlet space, we have the following extension [**25, 37**].

THEOREM 7.8. *Suppose* $\mathcal{M} \in Lat(S, \mathcal{D})$. *Then there is a sequence* $(\phi_n)_{n \geqslant 1} \subset \mathcal{M}$ *such that each* ϕ_n *is a multiplier of* $\mathcal{D}$ *and*

$$P_{\mathcal{M}} = \sum_{n=1}^{\infty} M_{\phi_n} M_{\phi_n}^*,$$

where the convergence above is in the strong operator topology. Furthermore,

$$\lim_{r \to 1^-} \sum_{n=1}^{\infty} |\phi_n(r\zeta)| = 1$$

for almost every $\zeta \in \mathbb{T}$.

We end this section with a discussion of the cyclic vectors for S on $\mathcal{D}$. Recall from Beurling's theorem (Theorem 7.1) that $f \in H^2$ is cyclic for S on H^2 if and only if f is an outer function. Suppose that f is cyclic for S on $\mathcal{D}$, then there is a sequence of polynomials $(p_n)_{n \geqslant 1}$ such that $p_n f \to 1$ in $\mathcal{D}$. But since the $\mathcal{D}$ norm dominates the H^2 norm, then $p_n f \to 1$ in the H^2 norm. Now apply Beurling's theorem to say the following.

PROPOSITION 7.9. *If* $f \in \mathcal{D}$ *is cyclic for* S, *then* f *is outer.*

Unfortunately there are outer functions which are not cyclic [**49**, Theorem 4.3] and so being an outer function does not guarantee cyclicity in the Dirichlet space. Indeed, one can find a set $E \subset \mathbb{T}$ so that $\mathcal{D}_E \neq \{0\}$ and such spaces always contain outer functions.

PROPOSITION 7.10. *Suppose* $f \in \mathcal{D}$ *and*

$$E_f := \left\{ \zeta \in \mathbb{T} : \lim_{r \to 1^-} f(r\zeta) = 0 \right\}.$$

If f *is cyclic for* S *on* $\mathcal{D}$, *then* E_f *must have capacity zero.*

Let us outline a proof of this result since it brings in the potential function f_μ mentioned earlier in eq.(4.5). Recall that with the inner product

$$(f, g) := f(0)\overline{g(0)} + \sum_{n=1}^{\infty} n a_n \overline{b_n},$$

where the a_n's are the Taylor coefficients of f and the b_n's are those for g, one can show that $f_\mu \in \mathcal{D}$, whenever $\mu \in \mathcal{E}$, and

$$(g, f_\mu) = \int g(\zeta) \, d\mu(\zeta).$$

Thus whenever $g \in \mathcal{D}$ and E_g has positive capacity, there is a compact subset $F \subset E_g$ with positive capacity and a measure $\mu \in \mathcal{E}$ supported on F. Notice that

$$(S^n g, f_\mu) = \int \zeta^n g(\zeta) d\mu(\zeta) = 0 \ \forall n \in \mathbb{N}_0$$

and so g is not cyclic.

It is also worth pointing out here that if E has zero capacity then certainly $\mathcal{D}_E = \mathcal{D}$. A construction of Brown and Cohn [10] says that if E is a closed set of capacity zero then there is an $f \in \mathcal{D}$ such that f is outer, $f \in A$, $E_f = E$, and f is cyclic. The main conjecture that has remained open for quite some time is the following:

CONJECTURE: A function $f \in \mathcal{D}$ is cyclic for S if and only if f is outer and E_f has capacity zero.

There are several partial results here (see [9, 11, 47, 48] for some examples) that support this conjecture. We mention a two of them.

THEOREM 7.11.   (1) *Suppose $f, g \in \mathcal{D}$ and $|f(z)| \geq |g(z)|$ for all $z \in \mathbb{D}$.*
Then $[f]_S \supset [g]_S$. In particular if g is cyclic, then f is cyclic.
(2) If f and $1/f$ belong to $\mathcal{D}$, then f is cyclic.

REMARK 7.12. Notice how when we replace $\mathcal{D}$ by H^2 in the above theorem how f must be an outer function, and hence cyclic for S on H^2.

Suppose that $f \in \mathcal{D}$ is univalent and cyclic. Then f can have no zeros on $\mathbb{D}$ and so by [22, Theorem 3.17] f is outer. Moreover [6], E_f has logarithmic capacity zero. One can prove [48] this definitive result.

COROLLARY 7.13 (Richter-Sundberg). *If $f \in \mathcal{D}$ is univalent, then f is cyclic for S on $\mathcal{D}$ if and only if $f(z) \neq 0$ for all $z \in \mathbb{D}$.*

We began this section asking whether or not every $\mathcal{M} \in \mathrm{Lat}(S, \mathcal{D})$ is of the form $\mathcal{M}(E, \vartheta)$. This next theorem [48] very much supports this conjecture.

THEOREM 7.14 (Richter-Sundberg). *Let $\mathcal{M} \in Lat(S, \mathcal{D}) \setminus \{0\}$ and ϑ be the greatest common divisor of the inner factors of $\mathcal{M}$. Then, there is an outer function $F \in \mathcal{D}$ such that $F, \vartheta F$ are multipliers of $\mathcal{D}$ and*

$$\mathcal{M} = [\vartheta F]_S = [F]_S \cap \vartheta H^2.$$

For any outer function $f \in \mathcal{D}$, is $[f]_S = \mathcal{D}_{E_f}$?

8. Backward shift invariant subspaces

The backward shift operator

$$Bf = \frac{f - f(0)}{z}$$

is a well-studied operator on H^2 and its invariant subspaces and cyclic vectors are known [21]. We will be more specific in a moment. For the Dirichlet space however, there is much work to be done. Suppose that f is non-cyclic for the backward shift on $\mathcal{D}$, that is to say,

$$[f]_B = \bigvee \{B^n f : n \in \mathbb{N}_0\} \neq \mathcal{D}.$$

Choose an

$$L \in [f]_B^\perp \setminus \{0\}.$$

Let us say a brief word about notation. Certainly $\mathcal{D}$ is a Hilbert space and so, via the Riesz representation theorem, all linear functionals are identified with unique elements of $\mathcal{D}$. However, the approach we are taking here from [51] to examine non-cyclic vectors, works in the general setting of Banach spaces of analytic functions where identifying the dual space is more complicated. With the annihilating L from above, form the meromorphic function

$$f_L(z) := L\left(\frac{f}{w-z}\right) \Big/ L\left(\frac{1}{w-z}\right), \quad z \in \mathbb{D}_e := \widehat{\mathbb{C}} \setminus \mathbb{D}^-.$$

In the setting of H^2, the above function f_L can be written as the quotient of two Cauchy integrals and so, via some well-known facts such as Fatou's jump theorem and the F. and M. Riesz theorem, one can show that the non-tangential limits of f (from $\mathbb{D}$) and f_L (from $\mathbb{D}_e$) both exist and are equal almost everywhere. One says that f and f_L are *pseudocontinuations* of each other [5] [18, p. 85] (see below). In fact, f_L is a pseudocontinuation of f whenever f is a non-cyclic vector for B on many of the Bergman-type spaces [5].

For meromorphic functions g on $\mathbb{D}$ and G on $\mathbb{D}_e$ we say they are *pseudocontinuations* of each other if the non-tangential limits of g and G exist and are equal almost everywhere. The following theorem of Privalov [19] implies that if g has a pseudocontinuation G, it must be unique.

THEOREM 8.1 (Privalov's uniqueness theorem). *Suppose that f is meromorphic on $\mathbb{D}$ and that the non-tangential limits of f vanish on a set of positive measure in $\mathbb{T}$. Then f must be identically zero.*

Thus, in the Hardy space setting, f_L is a pseudocontinuation of f and is independent of the annihilating L. Douglas, Shapiro, and Shields [21] completely characterize the non-cyclic vectors for B on H^2.

THEOREM 8.2 (Douglas-Shapiro-Shields). *A necessary and sufficient condition that $f \in H^2$ be non-cyclic for B is that f has a pseudocontinuation that can be written as the quotient of two bounded analytic functions on $\mathbb{D}_e$.*

Though the existence of a pseudocontinuation may seem somewhat mysterious, the fact that they are unique does give some specific information. For example, functions in H^2 which have isolated winding points, something like $f(z) = \sqrt{1-z}$, must be cyclic vectors for B on H^2. Indeed, if they were not, then f would have a pseudocontinuation f_L. However, f has an analytic continuation across any arc not meeting the point $z = 1$. By Privalov's uniqueness theorem, the analytic and pseudocontinuations must be one in the same, at least in some neighborhood of the arc. This would place a branch cut in the domain of analyticity of f_L, which is impossible (since f_L is meromorphic on $\mathbb{D}_e$). We refer the reader to [18, 21, 50] for more about the backward shift on the classical Hardy spaces.

In the Dirichlet space, the situation is very different. For one, the meromorphic function f_L is no longer a pseudocontinuation of f. In fact pseudocontinuations seem to have nothing to do with non-cyclic vectors on $\mathcal{D}$ [5].

THEOREM 8.3. *There is a non-cyclic vector f for B on $\mathcal{D}$ which does not have a pseudocontinuation across any set of positive measure in $\mathbb{T}$. More specifically, there is no set $E \subset \mathbb{T}$ of positive measure and no meromorphic G on $\mathbb{D}_e$ such that the non-tangential limits of f and G exist and are equal on E.*

The situation gets more complicated by the fact that, unlike the H^2 case (where f_L is a pseudocontinuation of f and as such, via Privalov's uniqueness theorem, is independent of the annihilating L), f_L depends on the annihilating L [50].

THEOREM 8.4. *There is a non-cyclic $f \in \mathcal{D}$ and $L_1, L_2 \in [f]_B^{\perp} \setminus \{0\}$ such that f_{L_1} is a pseudocontinuation of f while f_{L_2} is not.*

There are some positive results that seem to indicate that f_L can be regarded as a 'continuation' of f even though it is not a pseudocontinuation of f. We mention two of them. The first is from [5] while the second is from [50, 51].

PROPOSITION 8.5. *For non-cyclic $f \in \mathcal{D}$, and $L \in [f]_B^{\perp} \setminus \{0\}$, the non-tangential limit as $z \to \zeta$ of*

$$L\left(\frac{1}{w-z}\right)\{f_L(z) - f(\zeta)\}$$

is equal to zero for almost every $\zeta \in \mathbb{T}$.

In particular, this theorem says that if the function $z \to L((w-z)^{-1})$ has finite non-tangential limits almost everywhere, then f_L is a pseudocontinuation of f.

THEOREM 8.6. *Suppose $f \in \mathcal{D}$ is non-cyclic for B and has an analytic continuation to an open neighborhood U_ζ of $\zeta \in \mathbb{T}$. Then for any $L \in [f]_B^{\perp} \setminus \{0\}$, f_L agrees with f on U_ζ.*

A nice corollary, as was the case for H^2, is that the function $f(z) = (1-z)^{3/2}$ is a cyclic vector for B on $\mathcal{D}$.

COROLLARY 8.7. *Any $f \in \mathcal{D}$ with an isolated winding point on $\mathbb{T}$ must be cyclic for B on $\mathcal{D}$.*

We also wish to make some remarks about the possible linear structure on the set of non-cyclic vectors. Using the Douglas, Shapiro, Shields characterization of the non-cyclic vectors for B on H^2 (Theorem 8.2), one can prove that the sum of two non-cyclic vectors must be non-cyclic. Indeed, if f_1, f_2 are non-cyclic, then f_1 and f_2 have pseudocontinuations F_1/G_1 and F_2/G_2 respectively where F_j, G_j are bounded analytic functions on $\mathbb{D}_e$. The sum $f_1 + f_2$ will have $F_1/G_1 + F_2/G_2$ as a pseudocontinuation. For the Dirichlet space, we have the following curious pathology.

THEOREM 8.8. *There are two non-cyclic vectors f, g for B on $\mathcal{D}$, such that $f + g$ is cyclic.*

This phenomenon was originally discovered by S. Walsh [63]. Abakumov [1, 2] proved the some result in a more general setting by a gap series argument. There is another proof in [50] that uses some old spectral synthesis results of Beurling.

So far, we have discussed cyclic vectors for B on $\mathcal{D}$, at least as well as we could. What about a description of $\mathrm{Lat}(B, \mathcal{D})$? The B-invariant subspaces of H^2 are known. Indeed suppose $\mathcal{M} \in \mathrm{Lat}(B, H^2)$. Then, since $B = S^*$ on H^2, we know that $\mathcal{M}^{\perp}$ is S-invariant. By Beurling's theorem (Theorem 7.1) $\mathcal{M}^{\perp} = \vartheta H^2$, where ϑ is inner, and so $\mathcal{M} = (\vartheta H^2)^{\perp}$. A well-known characterization of Douglas, Shapiro, and Shields [21] better describes $(\vartheta H^2)^{\perp}$.

THEOREM 8.9 (Douglas-Shapiro-Shields). *For an inner function ϑ, the following are equivalent for $f \in H^2$.*

(1) $f \in (\vartheta H^2)^\perp$.

(2) f/ϑ has a pseudocontinuation G such that $G(1/z) \in H^2$ and vanishes at $z = 0$.

(3) There is a $g \in H^2$ with $g(0) = 0$ such that $f = \vartheta \overline{g}$ almost everywhere on $\mathbb{T}$.

Unfortunately, there is no similar type of theorem for $\text{Lat}(B, \mathcal{D})$. However, if one is willing to recast the problem in terms of approximation by rational functions, there is something to be said. Here is the set up. For each $n \in \mathbb{N}$, choose a finite sequence

$$E_n := \{z_{n,1}, \cdots, z_{n,N(n)}\}$$

of points of $\mathbb{D}$ (multiplicities are allowed) to create the tableau $\mathcal{S}$

$$z_{1,1}, z_{1,2}, \cdots, z_{1,N(1)}$$

$$z_{2,1}, z_{2,2}, \cdots, z_{2,N(2)}$$

$$\vdots$$

For each n, create the finite dimensional B-invariant subspace of $X = H^2$ (or $\mathcal{D}$)

$$(8.10) \qquad R_n := \bigvee \left\{ \frac{1}{(1 - \overline{z_{n,j}}z)^s} : j = 1, \cdots, N(n), s = 1, \cdots, \text{mult}(z_{n,j}) \right\},$$

where $\text{mult}(z_{n,j})$ is the number of times $z_{n,j}$ appears in E_n, the n-th row of the tableau. If $z_{n,j} = 0$ with multiplicity k, then the functions $1, z, z^2, \cdots, z^k$ are added to the spanning set for R_n. One can now form the 'liminf space' associated with the tableau $\mathcal{S}$ by

$$(8.11) \qquad R(\mathcal{S}) := \varliminf R_n = \left\{ f \in X : \lim_{n \to \infty} \text{dist}(f, R_n) = 0 \right\}.$$

When $X = H^2$, there is a condition that determines when $R(\mathcal{S}) \neq H^2$ [59, 62]: If

$$\beta(E_n) := \sum_{j=1}^{N(n)} (1 - |z_{n,j}|),$$

then

$$(8.12) \qquad R(\mathcal{S}) \neq H^2 \quad \Leftrightarrow \quad \lim_{n \to \infty} \beta(E_n) < \infty.$$

In the Dirichlet space, the quantity $\beta(E_n)$ is replaced by another quantity suitable for the Dirichlet space [39].

THEOREM 8.13. Let X be either H^2 or $\mathcal{D}$. If $\mathcal{M} \in \text{Lat}(B, X)$ with $\mathcal{M} \neq X$, then there is a tableau $\mathcal{S}$ so that

$$\mathcal{M} = R(\mathcal{S}).$$

The H^2 case was done by Tumarkin [59] while the $\mathcal{D}$ case was done recently by Shimorin [55]. A description of $\text{Lat}(B, \mathcal{D})$ in terms of the 'continuation' properties of the function, as was done with pseudocontinuations in the H^2 case, is very much an open problem worth of study.

We end this section with a final remark from [24] which says that the B-invariant subspaces of $\mathcal{D}$ have the F-property.

PROPOSITION 8.14. If $f \in \mathcal{D}$ and ϑ is inner with $f/\vartheta \in H^2$, then $f/\vartheta \in [f]_B$.

Notice from Theorem 8.9 how the B-invariant subspaces of H^2 have the F-property.

References

1. E. Abakumov, *Essais sur les operateurs de Hankel et capacité d'approximation des séries lacunaires*, Ph.D., L'Universite Bordeaux I, May 1994.

2. _____, *Cyclicity and approximation by lacunary power series*, Michigan Math. J. **42** (1995), no. 2, 277–299. MR **96f**:47056

3. D. Adams and L. Hedberg, *Function spaces and potential theory*, Grundlehren der Mathematischen Wissenschaften [Fundamental Principles of Mathematical Sciences], vol. 314, Springer-Verlag, Berlin, 1996. MR **1411441 (97j**:46024)

4. H. Aikawa, *Harmonic functions having no tangential limits*, Proc. Amer. Math. Soc. **108** (1990), no. 2, 457–464. MR **0990410 (90h**:31003)

5. A. Aleman, S. Richter, and W. T. Ross, *Pseudocontinuations and the backward shift*, Indiana Univ. Math. J. **47** (1998), no. 1, 223–276. MR **1631561 (2000i**:47009)

6. A. Beurling, *Ensembles exceptionnels*, Acta Math. **72** (1939), 1–13. MR 0001370 (1,226a)

7. _____, *On two problems concerning linear transformations in Hilbert space*, Acta Math. **81** (1948), 17. MR 0027954 (10,381e)

8. K. Bogdan, *On the zeros of functions with finite Dirichlet integral*, Kodai Math. J. **19** (1996), no. 1, 7–16. MR **1374458 (96k**:30005)

9. L. Brown, *Invertible elements in the Dirichlet space*, Canad. Math. Bull. **33** (1990), no. 4, 419–422. MR **1091345 (92e**:46054)

10. L. Brown and W. Cohn, *Some examples of cyclic vectors in the Dirichlet space*, Proc. Amer. Math. Soc. **95** (1985), no. 1, 42–46. MR **796443 (86j**:30043)

11. L. Brown and A. Shields, *Cyclic vectors in the Dirichlet space*, Trans. Amer. Math. Soc. **285** (1984), no. 1, 269–303. MR **748841 (86d**:30079)

12. L. Carleson, *On the zeros of functions with bounded Dirichlet integrals*, Math. Z. **56** (1952), 289–295. MR 0051298 (14,458e)

13. _____, *Sets of uniqueness for functions regular in the unit circle*, Acta Math. **87** (1952), 325–345. MR 0050011 (14,261a)

14. _____, *A representation formula for the Dirichlet integral*, Math. Z. **73** (1960), 190–196. MR 0112958 (22 #3803)

15. _____, *Selected problems on exceptional sets*, Van Nostrand Mathematical Studies, No. 13, D. Van Nostrand Co., Inc., Princeton, N.J.-Toronto, Ont.-London, 1967. MR 0225986 (37 #1576)

16. J. Caughran, *Two results concerning the zeros of functions with finite Dirichlet integral*, Canad. J. Math. **21** (1969), 312–316. MR 0236396 (38 #4692)

17. _____, *Zeros of analytic functions with infinitely differentiable boundary values*, Proc. Amer. Math. Soc. **24** (1970), 700–704. MR 0252649 (40 #5868)

18. J. A. Cima and W. T. Ross, *The backward shift on the Hardy space*, Mathematical Surveys and Monographs, vol. 79, American Mathematical Society, Providence, RI, 2000. MR **1761913 (2002f**:47068)

19. E. F. Collingwood and A. J. Lohwater, *The theory of cluster sets*, Cambridge University Press, Cambridge, 1966. MR 38 #325

20. J. Douglas, *Solution of the problem of Plateau*, Trans. Amer. Math. Soc. **33** (1931), no. 1, 263–321. MR 1501590

21. R. G. Douglas, H. S. Shapiro, and A. L. Shields, *Cyclic vectors and invariant subspaces for the backward shift operator.*, Ann. Inst. Fourier (Grenoble) **20** (1970), no. fasc. 1, 37–76. MR 42 #5088

22. P. L. Duren, *Theory of H^p spaces*, Academic Press, New York, 1970. MR 42 #3552

23. T. M. Flett, *A high-indices theorem*, Proc. London Math. Soc. (3) **7** (1957), 142–149. MR 0086914 (19,266c)

24. S. Garcia, *The backward shift on Dirichlet-type spaces*, Proc. Amer. Math. Soc. **133** (2005), no. 10, 3047–3056 (electronic). MR 2159784

25. D. Greene, S. Richter, and C. Sundberg, *The structure of inner multipliers on spaces with complete Nevanlinna-Pick kernels*, J. Funct. Anal. **194** (2002), no. 2, 311–331. MR **1934606 (2003h**:46038)

26. V. P. Havin and N. K. Nikolski (eds.), *Linear and complex analysis. Problem book 3. Part II*, Lecture Notes in Mathematics, vol. 1574, Springer-Verlag, Berlin, 1994. MR **1334346** (**96c**:00001b)

27. K. Hoffman, *Banach spaces of analytic functions*, Dover Publications Inc., New York, 1988, Reprint of the 1962 original. MR **92d**:46066

28. S. V. Hruščev, *Sets of uniqueness for the Gevrey classes*, Ark. Mat. **15** (1977), no. 2, 253–304. MR 0463443 (57 #3393)

29. J. Kinney, *Tangential limits of functions of the class S_α*, Proc. Amer. Math. Soc. **14** (1963), 68–70. MR 0143916 (26 #1466)

30. B. Korenblum, *Closed ideals of the ring A^n*, Funkcional. Anal. i Priložen. **6** (1972), no. 3, 38–52. MR 0324424 (48 #2776)

31. E. Landau, *Darstellung und begründung einiger neuerer ergebnisse der funktionentheorie*, Springer-Verlag, 1929.

32. J. Littlewood, *On a theorem of Fatou*, J. London Math. Soc. **2** (1927), 172–176.

33. A. J. Lohwater and G. Piranian, *The boundary behavior of functions analytic in a disk*, Ann. Acad. Sci. Fenn. Ser. A. I. **1957** (1957), no. 239, 17. MR 0091342 (19,950c)

34. O. Lokki, *Über analytische Funktionen, deren Dirichlet-integral endlich ist und die in gegebenen Punkten vorgeschriebene Werte annehmen*, Ann. Acad. Sci. Fennicae. Ser. A. I. Math.-Phys. **1947** (1947), no. 39, 57. MR 0022910 (9,277b)

35. P. Malliavin, *Sur l'analyse harmonique de certaines classes de séries de Taylor*, Symposia Mathematica, Vol. XXII (Convegno sull'Analisi Armonica e Spazi di Funzioni su Gruppi Localmente Compatti, INDAM, Rome, 1976), Academic Press, London, 1977, pp. 71–91. MR 0585689 (58 #28509)

36. A. Matheson, *Closed ideals in rings of analytic functions satisfying a Lipschitz condition*, Banach spaces of analytic functions (Proc. Pelczynski Conf., Kent State Univ., Kent, Ohio, 1976), Springer, Berlin, 1977, pp. 67–72. Lecture Notes in Math., Vol. 604. MR 0463926 (57 #3864)

37. S. McCullough and T. Trent, *Invariant subspaces and Nevanlinna-Pick kernels*, J. Funct. Anal. **178** (2000), no. 1, 226–249. MR **1800795** (**2002b**:47006)

38. A. Nagel, W. Rudin, and J. H. Shapiro, *Tangential boundary behavior of functions in Dirichlet-type spaces*, Ann. of Math. (2) **116** (1982), no. 2, 331–360. MR **672838** (**84a**:31002)

39. N. K. Nikol'skiĭ, *Treatise on the shift operator*, Springer-Verlag, Berlin, 1986. MR **87i**:47042

40. W. P. Novinger, *Holomorphic functions with infinitely differentiable boundary values.*, Illinois J. Math. **15** (1971), 80–90. MR 0269861 (42 #4754)

41. S. Richter, *Invariant subspaces in Banach spaces of analytic functions*, Trans. Amer. Math. Soc. **304** (1987), no. 2, 585–616. MR **911086** (**88m**:47056)

42. _____, *Invariant subspaces of the Dirichlet shift*, J. Reine Angew. Math. **386** (1988), 205–220. MR **936999** (**89e**:47048)

43. _____, *A representation theorem for cyclic analytic two-isometries*, Trans. Amer. Math. Soc. **328** (1991), no. 1, 325–349. MR **1013337** (**92e**:47052)

44. S. Richter, W. T. Ross, and C. Sundberg, *Hyperinvariant subspaces of the harmonic Dirichlet space*, J. Reine Angew. Math. **448** (1994), 1–26. MR **1266744** (**95e**:47045)

45. _____, *Zeros of functions with finite Dirichlet integral*, Proc. Amer. Math. Soc. **132** (2004), no. 8, 2361–2365 (electronic). MR **2052414** (**2005b**:30007)

46. S. Richter and A. Shields, *Bounded analytic functions in the Dirichlet space*, Math. Z. **198** (1988), no. 2, 151–159. MR **939532** (**89c**:46039)

47. S. Richter and C. Sundberg, *A formula for the local Dirichlet integral*, Michigan Math. J. **38** (1991), no. 3, 355–379. MR **1116495** (**92i**:47035)

48. _____, *Multipliers and invariant subspaces in the Dirichlet space*, J. Operator Theory **28** (1992), no. 1, 167–186. MR **1259923** (**95e**:47007)

49. _____, *Invariant subspaces of the Dirichlet shift and pseudocontinuations*, Trans. Amer. Math. Soc. **341** (1994), no. 2, 863–879. MR **1145733** (**94d**:47026)

50. W. T. Ross and H. S. Shapiro, *Generalized analytic continuation*, University Lecture Series, vol. 25, American Mathematical Society, Providence, RI, 2002. MR **1895624** (**2003h**:30003)

51. _____, *Prolongations and cyclic vectors*, Comput. Methods Funct. Theory **3** (2003), no. 1-2, 453–483. MR **2082029** (**2005h**:30001)

52. W. Rudin, *The closed ideals in an algebra of analytic functions*, Canad. J. Math. **9** (1957), 426–434. MR 0089254 (19,641c)

53. H. S. Shapiro and A. L. Shields, *On the zeros of functions with finite Dirichlet integral and some related function spaces*, Math. Z. **80** (1962), 217–229. MR 0145082 (26 #2617)

54. A. Shields, *Cyclic vectors in Banach spaces of analytic functions*, Operators and function theory (Lancaster, 1984), NATO Adv. Sci. Inst. Ser. C Math. Phys. Sci., vol. 153, Reidel, Dordrecht, 1985, pp. 315–349. MR **810450** (**87c:**47048)

55. S. Shimorin, *Approximate spectral synthesis in the Bergman space*, Duke Math. J. **101** (2000), no. 1, 1–39. MR **2001d:**47015

56. N. A. Shirokov, *Analytic functions smooth up to the boundary*, Springer-Verlag, Berlin, 1988. MR **90h:**30087

57. D. Stegenga, *Multipliers of the Dirichlet space*, Illinois J. Math. **24** (1980), no. 1, 113–139. MR **0550655** (**81a:**30027)

58. B. A. Taylor and D. L. Williams, *Zeros of Lipschitz functions analytic in the unit disc*, Michigan Math. J. **18** (1971), 129–139. MR 0283176 (44 #409)

59. G. C. Tumarkin, *Description of a class of functions admitting an approximation by fractions with preassigned poles*, Izv. Akad. Nauk Armjan. SSR Ser. Mat. **1** (1966), no. 2, 89–105. MR 34 #6123

60. J. B. Twomey, *Radial variation of functions in Dirichlet-type spaces*, Mathematika **44** (1997), no. 2, 267–277. MR **1600533** (**99d:**30033)

61. _____, *Tangential boundary behaviour of harmonic and holomorphic functions*, J. London Math. Soc. (2) **65** (2002), no. 1, 68–84. MR **1875136** (**2002k:**31006)

62. J. L. Walsh, *Interpolation and approximation by rational functions in the complex plane*, Amer. Math. Soc. Coll. Pub. (20), Providence, RI, 1935.

63. S. Walsh, *Noncyclic vectors for the backward Bergman shift*, Acta Sci. Math. (Szeged) **53** (1989), no. 1-2, 105–109. MR **1018678** (**90k:**47068)

64. Z. Wu, *Function theory and operator theory on the Dirichlet space*, Holomorphic spaces (Berkeley, CA, 1995), Math. Sci. Res. Inst. Publ., vol. 33, Cambridge Univ. Press, Cambridge, 1998, pp. 179–199. MR **1630650** (**2000a:**47061)

DEPARTMENT OF MATHEMATICS AND COMPUTER SCIENCE, UNIVERSITY OF RICHMOND, RICH-MOND, VIRGINIA 23173

E-mail address: wross@richmond.edu

Contemporary Mathematics
Volume **393**, 2006

Comparing topologies on the space of composition operators

Eero Saksman and Carl Sundberg

This paper is dedicated to Joseph Cima on the occasion of his 70th birthday.

ABSTRACT. Consider the set of analytic composition operators on the unit disc
$$\mathcal{C} := \{C_\phi \ : \ \phi : \mathbb{D} \to \mathbb{D} \text{ analytic}\}.$$
The set $\mathcal{C}$ is topologized in a natural way when it is considered as a subset of
the space of bounded linear operators on X, where X is a space of analytic
functions on $\mathbb{D}$ such that all the C_ϕ act boundedly on X. We compare the
different topologies so obtained in the case where X is a classical Hardy space,
Bergman space, or the Bloch space.

1. Introduction

For an analytic self-map $\phi : \mathbb{D} \to \mathbb{D}$ we let C_ϕ stand for the map $f \mapsto f \circ \phi$,
where f in any analytic function on $\mathbb{D}$. The set of all such composition operators is
denoted by
$$\mathcal{C} := \{C_\phi \ : \ \phi : \mathbb{D} \to \mathbb{D} \text{ analytic}\}.$$
It is also useful to consider separately symbols that vanish at zero:
$$\mathcal{C}_0 := \{C_\phi \ : \ \phi : \mathbb{D} \to \mathbb{D} \text{ analytic with } \phi(0) = 0\}.$$

Let X be a Banach space of analytic functions on $\mathbb{D}$, and let $L(X)$ stand for the space
of all bounded linear operators on X. Assume that $C_\phi \in L(X)$ for all $C_\phi \in \mathcal{C}$. The
set $\mathcal{C}$ is then topologized in a natural way by considering it as a subset of $L(X)$. Let
us call the induced topology τ_X. Given two such spaces X_1, X_2 a natural question
arises: are the topologies τ_{X_1} and τ_{X_2} comparable? In the present note we answer
this question in the case of most classical spaces, including the Hardy and Bergman
spaces.

This question is closely related to several other problems that have been studied previously in the theory of analytic composition operators. For instance, the
topological isolation of composition opeators was studied in [**3**] and [**21**], and the

2000 *Mathematics Subject Classification.* 47B33,32A35,32A36,30D55.

We thank Lund University and our collegues there for their hospitality during our very
pleasant stay there. We also thank Pekka Nieminen for his valuable comments on the manuscript.

component structure of $\mathcal{C}$ has been addressed in many works including [11], [12], [16], [6], [9], and [17]. Moreover, this study is closely related to the study of compactness of the differences of composition operators, as is seen in the papers already mentioned. Among other works in that direction we mention here [15] and [18].

In order to explain the results of the present note we first recall that if $\alpha > -1$ and $p \in [1, \infty)$ one denotes by $A_\alpha^p = A_\alpha^p(\mathbb{D})$ the weighted Bergman space with the norm given by $\|f\|_{A_\alpha^p}^p = \frac{(\alpha+1)}{\pi} \int_{\mathbb{D}} |f(z)|^p \, m(dz)(1 - |z|^2)^\alpha$, where m is the 2-dimensional Lebesgue measure. Actually, one obtains the classical Hardy spaces H^p as a natural limit when $\alpha \to -1^+$, whence we denote $A_{-1}^p := H^p$ for $p \in [1, \infty)$. In the unweighted case one applies the abbreviation $A_0^p = A^p$. As usual, $\|f\|_{H^\infty} = \sup_{|z|<1} |f(z)|$. Finally, the norm in the Bloch space $\mathcal{B}$ is $\|f\|_{\mathcal{B}} = |f(0)| + \sup_{|z|<1} |(1 - |z|^2)| |f'(z)|$.

It is useful to introduce special notation for the most important topologies considered here:

τ_H is the topology on $\mathcal{C}$ induced by $L(H^2)$,

τ_A is the topology on $\mathcal{C}$ induced by $L(A^2)$,

τ_∞ is the topology on $\mathcal{C}$ induced by $L(H^\infty)$,

$\tau_\mathcal{B}$ is the topology on $\mathcal{C}$ induced by $L(\mathcal{B})$.

For two topologies τ_1, τ_2 on the same set we write $\tau_1 \prec \tau_2$ if τ_2 is finer than τ_1. Our main result is the following.

THEOREM 1.1. **(i)** $\tau_A \prec \tau_H \prec \tau_\infty$. *None of these inclusions can be reversed.*

(ii) $\tau_A \prec \tau_\mathcal{B} \prec \tau_\infty$. *None of these inclusions can be reversed.*

(iii) *The topologies τ_H and $\tau_\mathcal{B}$ are not comparable.*

(iv) *For any $p \in [1, \infty)$ it holds that $\tau_{H^p} = \tau_H$.*

(v) *For any $p \in (1, \infty)$ and $\alpha > -1$ it holds that $\tau_{A_\alpha^p} = \tau_A$.*

For the symbols that fix the origin we prove the following sharp quantitative result.

THEOREM 1.2. *Let $T = \sum_{k=1}^m c_k C_{\phi_k}$ be an arbitrary (finite) linear combination of composition operators, where $C_{\phi_k} \in \mathcal{C}_0$ for all k. Then, if $-1 \leq \alpha < \beta$ and $p \in [1, \infty)$ it holds that*

$$(1.1) \qquad \|T\|_{A_\beta^p \to A_\beta^p} \leq \|T\|_{A_\alpha^p \to A_\alpha^p}.$$

Conversely, in the Hilbert space case for any $C_\phi, C_\psi \in \mathcal{C}_0$ the corresponding difference of composition operators satisfies

$$(1.2) \qquad \|C_\phi - C_\psi\|_{A_\alpha^2 \to A_\alpha^2} \leq c(\alpha, \beta)(\|C_\phi - C_\psi\|_{A_\beta^2 \to A_\beta^2})^{\frac{\alpha+1}{\beta+1}}.$$

The exponent $\frac{\alpha+1}{\beta+1}$ is optimal.

The previous Theorem has immediate corollaries to the study of Banach spaces generated by the composition operators. Thus, let X be a space of analytic functions on which all the C_ϕ operate boundedly. We define the Banach space E_X (resp. $E_{X,0}$) that is obtained by taking the closed linear span of the set $\mathcal{C}$ (resp. $\mathcal{C}_0$) inside $L(X)$. Theorem 1.2 shows immediately that the spaces $E_{A_\alpha^2,0}$ for $\alpha \geq -1$ are naturally ordered by strict inclusions.

The proofs of the above results are given in Section 2. Section 3 contains further comments and discussion on the topic.

2. Proof of Theorems 1.1 and 1.2

One can summarize our proof with four words: apply interpolation of operators! Let us first review the needed interpolation results in the simplest form that will suit our purposes. For further properties and basic notions of interpolation theory we refer to [5] or [4]. Let I be one of the intervals $(0,1)$, $[0,1)$, or $[0,1]$ and let $(X_s)_{s \in I}$ be an increasing (resp. decreasing) scale of Banach spaces, i.e. for $s < t$ (resp. for $s > t$) with $s, t \in I$ one has $X_s \subset X_t$ with continuous embedding, and X_s is dense in X_t. We say that the scale $(X_s)_{s \in I}$ is a *scale of interpolation* if for any $s, t \in I$ and $\theta \in (0,1)$ we have

$$\|T\|_{X_u \to X_u} \leq C(u,s,t)(\|T\|_{X_s \to X_s})^\theta (\|T\|_{X_t \to X_t})^{1-\theta},$$

where $u = \theta s + (1-\theta)t$ and T is any bounded linear operator on the larger of the spaces X_s, X_t that also restricts to a bounded linear operator on the smaller space.

LEMMA 2.1. *The following are scales of interpolation:*
(i) $(H^{1/s})_{s \in [0,1]}$ *with the understanding that* $1/0 = \infty$.
(ii) $(A_\alpha^{1/s})_{s \in (0,1]}$ *for any fixed* $\alpha > -1$.
(iii) $(A_{-1+ks}^2)_{s \in [0,1]}$ *for any fixed* $k > 0$.

PROOF. The claim of (i) is immediate, since it is known [10] that the spaces $H^{1/s}$, $0 \leq s \leq 1$ form a scale both in the real and complex interpolation methods. In turn, to obtain (ii) we fix $\alpha > -1$ and recall (see [8, Thm 1.10] that there is a fixed projection P that is a bounded projection from the weighted space $L^p(\mathbb{D}, \pi^{-1}(\alpha + 1)(1 - |z|^2)^\alpha m(dz))$ onto A_α^p for any $p \in [1, \infty)$. If T is the given operator between the weighted Bergman spaces, the claim follows by considering the operator TP and applying a standard interpolation result for weighted L^p spaces [4, Thm. 3.6]. Finally, for $\alpha \geq -1$ the norm of $f(z) = \sum_{k=0}^\infty a_k z^k$ is equivalent to

$$(2.1) \qquad \|f\|_{A_\alpha^2} \sim \left(\sum_{k=0}^\infty |a_k|^2 k^{-2(1+\alpha)} \right)^{1/2}$$

(see [8, Chapter 1, p. 4]). The stated claim is again a simple consequence of interpolation between weighted L^2-spaces. $\qquad \square$

We prove our Theorems in reverse order.

PROOF. *(Of Theorem 1.2)* Consider first the statement (1.2). Let $T = C_\phi - C_\psi$, where $C_\psi, C_\phi \in \mathcal{C}_0$. We apply Lemma 2.1 and interpolate between the weight indices -1 and β to obtain

$$\|T\|_{A_\alpha^2 \to A_\alpha^2} \leq C(\alpha, \beta) \|T\|_{H^2 \to H^2}^{\frac{\beta - \alpha}{\beta + 1}} \|T\|_{A_\beta^2 \to A_\beta^2}^{\frac{\alpha + 1}{\beta + 1}}.$$

The inequality (1.2) follows by observing that $\|T\|_{H^2 \to H^2} \leq 2$. In order to verify the optimality we renorm A_α^2 by the equivalent norm (2.1) and consider the the self-maps $\phi_m(z) = z^m$ and $\psi(z) = 0$. The space A_α^2 possesses an orthonormal basis $\{k^{1+\alpha} z^k\}_{k \geq 0}$. Observe that

$$(C_{\phi_m} - C_\psi)\left(\sum_{k=0}^n a_k k^{1+\alpha} z^k \right) = \sum_{k=1}^n m^{-1-\alpha} a_k (mk)^{1+\alpha} z^{mk}$$

whence clearly

$$(2.2) \qquad \|C_{\phi_m} - C_\psi\|_{A_\alpha^2 \to A_\alpha^2} = m^{-1-\alpha}$$

The optimality of the exponent in (1.2) follows immediately from this by considering large enough values of m.

Let us turn our attention to (1.1). At least for the case $p = 2$ one could prove the stated inequality by interpolating between the weight indices -1 and γ, letting $\gamma \to \infty$ and applying the fact that the norm of T stays uniformly bounded on A_γ^2 as $\gamma \to \infty$. However, in order to obtain the sharp estimate (1.1) we present here a self-contained argument that is based on a simple application of the maximum principle for vector-valued analytic functions (see (2.3) below).

Fix $p \in [1, \infty)$ and the parameters $-1 \leq \alpha < \beta$. We may as well assume that $\alpha > -1$, since the case $\alpha = -1$ is then obtained by considering the action only on polynomials and letting $\alpha \to -1^+$. Let $T = \sum_{k=1}^m c_k C_{\phi_k}$, where $C_{\phi_k} \in C_0$ for all k. For any given analytic function f on $\mathbb{D}$ and $r \in [0, 1]$ we define, as usual, $f_r(z) := f(rz)$. For self-maps ϕ with $\phi(0) = 0$ and $r \in (0, 1]$ it will also be useful to consider the map $\phi_{(r)}$, where $\phi_{(r)}(z) := r^{-1}\phi(rz)$ for $z \in \mathbb{D}$. The Schwarz lemma shows that $\phi_{(r)} \in C_0$. Analoguously, we set

$$T_{(r)} = \sum_{k=1}^m c_k C_{(\phi_k)_{(r)}}.$$

The crux of the proof is the following observation: for any $r \in (0, 1)$ it holds that

(2.3) $$\|T_{(r)}\|_{A_\alpha^p \to A_\alpha^p} \leq \|T\|_{A_\alpha^p \to A_\alpha^p}.$$

To verify this we extend the previous definition of $\phi_{(r)}$ by setting $\phi_{(w)}(z) := w^{-1}\phi(wz)$ if $w \in \overline{\mathbb{D}} \setminus \{0\}$, and $\phi_{(0)}(z) := \phi'(0)z$. By the Taylor expansion of ϕ we have analyticity also in $w \in \mathbb{D}$. This yields also the definition of $T_{(w)}$. Consider the $L(A_\alpha^p)$-valued map λ:

$$\lambda(w) = T_{(w)}, \quad w \in \overline{\mathbb{D}}.$$

Clearly λ is analytic inside $\mathbb{D}$. By considering the action on polynomials, we see that the norm $w \mapsto \|\lambda(w)\|_{A_\alpha^p \to A_\alpha^p}$ is continuous up to the boundary. By rotational symmetry one obviously has $\|\lambda(e^{it})\|_{A_\alpha^p \to A_\alpha^p} = \|T\|_{A_\alpha^p \to A_\alpha^p}$ for every $t \in [0, 2\pi]$. The inequality (2.3) is now a consequence of the maximum principle for vector-valued analytic functions.

Let us define the function $g_{\alpha,\beta}$ on $[0, 1]$ by the formula

$$g_{\alpha,\beta}(t) = \frac{2\Gamma(\beta + 2)}{\Gamma(\beta - \alpha)\Gamma(\alpha + 2)}(1 - t^2)^{\beta - \alpha - 1}t^{2\alpha + 3}.$$

By applying the basic properties of Euler's Beta-integral we see that $\int_0^1 t^{2u+1}(1 - t^2)^v \, dt = \Gamma(u + 1)\Gamma(v + 1)/2\Gamma(v + u + 2)$ for $u, v > -1$. This shows immediately that

(2.4) $$\int_0^1 g_{\alpha,\beta}(t) \, dt = 1.$$

We claim that for every positive and measurable h on $(0, 1)$ it holds that

(2.5) $$\int_0^1 g_{\alpha,\beta}(t) \left(2(\alpha + 1) \int_0^1 h(tr)r(1 - r^2)^\alpha \, dr \right) dt$$

$$= 2(\beta + 1) \int_0^1 h(r)r(1 - r^2)^\beta \, dt \, dr.$$

One may verify this by a change of variables and an application of the Fubini theorem. Instead, we give another argument. By approximation by polynomials it is enough to verify (2.5) in the case where $h(r) = r^{2k}$, with $k \in \{0, 1, \ldots\}$. With this choice the right hand side of (2.5) becomes $\Gamma(k+1)\Gamma(\beta+2)/\Gamma(k+\beta+2)$ and the left hand side breaks into a product of two Beta-type integrals:

$$
\frac{2(\alpha+1)\Gamma(\beta+2)}{\Gamma(\beta-\alpha)\Gamma(\alpha+2)} \int_0^1 t^{2k+2\alpha+3}(1-t^2)^{\beta-\alpha-1}\, dt \int_0^1 r^{2k+1}(1-r^2)^\alpha\, dr
$$

$$
= (\alpha+1)\left(\frac{\Gamma(\beta+2)}{\Gamma(\beta-\alpha)\Gamma(\alpha+2)}\right)\left(\frac{\Gamma(k+\alpha+2)\Gamma(\beta-\alpha)}{\Gamma(\beta+k+2)}\right)\left(\frac{\Gamma(k+1)\Gamma(\alpha+1)}{\Gamma(\alpha+k+2)}\right)
$$

$$
= \frac{\Gamma(k+1)\Gamma(\beta+2)}{\Gamma(k+\beta+2)}.
$$

This proves (2.5)

Let f be a polynomial. We will apply (2.5) twice, on functions h_1 and h_2. Here $h_1(r) = \frac{1}{2\pi}\int_0^{2\pi}|(Tf)(re^{i\theta})|^p\, d\theta$ and $h_2(r) = \frac{1}{2\pi}\int_0^{2\pi}|f(re^{i\theta})|^p\, d\theta$ Observe that for $r \in (0,1)$ we have

$$
h_1(rt) = \frac{1}{2\pi}\int_0^{2\pi}|(T_{(t)}f_t)(re^{i\theta})|^p\, d\theta.
$$

We keep in mind (2.3),(2.4) and estimate

$$
\begin{aligned}
\|Tf\|_{A_\beta^p}^p &= 2(\beta+1)\int_0^1 h_1(r)(1-r^2)^\beta r\, dr \\
&= \int_0^1 g_{\alpha,\beta}(t)\left(2(\alpha+1)\int_0^1 h_1(rt)(1-r^2)^\alpha r\, dr\right) dt \\
&= \int_0^1 g_{\alpha,\beta}(t)\|T_{(t)}f_t\|_{A_\alpha^p}^p\, dt \\
&\leq (\|T\|_{A_\alpha^p\to A_\alpha^p})^p \int_0^1 g_{\alpha,\beta}(t)\|f_t\|_{A_\alpha^p}^p\, dt \\
&= (\|T\|_{A_\alpha^p\to A_\alpha^p})^p \int_0^1 g_{\alpha,\beta}(t)\left(2(\alpha+1)\int_0^1 h_2(rt)(1-r^2)^\alpha r\, dr\right) dt \\
&= (\|T\|_{A_\alpha^p\to A_\alpha^p})^p 2(\beta+1)\int_0^1 h_2(r)(1-r^2)^\beta r\, dr \\
&= (\|T\|_{A_\alpha^p\to A_\alpha^p})^p\|f\|_{A_\beta^p}^p
\end{aligned}
$$

The proof of the Theorem is complete.   □

We next turn to the proof of our other result. We present the application of the interpolation method with full details in connection with part (i) and will not tease the reader by repeating them in analoguous applications. The reader is now asked to recall the well-known characterizations of compactness of C_ϕ on Hardy, Bergman, or Bloch spaces. A comprehensive exposition of these results are found in [20], [7], and [14].

PROOF. (Of Theorem 1.1) We will first show that $\tau_H \prec \tau_\infty$. For that end fix an operator $C_\phi \in \mathcal{C}$. In order to show that the topology at C_ϕ is finer in τ_∞ we choose a sequence $C_{\psi_k} \in \mathcal{C}$ with $\|C_{\psi_k} - C_\phi\|_{H^\infty\to H^\infty} \to 0$ as $k \to \infty$. We are to show that

(2.6) $\|C_{\psi_k} - C_\phi\|_{H^2 \to H^2} \to 0$ as $k \to \infty$.

By evaluating at $f(z) = z$ we observe that $\lim_{k\to\infty} |\psi_k(0) - \phi(0)| = 0$. This implies that $|\psi_k(0)| \le r_0 < 1$ for all k, whence standard estimates show that the norms of C_{ψ_k} and C_ϕ are uniformly bounded on H^1, say with bound A. We apply now interpolation between H^1 and H^∞ (see Lemma 2.1(i)) to obtain

$$\|C_{\psi_k} - C_\phi\|_{H^2 \to H^2} \le C(\|C_{\psi_k} - C_\phi\|_{H^\infty \to H^\infty})^{1/2} A^{1/2}.$$

This immediately yields (2.6).

We still must show that there is no reverse inclusion, i.e. that $\tau_H \ne \tau_\infty$. Consider an operator C_g that is compact on H^2 but not on H^∞; the choice

(2.7) $$g(z) = 1 - \frac{1}{4}\sqrt{1 - z}$$

will do. Let $K(H^\infty)$ stand for the subspace of $L(H^\infty)$ consisting of compact operators. Then obviously $\|C_g - C_{g_r}\|_{H^2 \to H^2} \to 0$ as $r \to 1^-$ but $\|C_g - C_{g_r}\|_{H^\infty \to H^\infty} \ge d(C_g, K(H^\infty)) > 0$ for any $r < 1$. This yields the claim.

The comparison $\tau_A \prec \tau_H$ is established in exactly the same way by applying Lemma 2.1(iii) to obtain A^2 as an interpolation space between H^2 and A_1^2. The same argument as above shows that the inclusion between the topologies is strict since there are compact operators on A^2 that are not compact on H^2 (see [13] or [20]).

We next consider part (ii). Again, to show that $\tau_B \prec \tau_\infty$ we fix C_ϕ and assume that $C_{\psi_k} \in \mathcal{C}$ with $\varepsilon_k := \|C_{\psi_k} - C_\phi\|_{H^\infty \to H^\infty} \to 0$ as $k \to \infty$. Fix z_0 and consider $f(z) := \dfrac{\phi(z_0) - z}{1 - \overline{\phi(z_0)}z}$. Since $\|f\|_\infty = 1$ we obtain

$$\varepsilon_k \ge |((C_{\psi_k} - C_\phi)f)(z_0)| = \left|\frac{\phi(z_0) - \psi_k(z_0)}{1 - \overline{\phi(z_0)}\psi_k(z_0)}\right| =: \rho(\phi(z_0), \psi_k(z_0)).$$

Here z_0 was arbitrary, so we infer that the pseudohyperbolic distance $\rho(\phi(z), \psi_k(z))$ tends to zero uniformly. Fix then a function $f \in \mathcal{B}$ with $\|f\|_\mathcal{B} = 1$. By the definition of the Bloch norm, f is 1-Lipschitz from the hyperbolic metric to the standard metric. Thus for large enough k we have $|f(\phi(z)) - f(\psi_k(z))| = O(\rho(\phi(z), \psi_k(z))) = O(\varepsilon_k)$, uniformly in z. Since the H^∞ norm dominates the Bloch norm we obtain

$$\|(C_{\psi_k} - C_\phi)f\|_\mathcal{B} \le C\|f \circ \psi_k - f \circ \phi\|_{H^\infty} = O(\varepsilon_k) \to 0   \text{ as } k \to \infty,$$

as was to be shown.

In order to treat the relation between the topologies τ_A and τ_B we apply interpolation in a more refined manner. Assume that ψ_k, ϕ are such that $\delta_k := \|C_{\psi_k} - C_\phi\|_{\mathcal{B} \to \mathcal{B}} \to 0$ as $k \to \infty$. Again by evaluating at the function z we see that $|\psi_k(0)| \le r_0 < 1$ for all k, and the norms of C_{ψ_k}, C_ϕ on the Bergman spaces A^2 and A^1 are uniformly bounded. Recall that with respect to the duality pairing

$$\langle f, g \rangle := \lim_{r \to 1^-} \frac{1}{\pi} \int_\mathbb{D} f_r(z)\overline{g}(z)m(dz)$$

we have the dualities $(\mathcal{B}_0)' = A^1$, $(A^1)' = \mathcal{B}$, and $(A^p)' = A^{p'}$ for $p \in (1, \infty)$ with $p^{-1} + p'^{-1} = 1$. Here $\mathcal{B}_0$ stands for the little Bloch space, see [8, Section 1] for these facts and the definition of $\mathcal{B}_0$. Fix $k \ge 1$, $r \in (0, 1)$ and denote $T_{k,r} := C_{(\psi_k)_r} - C_{\phi_r}$.

Then clearly $\|T_{k,r}\|_{\mathcal{B}\to\mathcal{B}} \le \delta_k$. Moreover, $T_{k,r}$ maps $\mathcal{B}_0$ boundedly into itself and we may consider the (Hermitean) adjoint $T'_{k,r}$. Then $T'_{k,r} : A^1 \to A^1$ with equivalent norm. By evaluating with polynomials we check that the restriction of $T'_{k,r}$ on A^p equals the adjoint of $T_{k,r} : A^{p'} \to A^{p'}$. In a similar vein the second adjoint $T''_{k,r}$ equals the operator $C_{(\psi_k)_r} - C_{\phi_r} : \mathcal{B} \to \mathcal{B}$. We are now ready to apply interpolation between A^1 and (say) A^4 to obtain

$$\|T'_{k,r}\|_{A^2 \to A^2} \le C(\|T'_{k,r}\|_{A^1 \to A^1})^{1/3} B^{2/3} \le B^{2/3}\delta_k^{1/3},$$

where $B = \sup \|T_{k,r}\|_{A^4 \to A^4} < \infty$. By duality it follows that $\|T_{k,r}\|_{A^2 \to A^2} \le B^{2/3}\delta_k^{1/3}$. Letting $r \to 1$ we obtain the same estimate for the norm $\|C_{\psi_k} - C_\phi\|_{A^2 \to A^2}$.

Let us now verify that the inclusion $\tau_\mathcal{B} \prec \tau_\infty$ is strict. By [14] there is a symbol g such that $\|g\|_\infty = 1$ but the induced map $C_g : \mathcal{B}_0 \to \mathcal{B}_0$ is compact. Since for functions $f \in \mathcal{B}_0$ it holds that $\|f - f_r\|_{\mathcal{B}_0} \to 0$ as $r \to 0$, it follows that $\lim_{r\to 1^-} \|C_g - C_{g_r}\|_{\mathcal{B}_0 \to \mathcal{B}_0} = 0$, whence by the duality described in the preceeding paragraph $\lim_{r\to 1^-} \|C_g - C_{g_r}\|_{\mathcal{B}\to\mathcal{B}} = 0$. However, $\|C_g - C_{g_r}\|_{H^\infty \to H^\infty} \ge c_0 > 0$ for all r since C_g is not compact on H^∞. An analoguous argument shows that τ_A is strictly coarser than $\tau_\mathcal{B}$ – one just considers C_g where g is defined through (2.7).

It will not be a big surprise for the reader that the proof of incomparability stated in part (iii) will be based on the corresponding incomparability of the sets of compact composition operators on the respective spaces. The fact that $\tau_\mathcal{B} \not\prec \tau_H$ follows just as in the preceeding paragraph by considering C_g with g given by (2.7). In turn, $\tau_H \not\prec \tau_\mathcal{B}$ is obtained by looking at C_ψ, where ψ is an inner function that induces a compact composition operator on $\mathcal{B}_0$ (the existence of such inner functions was shown in [22] and [1]). Thus on the little Bloch space, and, a fortiori, on the Bloch space the operator norm $\|C_\psi - C_{\psi_r}\|$ tends to zero as $r \to 1^-$, and this certainly cannot happen for the operator norm on H^2.

We are ready to proceed to the proof of part (iv). If $p \in (2, \infty)$, we interpolate as before between H^2 and H^{2p} to verify that $\tau_{H^p} \prec \tau_{H^2}$. Another interpolation between H^p and $H^{3/2}$ yields the converse inclusion. The case $p \in (1, 2)$ is similar. Next, an interpolation between H^1 and H^3 shows that $\tau_{H^2} \prec \tau_{H^1}$. To deduce the converse inclusion we could apply interpolation for the quasi-Banach couple $(H^{1/2}, H^2)$. However, a more direct argument is obtained as follows (compare the proof of [18, Prop. 2]). It is standard that any given function $f \in H^1$ with $\|f\|_{H^1} = 1$ can be written as a sum $f = f_1^2 + f_2^2$ where $\|f_1\|_{H^2} = \|f_2\|_{H^2} = 1$. In order to estimate the norm of an arbitrary difference $C_\phi - C_\psi$ on H^1 it is thus enough to evaluate on functions of the form $f = g^2$, where $\|g\|_{H^2} = 1$. We obtain by Cauchy-Schwarz inequality that

$$\|(C_\phi - C_\psi)f\|_{H^1} = \|(C_\phi g - C_\psi g)(C_\phi g + C_\psi g)\|_{H^1}$$
$$\le (\|C_\phi - C_\psi\|_{H^2 \to H^2})^{1/2}(\|C_\phi\|_{H^2 \to H^2} + \|C_\psi\|_{H^2 \to H^2})^{1/2}\|g\|_{H^2}^2$$

Thus we have the inequality

$$\|C_\phi - C_\psi\|_{H^1 \to H^1} \le 2(\|C_\phi - C_\psi\|_{H^2 \to H^2})^{1/2}(\|C_\phi\|_{H^2 \to H^2} + \|C_\psi\|_{H^2 \to H^2})^{1/2}$$

and it is obvious how one uses this inequality to deduce the missing relation $\tau_{H^1} \prec \tau_H$.

In the last part (v) one first obtains $\tau_{A_\alpha^2} = \tau_A$ for all $\alpha > -1$ by interpolation according to Lemma 2.1(iii) just as in the proof of part (iv). Finally, interpolation with respect to p using Lemma 2.1(ii) with fixed α yields the rest of the claim.

The proof of Theorem 1.1 is complete.   □

3. Complementary remarks

Our results have obvious implications to the comparison of the component structure of $(\mathcal{C}, \tau_X)$ with various X.

Let X be a space of analytic functions on which all the C_ϕ operate boundedly. Recall that at the end of Introduction we defined the corresponding completion E_X of the normed space spanned by $\mathcal{C}$. Many natural questions arise here. For example, are there examples of $E_X = E_Y$ (under the natural map) where X and Y are essentially different?

Also, a closely related question is the estimation of the norm of the difference $C_\psi - C_\phi$ on various spaces. On H^∞ one easily shows (see [12], [9] or recall the argument we used in the proof of $\tau_\mathcal{B} \prec \tau_\infty$) that the $\|C_\psi - C_\phi\|_{H^\infty \to H^\infty}$ is comparable to the quantity $\sup_{z \in \mathbb{D}} \rho(\phi(z), \psi(z))$. Corresponding, more complicated estimates in the case of Bloch spaces are contained in the recent work of P. Nieminen [17]. Let us also remark here that in the case of weighted Bergman spaces one may modify slightly the arguments of J. Moorhouse [15] and check that for $C_\psi, C_\phi \in \mathcal{C}_0$ there is an estimate for $\|C_\psi - C_\phi\|_{A^2 \to A^2}$ in terms of the quantity

$$\sup_{z \in \mathbb{D}} \left(\frac{1 - |z|^2}{1 - |\phi(z)|^2} + \frac{1 - |z|^2}{1 - |\psi(z)|^2} \right) \rho(\phi(z), \psi(z)).$$

All the examples considered in this note satisfy the following rule of thumb: Let X_1, X_2 be Banach spaces of analytic functions on $\mathbb{D}$ on which all elements of $\mathcal{C}$ act boundedly. Then $\tau_{X_1} \prec \tau_{X_2}$ only if $X_2 \subset X_1$. Another rule of thumb says that the coarser the topology, the less compact composition operators there are. It would be of interest to find natural examples violating one of these rules, or (under suitable assumptions) a general result to the positive direction.

References

[1] A. B. Aleksandrov, J. M. Anderson, and A. Nicolau: *Inner functions, Bloch spaces and symmetric measures*, Proc. London Math. Soc. (3) 79 (1999), no. 2, 318–352.

[2] C. Barks: *Junior Woodchucks Guide Book.*

[3] E. Berkson: *Composition operators isolated in the uniform operator topology.* Proc. Amer. Math. Soc. 81 (1981), no. 2, 230–232.

[4] C. Bennett and R. Sharpley: *Interpolation of operators.* Academic Press 1988.

[5] J. Bergh and J. Löfström: *Interpolation spaces. An introduction* Grundlehren der Mathematischen Wissenschaften, No. 223. Springer 1976.

[6] P. Bourdon: *Components of linear-fractional composition operators*, J. Math. Anal. Appl. 279 (2003), 228–245.

[7] C. Cowen and B. D. MacCluer: *Composition Operators on Spaces of Analytic Functions.* CRC Press, Boca Raton, 1995.

[8] H. Hedenmalm, B. Korenblum, and K. Zhu: *Theory of Bergman spaces.* Graduate Texts in Mathematics, 199. Springer-Verlag, 2000.

[9] T. Hosokawa, K. Izuchi, and D. Zheng: *Isolated points and essential components of composition operators on H^∞*, Proc. Amer. Math. Soc. 130 (2002), no. 6, 1765–1773.

[10] P. W. Jones: *L^∞ estimates for the $\bar\partial$ problem in a half-plane*, Acta Math. 150 (1983), no. 1-2, 137–152.

[11] B. D. MacCluer: *Components in the space of composition operators*, Integral Equations Operator Theory **12** (1989), 725–738.

[12] B. D. MacCluer, S. Ohno, and R. Zhao: *Topological structure of the space of composition operators on H^∞*, Integral Equations Operator Theory 40 (2001), 481–494.

[13] B. D. MacCluer, J. H. Shapiro: *Angular derivatives and compact composition operators on the Hardy and Bergman spaces*, Canad. J. Math. 38 (1986), no. 4, 878–906.

[14] K. Madigan and A. Matheson: *Compact composition operators on the Bloch space*, Trans. Amer. Math. Soc. 347 (1995), no. 7, 2679–2687.

[15] J. Moorhouse: *Compact differences of composition operators*, J. Funct. Anal. 219 (2005), no. 1, 70–92.

[16] J. Moorhouse and C. Toews: *Differences of composition operators*. Trends in Banach spaces and operator theory (Memphis, TN, 2001), 207–213, Contemp. Math., 321, Amer. Math. Soc., Providence, RI, 2003.

[17] P. J. Nieminen: *Compactness of the difference of composition operators on Bloch or Lipschitz spaces*. Manuscript in preparation.

[18] P. J. Nieminen and E. Saksman: *On compactness of the difference of composition operators*, J. Math. Anal. Appl. 298 (2004), no. 2, 501–522.

[19] J. E. Shapiro: *Aleksandrov measures used in essential norm inequalities for composition operators*, J. Operator Theory 40 (1998), 133–146.

[20] J. H. Shapiro: *Composition operators and classical function theory*. Universitext: Tracts in Mathematics. Springer, 1993.

[21] J. H. Shapiro and C. Sundberg: *Isolation amongst the composition operators*, Pacific J. Math. 145 (1990), no. 1, 117–152.

[22] W. Smith: *Inner functions in the hyperbolic little Bloch class*, Michigan Math. J. 45 (1998), no. 1, 103–114.

DEPARTMENT OF MATHEMATICS AND STATISTICS, UNIVERSITY OF JYVÄSKYLÄ, P.O. BOX 35 (MaD), FIN-40014 UNIVERSITY OF JYVÄSKYLÄ, FINLAND
E-mail address: `saksman@maths.jyu.fi`

DEPARTMENT OF MATHEMATICS, UNIVERSITY OF TENNESSEE, KNOXVILLE, TENNESSEE 37996, U.S.A.
E-mail address: `sundberg@math.utk.edu`

Contemporary Mathematics
Volume **393**, 2006

Brennan's conjecture for weighted composition operators

Wayne Smith

This paper is dedicated to Joseph Cima on his 70th birthday.

ABSTRACT. We show that Brennan's conjecture is equivalent to the existence of self-maps of the unit disk that make certain weighted composition operators compact.

1. Introduction

1.1. Brennan's conjecture. Brennan's conjecture concerns integrability of the derivative of a conformal map g of a simply connected planar domain G onto the unit disk $\mathbb{D}$. The conjecture is that, for all such G and g,

$$(1.1) \qquad \int_G |g'|^p dA < \infty$$

holds for $4/3 < p < 4$. Here dA is area measure on the plane, normalized so that $A(\mathbb{D}) = 1$. It is an easy consequence of the Koebe distortion theorem that (1.1) holds when $4/3 < p < 3$. Brennan [**2**] extended this to $4/3 < p < 3 + \delta$ where $\delta > 0$, and conjectured it to hold for $4/3 < p < 4$. The example $G = \mathbb{C} \setminus (-\infty, -1/4]$ shows that this range of p cannot be extended. The upper bound of those p for which (1.1) is known to hold has been increased by Ch. Pommerenke to $p \leq 3.399$, and then to $p \leq 3.421$ by D. Bertilsson. These results and more information can be found in [**5**, §8.3] and [**1**].

Brennan's conjecture can also be formulated for analytic and univalent maps of $\mathbb{D}$ by setting $\tau = g^{-1}$. The conjecture becomes

$$\int_{\mathbb{D}} (1/|\tau'|)^p dA < \infty$$

holds for $-2/3 < p < 2$ and for all conformal maps τ of $\mathbb{D}$, and this is known for the range $-2/3 < p \leq 1.421$.

2000 *Mathematics Subject Classification.* Primary 47B33, Secondary 30C35.
Key words and phrases. Composition operator, compact operator, Hardy space, Brennan's conjecture.

1.2. The conjecture for weighted composition operators. Fix a conformal map τ from the unit disk onto a domain G and denote by $H(\mathbb{D})$ the space of all holomorphic functions on $\mathbb{D}$. For φ an analytic self-map of $\mathbb{D}$ and $p \in \mathbb{R}$, define a linear operator on $H(\mathbb{D})$ by

$$(1.2) \quad (A_{\varphi,p}f)(z) = (Q_\varphi(z))^p\, f(\varphi(z)), \quad \text{where} \quad Q_\varphi(z) = \frac{\tau'(\varphi(z))}{\tau'(z)} \quad (z \in \mathbb{D}).$$

We are interested in the action of $A_{\varphi,p}$ on the Hilbert space $L_a^2(\mathbb{D})$ of analytic and square integrable functions on the unit disk. In general, $A_{\varphi,p}$ may not be bounded on $L_a^2(\mathbb{D})$. But for any τ and any p, the operator induced by the identity map is the identity operator. That is, if $I(z) = z$, then $A_{I,p}f = f$ for all $f \in L_a^2(\mathbb{D})$. Thus there are always some choices of φ that make $A_{\varphi,p}$ *bounded*. Our main result is that Brennan's conjecture is equivalent to the problem of determining for which p there is a choice of φ that makes $A_{\varphi,p}$ *compact*:

THEOREM 1.1. *Let τ be univalent and holomorphic on $\mathbb{D}$ and let $p \in \mathbb{R}$. $(1/\tau')^p \in L_a^2(\mathbb{D})$ if and only if there exists an analytic self-map φ of $\mathbb{D}$ such that $A_{\varphi,p}$ is compact on $L_a^2(\mathbb{D})$.*

From the results discussed above it follows that for any univalent map τ of $\mathbb{D}$ and any p in the range $-1/3 < p \le .7105$, there exists a φ for which $A_{\varphi,p}$ is compact on $L_a^2(\mathbb{D})$. Brennan's conjecture is equivalent to the extension of this range to $-1/3 < p < 1$.

One implication in Theorem 1.1 is trivial: For $b \in \mathbb{D}$ let ψ_b be the constant map $\psi_b(z) = b$, so that $A_{\psi_b,p}f = \tau'(b)^p f(b)/(\tau')^p$. If $(1/\tau')^p \in L_a^2(\mathbb{D})$, then $A_{\psi_b,p}$ is bounded and rank one, and hence compact on $L_a^2(\mathbb{D})$. The main content of the theorem is the reverse implication, which can be restated as follows: If *some* map φ induces a weighted composition operator $A_{\varphi,p}$ that is compact, then so do the constant maps ψ_b.

It might seem that more information can be obtained by considering the action of weighted composition operators on the Bergman spaces $L_a^r(\mathbb{D})$ for $r \ne 2$. We show in the next section that nothing is lost by considering just one value of r. The choice $r = 2$ was made in the statement of Theorem 1.1 so that Hilbert space methods may be used in the proof.

The methods we use to prove the main theorem were developed by the author and J. H. Shapiro in [7], for weighted composition operators acting on Hardy spaces or Bergman spaces. The starting point in that paper was an unweighted composition operator acting on $H^2(G)$ or $L_a^2(G)$, where $G = \tau(\mathbb{D})$ and τ is a conformal map of $\mathbb{D}$. A change of variable shows that such an operator is similar to $A_{\varphi,-1/2}$ acting on $H^2(\mathbb{D})$ or to $A_{\varphi,-1}$ acting on $L_a^2(\mathbb{D})$. It was shown in [7] that there exists φ for which the operator $A_{\varphi,-1/2}$ is compact on $H^2(\mathbb{D})$ if and only if the boundary of G has finite one-dimensional Hausdorff measure, and there exists φ for which the operator $A_{\varphi,-1}$ is compact on $L_a^2(\mathbb{D})$ if and only if G has finite area. This last result is just the case $p = -1$ in Theorem 1.1. In §3 we show the methods used in [7] work for all $p \in \mathbb{R}$, giving Theorem 1.1.

2. Weighted Composition Operators

Let $0 < r < \infty$. We are interested in the action of a weighted composition operator on the Bergman space $L_a^r(\mathbb{D})$ of functions $f \in H(\mathbb{D})$ for which

$$\|f\|_{L_a^r(\mathbb{D})}^r = \int_{\mathbb{D}} |f|^r dA < \infty.$$

Let $Q(z)$ be a non-vanishing holomorphic function on $\mathbb{D}$ and let $r > 0$. The weighted composition operator $W_{Q,r}$ is the linear operator defined on $H(\mathbb{D})$ by

$$(W_{Q,r}f)(z) = (Q(z))^{1/r} f(\varphi(z)).$$

Carleson measure methods introduced in [4] for unweighted composition operators provide a characterization of when $W_{Q,r}$ is bounded or compact on $L_a^r(\mathbb{D})$. Briefly, a change of variable formula from measure theory gives

$$\|W_{Q,r}f\|_{L_a^r(\mathbb{D})}^r = \int_{\mathbb{D}} |Q(z)||f(\varphi(z))|^r dA(z) = \int_{\mathbb{D}} |f|^r d\mu\varphi^{-1},$$

where $d\mu(z) = |Q(z)|dA(z)$ and $\mu\varphi^{-1}$ is the measure defined by $\mu\varphi^{-1}(E) = \mu(\varphi^{-1}E)$ for a Borel set E. This results in the following characterization of when $W_{Q,r}$ is bounded or compact.

LEMMA 2.1. *$W_{Q,r}$ is bounded (compact) on $L_a^r(\mathbb{D})$ if and only if $\mu\varphi^{-1}$ is a bounded (compact) Carleson measure.*

The definitions of bounded and compact Carleson measures can be found in [4, §4], where the case $Q(z) \equiv 1$ of the lemma was proved. In general, it is not easy to check if $\mu\varphi^{-1}$ is a bounded or compact Carleson measure. Our interest in Lemma 2.1 is that the characterization is independent of r, giving the next result.

PROPOSITION 2.2. *If W_{Q,r_0} is bounded (compact) on $L_a^{r_0}(\mathbb{D})$ for some r_0, $0 < r_0 < \infty$, then $W_{Q,r}$ is bounded (compact) on $L_a^r(\mathbb{D})$ for all r, $0 < r < \infty$.*

In view of this proposition, we may restrict our attention to the case $r = 2$ and only consider weighted composition operators acting on the Hilbert space $L_a^2(\mathbb{D})$.

3. Proof of the Main Theorem

Fix a conformal map τ from the unit disk onto a domain G. For φ an analytic self-map of $\mathbb{D}$ and $p \in \mathbb{R}$, let $A_{\varphi,p}$ be the linear operator on $H(\mathbb{D})$ defined by (1.2). Since τ' is non-vanishing on $\mathbb{D}$, $A_{\varphi,p}$ is a weighted composition operator. Before proving Theorem 1.1 we need some preliminary results. The first is that no automorphism of the disk induces a compact operator $A_{\varphi,p}$.

LEMMA 3.1. *If φ is an automorphism of $\mathbb{D}$, then $A_{\varphi,p}$ is not compact on $L_a^2(\mathbb{D})$.*

PROOF. Let φ be an automorphism of $\mathbb{D}$ with φ^{-1} its inverse. Then $A_{\varphi,p}C_{\varphi^{-1}}$ is the multiplication operator with symbol $(Q_\varphi)^p$. It is well known that the only compact multiplication operator on $L_a^2(\mathbb{D})$ is the one with symbol identically 0 (see for example [8, Corollary 6.1.5]), so $A_{\varphi,p}C_{\varphi^{-1}}$ is not compact. Since $C_{\varphi^{-1}}$ is bounded on $L_a^2(\mathbb{D})$, $A_{\varphi,p}$ is not compact. $\square$

Next we introduce the reproducing kernels for $L_a^2(\mathbb{D})$, which will serve as test functions for the action of $A_{\varphi,p}$.

3.1. Reproducing kernels. The $L_a^2(\mathbb{D})$-reproducing kernel K_a for the point $a \in \mathbb{D}$ is the function

$$K_a(z) = \frac{1}{(1 - \bar{a}z)^2},$$

which satisfies

$$\langle f, K_a \rangle = \int_{\mathbb{D}} f \overline{K_a} \, dA = f(a)$$

for all $f \in L_a^2(\mathbb{D})$; see [**8**, §4.1]. In particular,

(3.1) $\|K_a\| = \langle K_a, K_a \rangle^{1/2} = \sqrt{K_a(a)} = \dfrac{1}{1 - |a|^2}$ $(a \in \mathbb{D})$.

We need to calculate the action of the adjoint of $A_{\varphi,p}$ on a reproducing kernel K_a.

LEMMA 3.2. *If the operator $A_{\varphi,p}$ is bounded on $L_a^2(\mathbb{D})$, then for each $a \in \mathbb{D}$,*

$$A_{\varphi,p}^* K_a = \left(\overline{Q_\varphi(a)} \right)^p K_{\varphi(a)}.$$

PROOF. For each $f \in A^2$ we have

$$\langle f, A_{\varphi,p}^* K_a \rangle = \langle A_{\varphi,p} f, K_a \rangle = (A_{\varphi,p} f)(a) = Q_\varphi(a)^p \, f(\varphi(a))$$

$$= Q_\varphi(a)^p \, \langle f, K_{\varphi(a)} \rangle = \langle f, \left(\overline{Q_\varphi(a)} \right)^p K_{\varphi(a)} \rangle.$$

Since f was arbitrary, the result follows.   □

The final preliminary result needed for the proof of Theorem 1.1 is that φ must have a fixed point in $\mathbb{D}$ for $A_{\varphi,p}$ to be compact.

PROPOSITION 3.3. *If $A_{\varphi,p}$ is compact on $L_a^2(\mathbb{D})$, then φ has a fixed point in $\mathbb{D}$.*

PROOF. For $a \in \mathbb{D}$, let $k_a = K_a/\|K_a\|$, so that $\{k_a : |a| < 1\}$ is a family of unit vectors in $L_a^2(\mathbb{D})$. We need the growth estimate that if $f \in L_a^2(\mathbb{D})$, then

$$f(a) = o\left((1 - |a|^2)^{-1} \right) \quad \text{as } |a| \to 1-;$$

see [**3**, Theorem 3.1]. From this and (3.1) we get that if $f \in L_a^2(\mathbb{D})$, then

$$< f, k_a > = f(a)/\|K_a\| = f(a)(1 - |a|^2) \to 0 \quad \text{as } |a| \to 1-;$$

i.e. $\{k_a\}$ converges weakly to zero as $|a| \to 1-$.

Now assume that $A_{\varphi,p}$ is compact. Since compact operators take weakly convergent sequences to norm convergent sequences and have compact adjoints,

(3.2) $\|A_{\varphi,p}^* k_a\| \to 0$ as $|a| \to 1 -$.

To complete the proof, we must prove that this implies φ has a fixed point in $\mathbb{D}$.

We will prove the contrapositive, so assume that φ has no fixed point in $\mathbb{D}$. We must show that (3.2) fails. Since φ has no fixed point in $\mathbb{D}$, the Denjoy-Wolff Theorem asserts that φ has a unique boundary fixed point η where the angular derivative of φ exists and satisfies $0 < \varphi'(\eta) \le 1$; see [**6**, Chapter 5]. Without loss of generality we may assume that $\eta = 1$.

From Lemma 3.2 and (3.1), for $0 \le r < 1$ we have

(3.3) $\|A_{\varphi,p}^* k_r\| = \left| \dfrac{\tau'(\varphi(r))}{\tau'(r)} \right|^p \dfrac{\|K_{\varphi(r)}\|}{\|K_r\|} = \left(\dfrac{\delta[\tau](\varphi(r))}{\delta[\tau](r)} \right)^p \left(\dfrac{1 - r^2}{1 - |\varphi(r)|^2} \right)^{1+p},$

where $\delta[\tau](z) = |\tau'(z)|(1 - |z|^2)$ is the invariant derivative of τ. We now use (3.3) to get a lower bound for $\|A^*_{\varphi,p}k_r\|$ as $r \to 1-$. The last factor on the right side of (3.3) is easily handled. Since the angular derivative of φ exists at 1, it follows (see [**6**, Chapter 4]) that

$$(3.4) \qquad \lim_{r \to 1} \frac{1 - r^2}{1 - |\varphi(r)|^2} = \frac{1}{\varphi'(1)}.$$

The first factor on the right side of (3.3) can be estimated using the inequalities

$$(3.5) \qquad \left(\frac{1 - |\alpha_w(z)|}{1 + |\alpha_w(z)|}\right)^2 \leq \frac{\delta[\tau](z)}{\delta[\tau](w)} \leq \left(\frac{1 + |\alpha_w(z)|}{1 - |\alpha_w(z)|}\right)^2 \qquad (z, w \in \mathbb{D}),$$

where $\alpha_w(z) = (w - z)/(1 - \bar{w}z)$. When $w = 0$ this is just the classical Koebe Distortion Theorem, and the general case results from conformal invariance of $\delta[\tau]$; see [**7**, §3.5] for details. Since φ has an angular derivative at 1 with $0 < \varphi'(1) \leq 1$,

$$\alpha_r(\varphi(r)) = \frac{r - \varphi(r)}{1 - r\varphi(r)} = \frac{\left(\frac{1 - \varphi(r)}{1 - r}\right) - 1}{r\left(\frac{1 - \varphi(r)}{1 - r}\right) + 1} \to \frac{\varphi'(1) - 1}{\varphi'(1) + 1} \leq 0 \quad \text{as} \quad r \to 1 - .$$

With (3.5) this gives

$$\liminf_{r \to 1-} \frac{\delta[\tau](\varphi(r))}{\delta[\tau](r)} \geq \lim_{r \to 1-} \left(\frac{1 - |\alpha_r(\varphi(r))|}{1 + |\alpha_r(\varphi(r))|}\right)^2 = \left(\frac{1 - \left(\frac{1 - \varphi'(1)}{1 + \varphi'(1)}\right)}{1 + \left(\frac{1 - \varphi'(1)}{1 + \varphi'(1)}\right)}\right)^2 = \varphi'(1)^2.$$

Using (3.3) and (3.4) we see that

$$\liminf_{r \to 1-} \|A^*_{\varphi,p}k_r\| \geq \varphi'(1)^{p-1} > 0,$$

which shows (3.2) fails and thus completes the proof. $\qquad \square$

We can now give the proof of the main theorem.

3.2. Proof of Theorem 1.1. As remarked at the end of §1, one implication in the proof is trivial: If $(1/\tau')^p \in L^2_a(\mathbb{D})$, then the operator $A_{\varphi,p}$ induced by a constant map φ is bounded and rank one, and hence compact on $L^2_a(\mathbb{D})$.

Now assume there exists φ such that $A_{\varphi,p}$ is compact on $L^2_a(\mathbb{D})$. Proposition 3.3 then tells us that φ has a fixed point in $\mathbb{D}$, i.e. there exists $b \in \mathbb{D}$ such that $\varphi(b) = b$. Since $\varphi(b) = b$ and $Q_\varphi(b) = \tau'(\varphi(b))/\tau'(b) = 1$, from Lemma 3.2 we see

$$A^*_{\varphi,p}K_b = \left(\overline{Q_\varphi(b)}\right)^p K_{\varphi(b)} = K_b.$$

Hence the number 1 is an eigenvalue of $A^*_{\varphi,p}$ and so is in its spectrum and also in the spectrum of $A_{\varphi,p}$. The assumption that $A_{\varphi,p}$ is compact now tells us that 1 is an eigenvalue of $A_{\varphi,p}$; see for example [**6**, p. 95]. Thus there exists $f \in L^2_a(\mathbb{D})$ not identically 0 such that $A_{\varphi,p}f = f$, or equivalently the function $g = (\tau')^p f$ satisfies $g \circ \varphi = g$. From Lemma 3.1 we know that φ is not an automorphism, so its iterates φ_n converge pointwise to its fixed point b [**6**, Proposition 5.2.1]. Hence, for all $z \in \mathbb{D}$, $g(z) = g(\varphi(z)) = g(\varphi_n(z)) \to g(b)$ as $n \to \infty$. Thus g is the constant function $g(z) \equiv c$ where $c = g(b)$, and $c \neq 0$ since f is not identically zero. Hence $(1/\tau')^p = c^{-1}f \in L^2_a(\mathbb{D})$, which completes the proof. $\qquad \square$

References

[1] D. Bertilsson, *On Brennan's conjecture in comformal mapping*, Doctoral Thesis, Royal Institute of Technology, Stockholm, Sweden 1999.

[2] J. E. Brennan, *The integrability of the derivative in comformal mapping*, J. London Math. Soc. (2), 18(1978), 261-272.

[3] P. Duren and A. Schuster, *Bergman Spaces*, American Mathematical Society 2004.

[4] B. D. MacCluer and J. H. Shapiro, *Angular derivaties and compact composition operators on the Hardy and Bergman spaces*, Canadian J. Math. 38(1986), 878-906.

[5] Ch. Pommerenke, *Boundary Behavior of Conformal Maps*, Springer-Verlag 1992.

[6] J. H. Shapiro, *Composition Operators and Classical Function Theory*, Springer Verlag 1993.

[7] J. H. Shapiro and W. Smith, *Hardy spaces that support no compact composition operators*, J. Functional Analysis 205(2003), 62-89.

[8] K. Zhu, *Operator Theory in Function Spaces*, Marcel Dekker 1990.

UNIVERSITY OF HAWAII, HONOLULU, HI 96822, USA

E-mail address: wayne@math.hawaii.edu

Titles in This Series

For a complete list of titles in this series, visit the
AMS Bookstore at **www.ams.org/bookstore/**.